American Environmental Leaders

Volume II

American Environmental Leaders

From Colonial Times to the Present

Volume II
L–Z

Anne Becher

with Kyle McClure,
Rachel White Scheuering,
and Julia Willis

ABC-CLIO

Santa Barbara, California
Denver, Colorado
Oxford, England

Library of Congress Cataloging-in-Publication Data
Becher, Anne.
    American environmental leaders / Anne Becher with Kyle McClure, Rachel
White Scheuering, and Julia Willis.
        p.   cm.
Includes bibliographical references and index.
    ISBN 1-57607-162-6 (acid-free paper)
    1. Environmentalists—United States—Biography.   I. Title.
    GE55 .B43   2000
    363.7'0092'273—dc21

                                                                                    00-011296

06   05   04   03   02   01   00   10 9   8   7   6   5   4   3   2   1

ABC-CLIO, Inc.
130 Cremona Drive, P.O. Box 1911
Santa Barbara, California 93116-1911

This book is printed on acid-free paper ∞.
Manufactured in the United States of America.

Contents

List of American Environmental Leaders

LaBudde, Samuel

(July 3, 1956–)
Founder and Director of the Endangered Species Project

Samuel LaBudde is the founder and director of the Endangered Species Project (ESP), an organization that endeavors to curb species extinction through investigations of the illegal wildlife trade, public education campaigns, and wilderness preservation. The recipient of a 1991 Goldman Environmental Award, LaBudde undertook a risky 1988 undercover investigation of dolphin slaughter by the tuna industry. With videotape of this practice in hand, LaBudde helped organize a domestic campaign that resulted in industry reforms and protective legislation, which eventually led to a 95 percent reduction in dolphin kills. LaBudde is also credited for exposing the ecological destruction caused by driftnet fishing. His work on this problem resulted in a 1992 United Nations resolution banning the use of driftnets.

Born in Madison, Wisconsin, on July 3, 1956, to Bessie Freeman and John LaBudde, Samuel Freeman LaBudde (pronounced la-buddy) grew up in southern Indiana. Of mixed French Norwegian and Scotch Cherokee heritage, he was exposed to the regional influences of Hoosier pride and Bible Belt principles throughout his childhood and elementary school years. But in his teens, he transformed into a more rebellious young man: "I think it was an awareness of my own heritage and what happened to the American Indian in this country," LaBudde related in an interview with *The Atlantic Monthly.* "It's essentially the same thing that's been done to the land."

After graduating from high school LaBudde set off on a classic coming-of-age saga of college courses, travel, and outdoor jobs. He spent four years working in the West: He planted trees in the Pacific Northwest, was a technician on a seismic crew, and was a machinist in Alaska. By 1984 he had become increasingly concerned about the deforestation of tropical rain forests, and felt that if he earned a degree in biology he might be able to do something to stop it. He left Alaska and motorcycled to Indiana, where he returned to college and earned a B.S. in biology in 1986. After graduating, LaBudde took a job with the National Marine Fisheries Service (NMFS) as a fisheries biologist on a Japanese trawler in the Bering Sea. Back from his first tour, he was offered a second one, on an American tuna boat based in San Diego. He refused this offer, however, hoping to find a position with an environmental organization.

A few weeks later, while he was waiting in the offices of the Earth Island Institute (EII) for an interview with the Rainforest Action Network (which was than an affiliate of EII), he read in a recent issue of *Earth Island Journal* about how dolphin slaughter by the tuna industry had resulted in a 50- to 75-percent drop in population of the three principal dolphin species. Because tuna frequently school beneath dolphins, commercial tuna fisherman would set the nets around the dolphins in order to catch the tuna. Dolphins would die from exhaustion and muscle fatigue after the chase, or from being caught in the net and suffocated, or when they were pulled through the machinery used to haul in the net. LaBudde learned from EII staff that the biggest impediment to stopping the dolphin killings was the absence of documentation about the problem. He recontacted NMFS to ask for the job on a tuna boat, but learned that NMFS biologists had to sign a nondisclosure statement prohibiting them from speaking about what they saw on board, and that further, no cameras were allowed. So LaBudde sought out Stan Minasian, the director of the Marine Mammal Fund (MMF), to talk about going undercover on a foreign tuna boat. Armed only with his own determination and a promise from the Marine Mammal Fund to supply him with a video camera, LaBudde headed across the Mexican border in September 1987 and got a job aboard a tuna boat. His intent was to obtain graphic footage of the massacre during tuna catches that was needed to raise public consciousness and end the slaughter. Risking exposure and retribution, LaBudde—who says that from his experience waiting tables during college he knew what food was supposed to look like but had never formally prepared a meal for anyone other than himself—worked for 5 months as the cook aboard the Panamanian fishing boat *Maria Luisa*, ultimately securing the film he needed.

After debarking in January, 1988, LaBudde—with support from MMF and EII—set out on a major media campaign on the tuna industry's slaughter of dolphins. LaBudde's footage appeared on most every network's nightly news broadcast and news shows, and a massive boycott of tuna fish was organized and eventually focused on H.J. Heinz corporation, which controlled half of the United States market share of tuna. By April 1990, Heinz announced its decision not to buy or sell tuna that was not dolphin-safe. Congress subsequently passed legislation to heighten dolphin protection and strengthened the dolphin protection clauses of the 1972 Marine Mammal Protection Act in the Act's 1988 and 1994 reauthorizations.

Parallel to his work to protect dolphins, LaBudde also became involved in a fight to ban driftnet fishing. Driftnets (also called gillnets) are ten-meter-deep walls of net made from monofilament fishing line with mesh from two to eight inches wide. They ensnare anything that drifts into them and is large enough to get stuck: squid, billfish, tuna, cetaceans, seabirds, turtles, etc. This practice had been encouraged in East Asia after World War II, but by the late 1980s, when there were more than 1000 vessels each dropping 30 to 40 miles of the net every night in the north Pacific, that marine ecosystem was in extreme danger of collapse. With support from the Hawaiian organization Earthtrust, LaBudde helped organize and lead an expedition on the high seas to document the practice, and wrote a technical paper compiling data about the potential ecological impact of the driftnet fisheries. He provided the report and video to the governments of two dozen southern Pacific nations whose fisheries-dependent economies were most affected by driftnet fishing. Alarmed at the danger to their economies, these governments successfully lobbied the United Nations to pass a 1992 resolution banning driftnet fishing in international waters, and the same year, Pres. Bush signed into law legislation that severely limited the use of coastal driftnets, which essentially made driftnetting economically untenable. LaBudde then spent two years in Europe with Humane Society International on a successful campaign to ban the import of dolphin-deadly tuna and limit the use of driftnets by the European fishing industry.

In recognition of his courageous investigative work and effective public education campaigns, LaBudde was nominated by a network of internationally known environmental organizations and a panel of environmental experts for the 1991 Goldman Prize. He used the $60,000 monetary award to found the Endangered Species Project (ESP), which he still directs. ESP works with other domestic and international non-governmental organizations on investigations and campaigns to protect wildlife and wilderness, providing local activists with video cameras to obtain necessary documentation. ESP has financed extensive field investigations throughout Southeast Asia to document illegal trade in tigers and other endangered wildlife. It has exposed Vietnam as a center for the Southeast Asian wholesale trade in wildlife and has secured international resolutions condemning involvement of China and Taiwan in illegal wildlife trade. Thanks to pressure from ESP and other concerned entities, the U.S. implemented trade sanctions against Taiwan in 1994—the first economic sanctions in history that have been imposed against a country for violations of an international conservation accord. This pressure prompted China, Singapore, South Korea, and Taiwan to pass laws against trade in tigers. *Crime Against Nature*, a report and video overview that ESP pro-

duced in 1994, documents the involvement of organized crime in the illegal wildlife trade and the inability of the Convention on International Trade in Endangered Species (CITES) to regulate it. The report is available directly from ESP (espnotes@aol.com).

In addition to its research and work on policy and legislative matters, ESP has contributed to concrete conservation efforts. It helped establish the International Siberian Tiger Sanctuary in Eastern Russia, and is currently working to set up a bioreserve for gorillas in West Africa.

Despite the positive achievements LaBudde can claim, species conservation work can never be considered complete. In 1999, five years after the amendment of the Marine Mammal Protection Act, LaBudde found himself engaged in a lawsuit with other species protection activists—including Humane Society of the United States, Defenders of Wildlife, American Society for the Prevention of Cruelty to Animals, Animal Welfare Institute, and EII—against the United States government for its proposed weakening of dolphin-safe standards. The United States had come under fire by the World Trade Organization for its refusal to import dolphin-deadly tuna from Mexico; this was considered a contravention of international trade rules. A proposal by U.S. Secretary of Commerce William Daley to loosen the definition of "dolphin safe" tuna would have totally undermined much of the earlier work done by LaBudde and his fellow activists. In April 2000, Daley's proposal was struck down.

LaBudde makes his home and ESP work base in San Francisco. Copies of his video footage showing the devastating effects of tuna nets on dolphins are available through EII; video footage of drift-netting is available through Earthtrust.

BIBLIOGRAPHY

Brower, Kenneth, "The Destruction of Dolphins," *Atlantic Monthly*, 1989; "Earth Island Insti-tute," http://www.earthisland.org; "Earthtrust," http://www.earthtrust.org; "Goldman Prize: Re-cipients," http://www.goldmanprize.org/recipi-ents/recipients.html; Wallace, Aubrey, *Eco-Heroes: Twelve Tales of Environmental Victory*, 1993.

LaDuke, Winona

(August 18, 1959–)
Community Restoration Worker, Cofounder and Cochair of Indigenous Women's Network

Native American economist, writer, and activist Winona LaDuke (An-ishinabe) is an advocate for in-digenous people throughout the world, promoting indigenous control of tradi-tional homelands. Native peoples have honed sustainable lifestyles that have al-lowed them to live for thousands of years without damaging the ecology of their homelands, and LaDuke works to nur-ture that specialized knowledge. She co-founded and cochairs the Indigenous Women's Network, a coalition of Native women who apply traditional knowledge to resolve contemporary problems.

Winona LaDuke, born on August 18, 1959 in California, spent her childhood in Los Angeles and in Ashland, Oregon. Her father, Vincent LaDuke, also known as Sun Bear, was an Anishinabe (or Ojibwe, Chippewa) spiritual thinker, writer, actor, and activist who enrolled his daughter as a member of his White Earth reservation and took her with him to powwows and other Native American functions. Both Winona and her father remember that powwow dancing made a deep impres-sion on her. When Winona was five, her parents split up, and she moved with her mother, artist and political activist Betty LaDuke, to Ashland, Oregon. She visited White Earth frequently, and her mother, though not Native American herself, en-couraged her to spend summers living with Native peoples in order to learn more about her heritage and Native struggles. Winona LaDuke was recruited by Harvard University, studied economic development there, and graduated in 1982. Upon graduation, she was invited to become high school principal on the White Earth Reservation. Returning to White Earth, she told *People Magazine*, was like coming home for the first time.

LaDuke works to secure Native land tenancy at White Earth, where 90 percent of the 830,000 acres in northwestern Min-nesota that were originally declared White Earth Reservation in 1867 are now owned by non-Anishinabe people. With the $20,000 the Reebok Foundation awarded to her in 1988 in recognition of her work championing human rights, LaDuke founded the White Earth Land Recovery Project. That seed money was used to purchase 1,000 acres of White Earth land, and the project hopes to ac-quire hundreds of thousands more.

Winona LaDuke (Dilip Mehta/Contact Press Images/PictureQuest)

In addition to the White Earth Land Recovery Project, LaDuke has worked on many other fronts to fortify Anishinabe culture. She has helped establish an artisan cooperative for traditional handicrafts, set up an Ojibwe language program for children and adults, arranged for all road signs in White Earth to be in the Ojibwe language, and taken legal action to prevent unsuitable development on sacred burial sites. She has written two books: *Last Standing Woman* (1997), a novel tracing the history of the Anishinabe people, and *All Our Relations: Native Struggles for Land and Life* (1999), which describes how eight Native nations across the United States are struggling to maintain self-determination and community despite threats from outside.

LaDuke's affiliations and commitments are numerous. She cochairs the Indigenous Women's Network, which was founded in 1985 by a coalition of Native women to revitalize indigenous languages and cultures, protect religious and cultural practices, protect the environment, and work toward recovery of indigenous lands. She is a member of the editorial collective that publishes *Indigenous Woman* magazine. Since 1993 LaDuke has organized biannual concert tours with the musical duo Indigo Girls to educate youth about Native issues and raise money for Native environmental causes. She ran for vice president in 1996 and 2000 on the Green Party ticket with Ralph Nader and serves on the board of Greenpeace. She was recognized by *Time* magazine in 1995 as one of the Fifty Leaders for the Future and in *Ms.* magazine in 1998 as one of their Women of the Year.

LaDuke has two children from a marriage to Randy Kapashesit, a Cree from Moose Factory, Ontario, and now lives with her companion, Kevin Gasco, an Odawa coffee roaster from Northern Michigan.

Bibliography

Barsamian, David, *Winona LaDuke: Activism on and off the Reservation*, audio recording, 1998; Bowermaster, Jon, "Earth of a Nation," *Harpers Bazaar*, 1993; "Indigenous Women's Network," http://www.honorearth.com/iwn/; LaDuke, Winona, *Last Standing Woman*, 1997; Paul, Sonya, and Robert Perkinson, "Winona LaDuke," *The Progressive*, 1995; Rosen, Marjorie, and Margaret Nelson, "Friend of the Earth," *People*, 1994.

Lammers, Owen

(March 20, 1963–)
Executive Director of the Glen Canyon Action Network, River Advocate

Owen Lammers is the executive director of the Glen Canyon Action Network (GCAN), a Moab, Utah–based organization that seeks to restore Glen Canyon by draining Lake Powell. Prior to joining GCAN in 2000, Lammers spent 12 years leading the river advocacy organization International Rivers Network (IRN). IRN is a Berkeley-based group that promotes river management practices that respect both human rights and environmental protection. As head of program development at IRN, Lammers mobilized international campaigns to challenge destructive river development projects around the world and to discourage governments and businesses from financing such projects as the Three Gorges Dam in China.

Owen Thomas Lammers was born on March 20, 1963, in Los Angeles and raised in Walnut Creek and Berkeley, California. His father, Thomas Lammers, is a civil engineer, and his mother, Mary Josephine (Coor) Lammers, is a teacher and moving consultant. Lammers graduated from the University of California at Berkeley in 1986 with a B.S. in natural resources economics. After working for a few years as a real estate appraiser, in 1987 Lammers joined IRN, which at that time was an all-volunteer organization. Lammers plunged into fundraising, and within three months, he was a paid staff member. Over the next 12 years, Lammers worked his way up from administrative director to executive director, eventually becoming vice president for program development. Under his direction, IRN grew from a volunteer organization to one with 21 staff members, three offices, and a budget of over a million dollars.

In the 1990s, Lammers coordinated international efforts to challenge some of the most destructive river development projects in the world, including the Three Gorges Dam in China and the Maheshwar Dam in India. IRN scored a success by targeting Morgan Stanley Dean Witter because the investment firm had been underwriting securities for the State Development Bank of China, which was helping to finance the Three Gorges project. In response to pressure from environmental groups, including IRN, the shareholders voted to withdraw support for the project and to set guidelines that could prevent the financing of similar projects in the future.

In 1998, IRN was one of three environmental organizations to receive $1 million from the San Francisco–based Richard and Rhoda Goldman Fund, administrator of the prestigious Goldman Environmental Prize. IRN used the money to further support projects in Latin America and Southern Africa. One of the projects supported by IRN is the International Meeting of People Affected by Dams, which held its first meeting in 1997 in Brazil and which works to build an international network of antidam activists.

Lammers, along with Dr. Robert Haas, U.S. poet laureate (1995–1997), helped to create the River of Words Project, a nonprofit educational program established in 1993. The respected environmental education program distributes curriculum guides that teachers may use to develop activities to teach children about their local environment; students then create artwork or poetry to submit to an international environmental poetry and art contest. River of Words also offers teacher training workshops and annually honors a Teacher of the Year.

Lammers's strength as an environmental leader lies in his skill as a campaigner and his ability to mobilize public support. By 1999, IRN had developed from a grassroots activist organization into a well-funded service organization. Lammers felt that it was time to move on. He resigned from IRN in August and within a few months had left Berkeley and moved to Utah. The move was Lammers's response to a growing sense that he could not be a credible environmental leader while spending little time engaged with the natural world. He had been introduced to the red rock country of southern Utah in 1993 when he joined the board of the Four Corners School of Outdoor Education based in Monticello. (He served as a consultant to the school until 1998.) Lammers decided Moab, Utah, was a good place to reestablish his connection to the environment after years of living in the city.

Upon his arrival in Moab, area residents with an interest in restoration of the Colorado River seized the opportunity to organize a group with Lammers at the helm. The idea for GCAN arose out of an informal discussion between Lammers and local river guides and outfitters in a Moab restaurant in late 1999. The group's initial objective is to build public support for the decommissioning of the Glen Canyon dam, with the end goal of draining mammoth Lake Powell. Mobilizing with lightning speed, the group was up and running by January 2000. Lammers enlisted the support of prominent environmentalist DAVID BROWER and set about developing a broad-based membership that included activists, river outfitters, educators, and writers. GCAN began a program of monthly lectures in Moab and established a nonprofit ice cream shop in downtown Moab, with all proceeds going toward restoring the Colorado River watershed.

Three months after GCAN was created, it hosted a Restoration Celebration and Rendezvous in Page and Flagstaff, Arizona. The date of the Rendezvous, March 13, was set to coincide with the 37th anniversary of the start of the filling of Lake Powell and the international Third Annual Day of Action Against Dams and for Rivers, Water, and Life. Eighty-seven-year-old David Brower was the keynote speaker, and Secretary of the Interior BRUCE BABBITT made an appearance wielding a sledgehammer. Two hundred individuals and representatives of 45 organizations signed the Glen Canyon Declaration calling for the decommissioning of Glen Canyon Dam and the draining of Lake Powell.

Lammers lives in Moab, Utah.

BIBLIOGRAPHY

"Glen Canyon Action Network," www.drainit.org; "International Rivers Network," www.irn.org; "The Rivers Movement," *New Internationalist*, 1995.

Lappé, Frances Moore

(February 10, 1944–)
Writer, Cofounder of Institute for Food and Development Policy and Center for Living Democracy

Through her writing and activism, Frances Moore Lappé has fought to dismiss popular myths about the causes of hunger, poverty, and environmental degradation. She has written many books on these subjects, including *Diet for a Small Planet*, the best seller that linked hunger to economic and political issues and urged developed countries to shift from meat production to vegetable production. This groundbreaking book revealed the vast waste of resources incurred by meat production and changed the way millions of people thought about how food choices affect the environment. Lappé was also founder and director of the Institute for Food and Development Policy and more recently launched the Center for Living Democracy to involve citizens in public dialogue in political and economic issues.

Frances Moore was born on February 10, 1944, in Pendleton, Oregon, to John Gilmer Moore, a forecaster for the U.S. Weather Service, and Ina (Skrifvars) Moore, a transportation agent for the Corps of Engineers. Her parents' professions made less of an impression on her than their open-minded and progressive home life and volunteer efforts. The family moved to Fort Worth, Texas, when Frances was just a few years old, and with the help of some close friends they founded the First Unitarian Church of Fort Worth. Lappé loved growing up against the backdrop of the church, which provided a forum for discussing and participating in social issues. She attended American University in Washing-ton, D.C., from 1962 to 1963 and then finished up at Earlham College in Richmond, Indiana, earning her B.A. in 1966. In the late 1960s, Lappé headed to Berkeley, California, to study social work in graduate school. But she found herself delving deeper and deeper into the shelves at the campus agricultural library, where she discovered more and more links between food and politics. She began trying to figure out why millions were going hungry despite an apparent bounty of food. In 1967 she married Marc Alan Lappé, a biology research associate, with whom she would have two children (their marriage ended in divorce ten years later). She moved to Philadelphia, Pennsylvania, that year in order to work as a community organizer at the Philadelphia Neighborhood Renewal Program, where she stayed two years.

As her interest in agriculture and world hunger grew, Lappé continued to educate herself, taking notes and tacking messages on health food store bulletin boards. She compiled her findings, and in 1971 she published *Diet for a Small Planet*, a book that created a nutrition revolution and changed the way millions of Americans thought about food and world hunger. She pointed out that most of the grain grown on harvested agricultural land is fed to livestock, an extremely inefficient use of energy and farming resources, since a cow must be fed 21 pounds of protein to produce one pound of protein for human consumption. Her book tapped into a powerful well of concern across the country when it showed how this pattern of waste contributes to the overconsumption of resources that the United States had come to be known for. However, Lappé, having grown up eating meatloaf and hamburgers herself, knew that the fixed cultural attitude of the time frowned on vegetarianism and that people were concerned about getting enough protein. So her book explained how to get enough protein without centering one's diet around meat by combining grains with legumes in various combinations, such as beans and rice, beans and tortillas, or soy and rice. This produces a healthy diet rich in protein and makes much better use of the earth's productivity. Lappé accomplished her goal of establishing a sense of the direct impact food choices have on the earth, and her book, which has sold over three million copies, spawned a new wave of environmentally sensitive eating.

In 1975, Lappé cofounded with Joseph Collins the Institute for Food and Development Policy, also known as Food First, in San Francisco, California. The guiding principle of the institute was that the world hunger problem comes not from a scarcity of food, but a scarcity of democracy. The institute studied economics and politics through field research and published reports, articles, pamphlets, school curricula, and study guides. One of the institute's seminal works, *Food First: Beyond the Myth of Scarcity* (1977), written by Lappé and Collins, was internationally recognized as a major achievement in its analysis of global politics surrounding world hunger. They presented the argument that starvation in the world's population is caused not by a scarcity of food supplies or by outmoded farming methods used in developing countries but by political and economic

problems such as centralized control of farmland and the colonization of Third World countries by western nations. Lappé and Collins wrote that communities should make it their top priority to become self-reliant in production of food, tools, and fertilizers, though this would require land reform on a grand scale and a return of ownership of local farmland to the people who live on it.

Lappé lectured widely on these issues through the 1980s but eventually began feeling as if public speaking was not enough. She developed the idea that public life must involve dialogue and discourse about the values that inform political views. In 1990 she cofounded with Paul Du Bois the Institute for the Arts of Democracy (now called the Center for Living Democracy) in San Rafael, California, to serve as a catalyst, offering learning tools and training to incorporate a practical vision of democracy into people's lives and get them directly involved in economic and political matters. The center relocated to Brattleboro, Vermont, in 1993 to continue its development, and in 1994 one of its major efforts, the book *The Quickening of America: Rebuilding Our Nation, Remaking Our Lives*, was published. This interactive book serves as a guide for the importance of democracy and gives advice on participating in public life.

Lappé has received 15 honorary doctorates from distinguished institutions. In 1987 she became the fourth American to receive the prestigious Right Livelihood Award, sometimes called the "alternative Nobel." In January 2000, she began a year's appointment as a visiting scholar at the Massachusetts Institute of Technology, where she is writing a 30th-anniversary sequel to *Diet for a Small*

Planet. She and her second husband, Paul Du Bois, also continue their work at the Center for Living Democracy, which has become recognized as a leading resource for community innovations.

BIBLIOGRAPHY

Blanchard, Bob, and Susan Watrous, "Frances Moore Lappé: 'Something New is Possible under the Sun,'" *The Progressive*, 1990; Lappé, Frances Moore, *Diet for a Small Planet*, 1971; Lappé, Frances Moore, and Joseph Collins, *Food First: Beyond the Myth of Scarcity*, 1977; Lappé, Frances Moore, and Paul Martin Du Bois, *The Quickening of America: Rebuilding Our Nation, Remaking Our Lives*, 1994; Turner, Tom, "The World According to Frances Moore Lappé: Food and Democracy Go Hand in Hand," *E Magazine*, 1992.

Leopold, Aldo

(January 11, 1886–April 21, 1948)
Writer, Professor of Wildlife Management, Cofounder of the Wilderness Society

Aldo Leopold is known as the father of environmental ethics and game management. His more than 300 articles record his gradual definition of these two fields, but he is best remembered by the general public as author of *A Sand County Almanac and Sketches Here and There* (1949). This lean, poetic volume, a model for all subsequent nature writers, advocated a change in "the role of *Homo sapiens* from conqueror of the land-community to plain member and citizen of it." Leopold's science and writings have served as inspiration for many environmentalists and conservation organizations, most directly the Wilderness Society, which Leopold helped found.

Born on January 11, 1886, in Burlington, Iowa, Aldo Leopold grew up hunting ducks and partridges in the marshes and woods of Iowa. With the gift of his first rifle at a young age, his father imposed strict hunting rules on the boy, for example, that he could shoot birds only when they were in the air. Leopold went east for preparatory high school and studied forestry at Yale University. After graduating in 1909, he took a job with the U.S. Forest Service and moved to New Mexico, where he worked his way up to supervisor of Carson National Forest.

Leopold quickly began to form his own ideas on conservation. He did not like what he saw happening in national parks in the 1920s. He compared the multitudes of tourists in Yosemite to "wedding guests" consuming "wedding-cake"—the spectacular features like Half Dome, Bridal Veil Falls, and El Capitan. Leopold advocated preservation of wild lands that lacked such impressive scenery but were nonetheless important natural resources and habitat. He worked with New Mexican stockmen to establish the Gila National Forest in 1924, to be used as grazing country by their cattle but to remain undeveloped and inaccessible to tourists.

Once the Gila National Forest was established, Leopold participated eagerly in the extermination of cattle and deer predators such as wolves, but an encounter with a wolf he had just shot awakened him to aspects of ecology that would later help him formulate the science of game management. As Leopold watched the wolf's fading green eyes, he suddenly realized that he had been mistaken to think that predators were exclusively bad and dangerous. As the wolf died, it occurred to him that they belonged to their habitat and that they must play an important role there. This epiphany was promptly confirmed when the deer population surged out of control after Leopold and his Forest Service team had successfully eradicated wolves from the Gila.

Leopold and his wife, Estella Luna Bergere, whom he married in 1912 in New Mexico, moved to Madison, Wisconsin, in 1924. There he continued ruminating on the epiphany he had had with the wolf. He developed a theory of game management and became the University of Wisconsin–Madison's first professor of wildlife management in 1933. His primer, *Game Management*, was also published in 1933 and has guided biologists in managing wildlife ever since. In that book, as well as in his articles written for foresters, sportsmen, conservationists, economists, and the general public, Leopold insisted that a respect for all life forms occurring in a given environment is fundamental for conservation work. This challenged the prevailing basis for conservation of that time, which was that the environment should be conserved primarily to assure future generations a ready supply of natural resources.

Though primarily an academician and scientist, Leopold became active in such conservation organizations as the Izaak Walton League and the Wilderness Society, of which he was a founding member. Leopold approached his conservation

Aldo Leopold (Courtesy of Anne Ronan Picture Library)

work with the sharp mind of a skeptic, constantly questioning the goals and projects of conservation groups. One of his friends called him a "living question mark."

Leopold enriched his academic career with his work at an old, depleted farm that his family bought in 1935 in Sand County, Wisconsin. Calling it a "living laboratory," the family tried to revive the land by planting trees and nurturing the soil. Leopold recorded the changes on the farm with the passage of seasons and years in *The Sand County Almanac*, now considered an environmental classic. This volume not only lovingly describes what happens in Sand County month by month (the arrival, nesting, and departure south of birds, the growth and blossoming of plants, the rise and fall of rivers and streams), but also proposes new paradigms for how humans can live in greater harmony with their natural environment. Near the end of the book, Leopold articulates his idea for a land ethic—taken for granted by traditional land-based communities, but new to the majority of people who consider themselves civilized:

> The land simply enlarges the boundaries of the community to include soils, water, plants, and animals, or collectively: the land.

> This sounds simple: do we not already sing our love for and obligation to the land of the free and the home of the brave? Yes, but just what and whom do we love? Certainly not the soil, which we are sending helter-skelter downriver. Certainly not the waters, which we assume have no function except to turn turbines, float barges, and carry off sewage. Certainly not the plants, of which we exterminate whole communities without batting an eye. Certainly not the animals, of which we have already extirpated many of the largest and most beautiful species. A land ethic of course cannot prevent the alteration, management, and use of these "resources," but it does affirm their right to continued existence, and at least in spots, their continued existence in a natural state.

> In short, a land ethic changes the role of *Homo sapiens* from conqueror of the land-community to plain member and citizen of it. It implies respect for his fellow-members, and also respect for the community as such.

Leopold died on April 21, 1948, while fighting a fire on a neighbor's farm. *A Sand County Almanac* was published posthumously, in 1949. His five children all became scientists.

BIBLIOGRAPHY

Callicott, J. Baird, and Eric T. Freyfogle, eds., *For the Health of the Land: Previously Unpublished Essays and Other Writings by Aldo Leopold*, 1999; Flader, Susan, *Thinking like a Mountain: Aldo Leopold and the Evolution of an Ecological Attitude toward Deer, Wolves, and Forests*, 1974; Leopold, Aldo, *Round River; from the journals of Aldo Leopold*, 1953; Loribiecki, Marybeth, *Aldo Leopold: A Fierce Green Fire*, 1996; McCabe, Robert A., *Aldo Leopold, the Professor*, 1987; Meine, Curt, *Aldo Leopold: His Life and Work;* 1988; Nash, Roderick, *Wilderness and the American Mind*, Revised Edition, 1973; Strong, Douglas H., *Dreamers & Defenders: American Conservationists*, 1988; Tanner, Thomas, ed., *Aldo Leopold: The Man and His Legacy*, 1933.

Lindbergh, Anne Morrow, and Charles Augustus Lindbergh

(June 22, 1906– ; February 4, 1902–August 26, 1974)
Novelist, Poet, Aviator; Aviator, Writer

In a dynamic and world-famous partnership, Charles and Anne Morrow Lindbergh made key contributions to pioneering aviation, exploration, writing, and conservation. In their exploratory flights around the globe, they had many opportunities to view the earth from above and grew concerned that technology and development were causing negative impacts on the environment. Anne was a productive and sensitive writer and used her writing to convey her belief that wilderness should be protected, while Charles used his fame and status to speak out on such conservation issues as saving endangered species. In honor of the Lindberghs' environmental vision, the Charles A. and Anne Morrow Lindbergh Foundation was created to help fund research and educational projects that contribute to a balance between technological advancement and environmental preservation.

Anne Spencer Morrow was born on June 22, 1906, in Englewood, New Jersey, to Dwight Whitney Morrow and Elizabeth Reeve (Cutter) Morrow. Her father, an attorney, later served as ambassador to Mexico from 1927 to 1929 and as Republican senator from New Jersey in 1930 and 1931. Her mother, a poet and educator, was acting president of Smith College from 1939 to 1940 and was an advocate for women's education. Anne grew up with her three siblings in the sheltered comfort of the upper class—traveling with her family on several European tours and attending Miss Chapin's School, a college preparatory school in New York City. Shy and introspective, she filled much of her time with writing. She began studying at Smith College in 1924, majoring in English and continuing her literary efforts. In 1928, the year she graduated with her B.A., she won a prize for the best essay on women of the eighteenth century and had a poem published in *Scribner's Magazine.*

Charles Augustus Lindbergh was born on February 4, 1902, in Detroit, Michigan, to Evangeline (Land) and Charles August Lindbergh and spent most of his childhood years on the family's farm near Little Falls, Minnesota. His parents had little respect for formal education, and when teachers complained that Charles was falling behind, he was pulled out of the school and placed elsewhere. Between the ages of 8 and 16, he attended at least 11 different institutions. After graduating from high school in 1918 in Little Falls, Charles Lindbergh enrolled at the University of Wisconsin in Madison to study engineering. But during his second year, he could no longer resist his growing attraction to flight and enrolled in an aviation school in Lincoln, Nebraska. He went on to become a stunt flier, learning wing walking and parachuting, and then in 1924 joined the army so he could attend the army flight school in San Antonio, Texas. After graduating first in his class, he became the first airmail pilot between Chicago and St. Louis and began plans to attempt the first solo nonstop flight between New York and Paris. On May 20, 1927, Charles Lindbergh left New York in his plane, *The Spirit of St. Louis,* and flew nonstop for 33½ hours to Paris, the first person ever to achieve this.

When news of his accomplishment spread, Charles Lindbergh stepped into the international spotlight. He was awarded the Congressional Medal of Honor by the U.S. government and began flying on tour to promote aviation and express goodwill to other countries. He was invited to the U.S. Embassy in Mexico, where Anne Morrow Lindbergh's father was serving as ambassador. Anne was at the embassy on Christmas break from her junior year of college and met Charles during his two-week stay there. He took Anne flying, which she found exhilarating, and the two of them began a courtship that they tried to shield from the ever-present press. On May 27, 1929, they succeeded in getting married in Englewood without the press's finding out, though photographers interrupted their honeymoon.

Anne Morrow Lindbergh caught on to her husband's passion for flying, and she herself soon learned to fly. In 1930 she became the first woman in the United States to obtain a glider pilot's license, and she earned her private pilot's license in 1931. Much of their early years of marriage was spent flying all over the globe, exploring and charting routes for commercial air travel. Flying provided the inspiration for Anne Morrow Lindbergh's first book, *North to the Orient* (1935), an account of a survey flight to Asia by way of Canada, Alaska, and Siberia.

In March 1932, the Lindberghs' 21-month-old baby boy, Charles, was kidnapped from their home in Hopewell, New Jersey, and though they paid the ransom, they learned after 72 days that their child had been killed the night of his abduction. The publicity that followed was incessant and unbearable. In 1935, after the birth of their second son and the conclusion of the trial of the man accused of the kidnapping, the Lindberghs moved to England for relative privacy. Anne Morrow Lindbergh continued writing, and in 1938 her next book, *Listen! the Wind*, again based on survey flights she took with her husband, was published. In 1938 the Lindberghs moved to France, and Charles began flying aviation intelligence missions for the U.S. military. In the years following, rumors that many of Charles Lindbergh's beliefs echoed Nazi dogma spread to the United States, and his popularity took an icy

plunge. After World War II the Lindbergh family returned to the United States and purchased a home in Darien, Connecticut. In 1955, Anne Morrow Lindbergh's next book appeared. *Gift from the Sea*, her most popular book, contains eight personal essays reflecting on nature, family, the passing of time, and the need to remove oneself occasionally from the routine of everyday life and seek the solitude vital for self-discovery. This book has been hailed as a testimony to the powerful emotional and spiritual value of an intimate relationship with the natural environment. Charles Lindbergh also authored a number of books in his lifetime and is probably best known for the 1954 Pulitzer Prize winner, *The Spirit of St. Louis*, a memoir of his famous flight.

During their travels, the Lindberghs had many opportunities to view the earth's landscapes from the sky. They began noticing changes in the land and became very concerned about the effects of pollution and the disappearance of wild places. Anne Morrow Lindbergh gave voice to these concerns in *Earth Shine* (1969), a book consisting of a pair of essays that illustrate her view that the roots of life are in wilderness. She speaks not only of her fear of the extinction of animal species but also her fear that if wilderness is lost, humans will lose an element vital to their being. In wilderness is renewal she writes, and through that renewal is the possibility of making connections with other life. After many years of privacy, Charles Lindbergh's abiding interest in preserving the environment brought him back into the public eye in the late 1960s, when he began speaking out for conservation. He campaigned to protect endangered species, especially humpback and blue whales. Appointed by the International Union for the Conservation of Nature, he attended the 1966 conference of the International Whaling Commission. He also joined the World Wildlife Fund's Committee of 100, a panel of internationally recognized figures who lobbied heads of state to save endangered wildlife and habitats. His speeches and writings from this time reinforced his belief in the need for both technology and conservation and the necessity of balancing the two. Having pushed the frontiers of aviation technology, Charles Lindbergh was aware of both the achievements of technology and its capacity for destruction. For example, while he supported advancements in aviation, he opposed the development of supersonic transport planes because he feared they would pollute the upper atmosphere. Anne Morrow Lindbergh also believed in balance, and at a conference at Smith College in February 1970, she made a rare public speech, arguing that human values are derived from earth values and therefore the earth must be protected.

The Lindberghs' vision, that the use of technology must be balanced to secure the long-term survival of the earth's life support systems, led to the creation of the Charles A. and Anne Morrow Lindbergh Foundation in 1977. The foundation presents grants each year to research projects that contribute to an equilibrium between technology and environmental preservation and that have included areas of special interest to the Lindberghs, such as exploration, conservation of natural resources, health and population sciences, and wildlife preservation.

Both of the Lindberghs have been presented with numerous honorary degrees and other awards. Charles Lindbergh died of cancer on August 26, 1974, on the island of Maui, Hawaii. Anne Morrow Lindbergh lives in Connecticut.

BIBLIOGRAPHY

"The Charles A. and Anne Morrow Lindbergh Foundation," http://lindberghfoundation.org; Hertog, Susan, *Anne Morrow Lindbergh: Her Life*, 1999; Lindbergh, Anne Morrow, *Earth Shine*, 1969; Lindbergh, Anne Morrow, *Gift from the Sea*, 1955; Milton, Joyce, *Loss of Eden: A Biography of Charles and Anne Morrow Lindbergh*, 1993.

Littletree, Alicia

(March 24, 1974–)
Musician, Environmental Organizer

Alicia Littletree is an organizer for Earth First! in northern California and is a main protagonist in the struggle to end corporate and government harassment of environmentalists practicing nonviolent direct action protests to protect wilderness.

Alicia Littletree was born on March 24, 1974, in Los Angeles, California, and was raised in Sacramento, the only child of a single mother. She was educated in the public schools, in an accelerated program that featured camping trips and outdoor classes with naturalists. She and the other students in that program were encouraged to rely on their own minds and to develop their imaginations. While in high school, Littletree founded Students for a Green Earth, the school's environmental club. During the first Gulf War in 1991, Littletree joined Sacramento's large protests against the bombing of Iraq, and through those activities, she met a group of Earth First! activists from Humboldt County, on the north California coast.

Littletree moved at the age of 17 to a small cabin on the wooded coast of Humboldt County. She enrolled in an independent high school program there, worked at the Institute for Sustainable Forestry, and attended local Earth First! meetings. Earth First! is a loosely organized group of environmental activists who use nonviolent civil disobedience and direct action to attempt to stop logging, mining, and road building in wild areas. Earth First! works in northern California against clear-cutting practiced by large multinational timber corporations in this country's last remaining temperate rain forests, including the last of the ancient redwoods.

It was at Earth First! meetings that Littletree met Earth First! and union organizer and fellow musician JUDI BARI, with whom she was to live and collaborate for the rest of Bari's life. Littletree celebrated her involvement with Earth First! by holding her first tree sit in the spring of 1992 during what is now referred to as the Albion Uprising, a two-month mass protest of Louisiana Pacific's clear-cutting. A tree sit is one of the varied tactics that nonviolent direct action environmentalists use to protect a forest from being cut. It involves climbing a tree and sitting on a branch or a platform, in order to prevent loggers from cutting that tree or any others near it. Littletree's first tree sit lasted nine days, during which she stayed alone aloft in the tree, receiving necessary food, water, and other supplies from a support team on the ground.

Littletree became more involved in Earth First! through the mid-1990s. She helped organize Earth First!'s major protests in the Headwaters forest in 1993 and helped coordinate a 7,000-person rally in the spring of 1996, during which 1,000 people were arrested. In the days following that action, Littletree worked with 13 "affinity groups," tight yet decentralized groups of people collaborating on a nonviolent direct-action protest. The affinity groups shut down the operations of clear-cut giant Pacific Lumber/Maxxam for a full day by blocking every access gate to the forest being cut, holding several tree sits concurrently, and rallying 200 protesters at the company headquarters. Littletree and hundreds more continued on the front lines of the 1996 Pacific Lumber/Maxxam protest for two months.

As well as working with Earth First! to defend the forest, Littletree worked closely with Judi Bari on her court case about the May 24, 1990, car-bombing that had almost killed Bari and injured her passenger Darryl Cherney, another Earth First! activist. Bari and Cherney sued the Oakland police and the Federal Bureau of Investigation (FBI) for false arrest, illegal search and seizure, and conspiracy to violate their first amendment rights by using the bombing to discredit them as terrorists. Littletree works with the organization set up to coordinate the case, the Redwood Summer Justice Project. She believes that the case is of great importance because it exposes the abuses of rights by the FBI. This is especially important now since violence against non-violent environmental activists has become more prevalent and dangerous in recent years.

In addition to her work on the bombing case, Littletree continues her work to protect North Coast forests from clear-cutting. She hosts a biweekly hour-long slot on community radio station KZYX called "Truth to Power," a talk show on activism, local to global. In fall 1999 she helped organize a national meeting of environmental activists suffering corporate and government harassment, which led to the founding of a National Clearinghouse on Intimidation and Disruption to collect reports of harassment from all participating environmental groups and disseminate it to all interested parties, in the hopes that a coordinated response might discourage it. Littletree resides in Humboldt County, California.

BIBLIOGRAPHY

Hansen, Brian, "Life on the Front Lines," *Colorado Daily*, 1999; Littletree, Alicia, *Curve of the Earth*, forthcoming (audio compact disk); Littletree, Alicia, *Uprise Singing: Songs of Redwood Nation Earth First!*, 1995; "Redwood Summer Justice Project," http://www.monitor.com/~bari.

Lockett, Jackie

(April 30, 1944–)
Environmental Activist, Cofounder of Border Information & Solutions Network

Former Brownsville, Texas, city commissioner and grassroots toxics activist Jackie Lockett established the Border Information & Solutions Network in 1994 to promote—primarily via the Internet—sustainable development in the Brownsville region of the U.S.-Mexico border.

Jackie Frazier Lockett was born on April 30, 1944, in Brownsville, Texas. Lockett studied chemistry at Southwest Texas University, earning a B.S. in 1966 and an M.A. in 1967. Lockett taught chemistry at Texas Southmost College (which has now merged with the University of Texas at Brownsville) during the late 1970s.

However, her true community environmental activism developed once she retired from teaching and had the time to help lead a grassroots challenge to incineration of toxic wastes on incinerator ships in the Gulf of Mexico. Lockett and other members of the Gulf Coast Coalition for Public Health worked for eight years until the U.S. Environmental Protection Agency (EPA) finally denied any permits and "de-designated" the site in the Gulf for such activities. A high point in the work was a 1983 public hearing that the group organized, in which more than 6,000 residents participated. This was the largest public hearing attendance ever recorded by the EPA at the time. In 1987, Lockett helped found the Cameron County Local Emergency Planning Committee (LEPC) in response to the congressional mandate through the

Superfund Amendments and Reauthorization Act (SARA) Title III (now referred to as the Emergency Planning and Community Right to Know Act). In 1990, the Cameron County LEPC and the Matamoros Comité Local de Ayuda Mutua (CLAM) organized the first full-scale, international hazardous material response exercise on the U.S.-Mexico border.

Lockett was elected Brownsville city commissioner in 1992 for a short term. She was reelected in 1993, and she retained that post until 1997. During her tenure as commissioner, Lockett was appointed to the Government Advisory Committee of the North American Commission on Environmental Cooperation, the Cameron County Appraisal District Board, and the Lower Rio Grande Valley Development Council. As a city commissioner she supported improved urban planning, more extensive public involvement, improved emergency response, and better natural resource conservation. She was also a booster of a 1995 plan to develop "eco-industrial parks" in the area, industrial parks where factories are grouped to share environmental management services (wastewater treatment, recycling, and so on) and, in the best of cases, where waste from one factory can be used as raw material by another. The Brownsville virtual eco-park project led to a computer model that now allows the Brownsville Economic Development Council to discuss coproduction and waste-sharing with local industries and those new to the area. The identification of usable wastes being discarded in the area is expected to lead to the development of new business opportunities. Already, a few local manufacturers have incorporated the concepts into their manufacturing operations.

In 1994, Lockett and several collaborators, including Dr. Genaro López and Juan Carlos Cuellar, founded the Border Information & Solutions Network (BISN), a nonprofit organization dedicated to promoting sustainable development of the U.S.-Mexico border, especially along the southeastern tip of Texas. Through BISN, Lockett has organized several conferences to discuss sustainability and binational cooperation toward that goal. BISN interns Jesús Verduzco and Polo Bañuelos carried out a survey in 1997 of local high school and college students, asking them their opinions on local industry. They discovered that although the students admitted having little knowledge about local industry, their opinion was overwhelmingly negative about the effects of industry on the local environment. Because there are several factories in Brownsville and Matamoros, across the border in Mexico, that exemplify environmentally conscientious industry (such as Deltrónicos de Matamoros [Delphi], a Matamoros *maquiladora* that in 1997 recycled 72 percent of its production waste, and Chem-Pruf Doors, Inc., a manufacturer in Brownsville that recycled 90 percent of its waste), BISN proposed a Borderplex Industrial Awards Program. The program would increase community knowledge about local industry; reward achievements in health, safety, and environment; and promote careers in the environmental field for students. The city of Brownsville officially lent its support to the Industry Award Program, but it is currently on hold pending funding for coordination.

In 1998, BISN together with University of Texas–Austin senior Javier del Castillo offered a summer workshop for local students on city management and urban planning. The workshop's intent was to interest young people in these fields. From February 1998 through January 1999, BISN developed the Borderplex Environmental Information Center for the city of Brownsville at the Brownsville Public Library. The binational project provided information about municipal solid wastes, natural resources, and emergency planning to a binational audience through monthly meetings and newsletters. BISN is currently developing a Public Information Campaign for Cameron County and the Cameron County LEPC. Three schools along an international hazardous material truck route will learn about materials being transported and plans for appropriate emergency response.

For her work individually and with BISN, Lockett has been the recipient of several awards. Region VI of EPA presented Lockett with the Unsung Hero Award in 1990 for her work in coordinating the Binational Joint Response Team conference in Brownsville and with a

Certificate of Recognition in 1992. The Texas Natural Resource Conservation Commission and the Texas General Land Office awarded her certificates for outstanding service to the environmental community and preservation of natural resources. And she was named in 1998 a Paul Harris Fellow by the Rotary Foundation of Rotary International.

Lockett has been married to Dr. Ford L. Lockett, D.D.S., since 1967. They have two adult children, Renee and Ford III.

BIBLIOGRAPHY

"Border Information & Solutions Network (BISN)," http://www.bisn.org/; Mader, Ron, "Eco-Industrial Park," *El Planeta Platica* (http://www.planeta.com), 1995; Piasecki, Bruce, and Peter Asmus, *In Search of Environmental Excellence, Moving beyond Blame*, 1990; United States Environmental Protection Agency, "Successful Practices," *Title III Implementation, Chemical Emergency Preparedness and Prevention Technical Assistance Bulletin*, 1991.

Lopez, Barry

(January 6, 1945–)
Writer

As a writer of both fiction and nonfiction, Barry Lopez incorporates elements of imagination, science, and history to create a means of report that many readers find more accurate than work written within the confines of a single discipline. He finds that fiction and nonfiction combined can create an atmosphere in which truth can reveal itself, and he aspires to have

his writing serve this shamanic purpose. His growing body of work includes *Desert Notes/River Notes* (1976), *Of Wolves and Men* (1978), *Crow and Weasel* (1990), and *Field Notes* (1994). His 1986 book, *Arctic Dreams*, a contemplative natural history, is widely regarded as his masterpiece.

Born in Port Chester, New York, on January 6, 1945, Barry Holstun Lopez is the son of John Edward Brennan and Mary Frances (Holstun) Brennan, both journalists. When he was three years old, his family moved to southern California; at five, his parents divorced; when he was ten years old, Barry Lopez's mother married Adrian Bernard Lopez. As a boy, Barry Lopez remembers encounters with coyotes, rattlesnakes, deer, and bear. His imagination was captured by the intensity and clarity of these animals, and he idealized their ascetic life and its low impact on the desert landscape. Lopez's prowess with academics earned him a formal education at a Jesuit prep school in New York City and at the University of Notre Dame, where he was awarded an A.B. in 1966 and an M.A. in teaching in 1968. In addition to studying Christian theology, he scoured the full range of Western philosophy and the philosophy of science. Lopez felt that his education benefited him greatly; it endowed him with an appreciation for the rigors of scholarship, the questioning of authority, and a sense of living an ethical life. Yet after completing another year of graduate study in the creative writing program at the University of Oregon in 1969, Lopez felt that his learning was only about to begin. In his essay "The Language of Animals," published in *Wild Earth*, Lopez wrote, "There were other epistemologies out there as valid as the ones I learned in school. Not convinced of the superiority of the latter, I felt ready to consider these other epistemologies, no matter how at odds." This essential curiosity for the unknown or unacknowledged led Lopez in 1970 to commit himself to the life of a nature writer and explorer.

The subject of his first book-length work, *Of Wolves and Men* (1978), was born of this curiosity. In that book he explains his motives: "Let's say there are 8,000 wolves in Alaska. Multiplying by 365, that's about 3 million wolf-days of activity a year. Researchers may see something like 75 different wolves over a period of 25 or 30 hours. That's about 90 wolf-days. Observed behavior amounts to about three one-thousandths of one percent of wolf behavior. The deductions made from such observations represent good guesses, and indicate how incomplete is our sense of worlds outside our own."

Lopez sought to make the wolf less incomplete in his mind by reading all he could about them, then journeying into their territory. While there he became obsessed with the North American polar regions, and repeated excursions into the Arctic tundra resulted in two best-selling books: *Of Wolves and Men* and perhaps his greatest work, *Arctic Dreams: Imagination and Desire in a Northern Landscape* (1986), which won a National Book Award.

His expedition into the North's natural history became a spiritual journey. Studying and hiking the North affected Lopez's imagination profoundly. Icebergs became cathedrals. The bear was the embodiment of intelligence and physical power. Facts about snow geese, seals, and other mammals that traverse the Bering Strait were presented in the book alongside Eskimo myths and folktales. The resourcefulness and physical attributes of the animals of the Arctic amazed Lopez. He traveled the tundra with Eskimos, drawing inspiration from the early documentary filmmaker Robert Flaherty, who made the first full-length documentary film, "Nanook of the North," in 1922. Lopez's study was not anthropological; he didn't learn indigenous languages or hunt with Eskimos, as an anthropologist might have. *Arctic Dreams* is more poetic than anthropological. The reader learns as much about the inner workings of Barry Lopez as about the natural history of the polar regions. However the concluding chapters return to the pure scholarship entailed in writing such a book. Lopez questioned the motives of the ambitious early European expeditions and of Robert Peary, the American who reached the North Pole in 1909. Lopez criticizes these missions after defining, throughout the book, assorted dignified relationships with the land.

Uniting Native American ecological sensibilities and Western science's disciplined inquiry into the natural world is the trademark of Lopez's literary production. He possesses a keen sense for the moment when the geography of land and the geography of mind are synchronized, and he charts those epiphanous "sacred encounters" with the land. *Giving Birth to Thunder, Sleeping with His Daughter* (1977) is a book of allegorical fables that uses American Indian folklore's archetypal trickster character, Coyote. In Lopez's tales, Coyote plays a role that readers can identify as a typically human role: interventionary, impulsive, and ever providing the impetus for situations to which the animal world must adapt or die. In his trilogy *Desert Notes* (1976), *River Notes* (1979), and *Field Notes* (1994), the characters Lopez develops are ones with which he has personal familiarity, aspects of himself perhaps, the transformed scholar. Eccentric academic characters are altered by experiences with nature. In *River Notes*, a visionary naturalist has spiritual insights revealed to him through nature. In the story "Homecoming" in *Field Notes*, the ambi-

tion-driven botanist has to stop his career in its tracks to renew his faith and devotion to the land itself. But the land, as much a character as anyone else in his books, is not always a benevolent guru. As in some superstitious tectonic beliefs of our indigenous ancestors, Lopez invokes earthquakes, floods, droughts, and general discord and disharmony to follow the sacrilege of broken promises to the land. In the intermingling of the natural and supernatural, Lopez's fiction can resemble the magical realism of Latin American writers Gabriel García Márquez and Carlos Fuentes.

Barry Lopez can be seen as an Emersonian transcendentalist in that he believes that natural facts reveal spiritual truths and that nature holds power to transform the individual. Like most contemplative nature writers, Lopez treats the mingling of the external landscape and the internal human mind at work. He has likened the awe and wonder that one feels in nature with the passions of falling in love.

Barry Lopez's travel writing appears in literary and environmental journals as well as *Harpers*, *National Geographic*, and the *New York Times Magazine*. His poetic coverage of diverse bioregions—the Great Plains of the Dakotas, Florida's Gulf Coast, the world beneath the ice of Antarctica—make him one of the most popular American nature writers and the recipient of numerous literary awards. Lopez lives near the Mackenzie River on the western slope of the Cascades in Oregon.

BIBLIOGRAPHY

Elder, John, ed., *American Nature Writers*, Vol. 1, 1996; Lopez, Barry, *About This Life: Journeys on the Threshold of Memory*, 1998; Lopez,

Barry, *Crossing Open Ground*, 1978; Lopez, Barry, "The Language of Animals," *Wild Earth*, 1998–1999; Lopez, Barry, *Winter Count*, 1976; Murray, John A., "About this Life: A Conversation with Barry Lopez," *The Bloomsbury Review*, 1998; Sherman, Paul, "Making the Turn: Rereading Barry Lopez," *For Love of the World: Essays on Nature Writers*, 1992.

Lovejoy, Thomas

(August 22, 1941–)
Tropical Ecologist, Counselor for Biodiversity and Environmental Affairs at the Smithsonian Institution

Thomas Lovejoy is known for his effective activism about tropical deforestation. During the 1980s, Thomas Lovejoy was one of the main voices raising public concern about the destruction of tropical rain forests. He also proposed what has become a successful way to combat tropical deforestation. Through what he called "debt-for-nature swaps," indebted countries are forgiven a portion of their foreign debt in return for preserving from development an ecologically rich area. Also a respected tropical ecologist, Lovejoy began his Brazil-based Biological Dynamics of Forest Fragments Project in the late 1970s to study the loss of biodiversity when only small patches of forest are left standing. The results of this experiment, which will continue well into the twenty-first century, will help determine how much biodiversity a small and isolated wilderness area can retain.

Thomas Eugene Lovejoy was born in New York City on August 22, 1941, into a wealthy family that owned the Manhattan Life Insurance Company. He was sent to the private boarding school Milbrook, where the founder of the school's zoo, Frank Trevor, encouraged Lovejoy to study field biology. Lovejoy was especially interested in birds. At Yale University, Lovejoy studied under the eminent ecologist Evelyn Hutchinson, who instructed him about the subtleties and complexities of field biology. Lovejoy received a B.S. from Yale in 1964 and a Ph.D. in 1971. His Ph.D. research was carried out in Belem, Brazil, near the mouth of the Amazon River.

Lovejoy joined the Academy of Natural Sciences upon finishing his Ph.D., serving as executive assistant to the science director and assistant to the vice president of resources and planning. In 1973, he moved to the World Wildlife Fund–U.S. (WWF-U.S.), where he remained until 1987, acting as program director (1973–1978), vice president for science (1978–1985), and executive vice president (1985–1987). Lovejoy has also served as chairman of the Wildlife Preservation Trust International and as a member of the Species Survival Commission of the International Union for the Conservation of Nature. Currently Lovejoy is counselor for biodiversity and environmental affairs at the Smithsonian Institution.

During the late 1970s, Lovejoy initiated what has become one of the largest and longest-term biology field experiments. He was interested in expanding on the island biogeography hypothesis of biologist E. O. WILSON and mathematician Robert MacArthur, which held that the number of species on oceanic islands was predictable and based on the size of the island and its distance from other landforms. Facilitated by a Brazilian law requiring land developers to leave 50 percent of their plots under forest cover, Lovejoy convinced Amazonian cattle ranchers to leave rain forest "islands" of certain sizes sprinkled through their cattle pastures. Working with biologist Rob Birregaard and several Brazilian colleagues, Lovejoy set up the Minimum Critical Size of Ecosystems Project (later renamed the Biological Dynamics of Forest Fragments Project), with 24 forest patches of one, ten, 100, and 200 hectares and a 10,000-hectare control site. Hundreds of researchers have used the plots, and 20 or 30 of them are there at any given time. Although the experiment will continue indefinitely, researchers have already found that the smaller the plot, the more severe the decline of species. Trees die along the edge of the plots, desiccated by the surrounding dry pastures. As they die, so do the animals that depend upon them for food and shelter. What has come to be called the "edge effect" has important implications for those who plan wildlife reserves.

During the mid-1980s, Lovejoy first raised the alarm about tropical deforestation, becoming what Congressman Timothy Wirth called "the Tom Paine of the Rainforest." With his social and political connections, as well as his firsthand knowledge about the political and economic workings of indebted tropical countries, Lovejoy invented and then proposed in the *New York Times* the idea for debt-for-nature swaps. What happened afterwards illustrates an underlying current in Lovejoy's career: he proposes grand ideas, then lets others fine-tune them and implement them. This happened with the Forest Fragments project and with debt-for-nature swaps, dozens of which have been arranged by groups such as Conservation International and the World Wildlife Fund and by several European governments.

Another of Lovejoy's particular abilities is to attract the rich and famous to rain forest conservation. Among those who have visited his Forest Fragments project in Brazil and/or become his influential allies are Sting, Robert Redford, Tom Cruise, Olivia Newton John, and many U.S. politicians. Lovejoy felt that if he could convince at least 15 senators of the importance of tropical conservation, that would be a critical mass for passing important conservation legislation. The late Republican senator H. John Heinz III of Pennsylvania said of Lovejoy that "he makes believers of skeptics," and biologist E. O. Wilson credits Lovejoy with building bridges between science and the public. Lovejoy resides outside of Washington, D.C.

BIBLIOGRAPHY

Gaither, Rowan, "The Natural," *New York*, 1991; Laurence, William F., "Fragments of the Forest," *Natural History*, 1998; Misler, Rachel, "Thomas E. Lovejoy," http://www.si.edu/resource/direct/lovejoy.htm; "Tom Lovejoy and the Last Crusade," *GQ*, 1989; Sun, Marjorie, "How Do You Measure the Lovejoy Effect?" *Science*, 1990.

Lovins, Amory, and Hunter Lovins

(November 13, 1947– ; February 26, 1950–)
Cofounders of Rocky Mountain Institute

Amory Lovins and Hunter Lovins are cofounders of Rocky Mountain Institute (RMI), an independent, nonprofit research and implementation organization that fosters the efficient and sustainable use of resources as a path to global security. Amory Lovins proclaimed that energy waste was a crippling economic and environmental problem in the early 1970s—before the oil embargo of 1972 woke the rest of the world up to that fact—and he innovated the conceptual, technical, and business framework for what has become the $5 billion electricity-saving (or "negawatt") industry in the United States. The Lovinses and their colleagues continue to insist that the solution is not to build more power plants to supply a growing appetite for energy, but rather to reduce society's use of energy through innovation and advanced energy efficiency. The Lovinses and the staff of Rocky Mountain Institute combine sophisticated yet cost-efficient technologies and practices to promote "elegant frugality" with better service and lower cost.

Amory Lovins was born on November 13, 1947, in Washington, D.C., to Gerald and Miriam Lovins, a scientist and social services administrator, respectively. He was a frequent participant in national and international science fairs during high school, and one experimental physics project, "Method and Means for Detecting Nuclear Magnetic Resonances," even received a U.S. patent in 1965. Lovins attended Harvard College as a Presidential Scholar for two years but then transferred in 1967 to Magdalen College at Oxford University as an advanced student in theoretical physics. At Oxford's Merton College, where Lovins became a don in 1969, he requested permission to do his doctoral research on energy and resource policy, but he was refused, because at that time energy was not considered worthy of academic study. Rather than pursue another related field, Lovins left Oxford in 1971 to work independently. Serving as the British representative to DAVID BROWER's Friends of the Earth, Lovins pulled off a successful conservation campaign. The mountains surrounding Snowdonia National Park in northern Wales were threatened with copper stripping by the world's largest mining company. Lovins took photographs and wrote text for the coffee-table volume *Eryri: The Mountains of Longing*, which was supplemented by a BBC television program. Public outrage about the threatened destruction, spurred by Lovins's work, halted the project.

Lovins remained headquartered in London throughout the 1970s. During that decade, he worked as an international consultant on energy and resource policy and its link to development, security, and the environment. A 1976 article in *Foreign Affairs* redefined the energy problem, with profound effect: he recommended choosing the cheapest ways (typically efficient use, then appropriate renewable resources) to provide energy in the right amount, quality, and scale to do each desired task. Lovins's 1977 book *Soft Energy Paths: Toward a Durable Peace* inspired a new generation of scien-

Amory and Hunter Lovins (Courtesy of Anne Ronan Picture Library)

tists and decision makers and steered energy companies away from over a trillion dollars in needless investments.

Hunter Sheldon Lovins was born in Middlebury, Vermont, on February 2, 1950, to Farley Hunter Sheldon and Paul Millard Sheldon and was raised in the mountains east of Los Angeles and in the Roaring Fork Valley of Colorado. She learned to ride horses before she could walk and began to ride in rodeos as a teenager, an activity she still enjoys. Her family embarked on camping trips through the western United States and Mexico for weeks at a time, and whenever they witnessed environmental problems, her parents encouraged their children to think about constructive solutions, rather than

complaining about the problems. Lovins attended Pitzer College in California, graduating in 1972 with a B.A. in political studies and sociology. She served as assistant director of the California Conservation Project, from 1973 to 1979, attending Loyola law school and receiving her J.D. in 1975. She was admitted to the California bar in 1975 and is still a member. While directing a Los Angeles–based conservation organization called TreePeople in 1976, Lovins came across the energy-efficiency work of Amory Lovins and began to refine its presentation so that it would be more accessible to laypeople. The chief economist of the Atlantic Richfield oil company introduced the two in 1977, and they decided to marry and to integrate careers in 1979. The Lovinses became policy advisers for Friends of the Earth, posts they held until they moved to the rural valley of Old Snowmass, Colorado, to create Rocky Mountain Institute in 1982.

Housed in a superefficient building, which the Lovinses designed themselves to showcase energy conservation design and which was built by a group totaling over 100 volunteers, RMI is an organization with about 45 staff members working on issues of natural capitalism; superefficient use of energy, water, and transportation; profitable climate protection; green real estate development; community economic renewal; global security; and related issues. In all of these fields, RMI's focus is on profitable routes to sustainability. Because its research has immediate and recognized value, RMI is able to provide for half of its budget by selling information and consulting services to industry and many levels of government. RMI has formed and spun off four for-profit companies that sell the information it generates. E source, for example, provides technical information on advanced electric efficiency; it was sold in 1999 to the *Financial Times* group. Private donations and research grants provide the rest of RMI's income.

RMI has confidence in the ability of a savvy and efficient private sector to embrace the goal of sustainability; business has been an eager client of RMI's money-saving advice for energy efficiency. RMI's Hypercar Center® and its spinoff, Hypercar, Inc., for example, have invented the ultralight hybrid-electric Hypercar™ that would use four to eight times less energy than today's cars, emit nothing but hot drinking water, be superior in all respects, and sell for a competitive price.

Such cars differ from currently existing electric cars in that they lack heavy batteries and instead are driven by an electric motor powered by a small engine, turbine, or fuel cell on board. Amory Lovins, who first conceived the Hypercar™ concept, envisions stations where Hypercars™ could be parked with their fuel cell left running and plugged in, so they send back to the utility grid enough electricity to repay up to half their lease cost. Hunter and Amory Lovins shared the 1993 Nissan Prize for the Hypercar™, and by 1999 the automobile industry had already invested five billion dollars in Hypercar™ research and development.

RMI is working on many other technical innovations as well. Amory Lovins told interviewer David Kupfer of *The Progressive* about "an inexpensive, easy-to-make, easy-to-fix photovoltaic ultraviolet disinfectant for water. Water circulates through a trough in thin layers and gets exposed to hard ultraviolet light made by a solar-powered lamp. The water that comes out is free of disease-causing bacteria and viruses. Once it's deployed in a few years, this gadget will save millions of babies now dying of dysentery and similar water-borne diseases." And he also described "superwindows . . . that can let in light without unwanted heat, so that in very hot or cold climates, we can eliminate heating or cooling equipment, and therefore save enormous amounts of energy, pollution, and capital costs. In RMI's headquarters, we're growing our eighteenth passive-solar banana crop with no furnace despite outdoor temperatures as low as minus-forty-seven degrees Fahrenheit, and it was cheaper than usual to construct." As of late 1999, they were growing the 27th banana crop.

The Lovinses together with RMI staff and their collaborators continue to generate unorthodox ideas that make perfect sense for solving world problems. Amory Lovins told *Whole Earth's* PETER WARSHALL in 1998 that the challenge of new solutions usually involves defying status quo, be it in the form of bureaucracy, rigid specifications for technology, or accepted yet inefficient ways of doing things. The Lovinses have authored—individually, together, and with other writers—several hundred articles and 27 books. Their most recent book, *Natural Capitalism* (1999), written with sustainability expert PAUL HAWKEN, explains new business practices, being rapidly adopted, that restore the earth while providing striking profitability and competitive advantage. The book has received widespread praise for its scope and innovative suggestions.

Amory Lovins received the 1999 World Technology Award (Environment), the 1997 Heinz Award, the Onassis Foundation's first Delphi Prize, and a 1993 MacArthur Fellowship. The *Wall Street Journal* predicted that he would be one of 28 people worldwide "most likely to change the course of business in the '90s"; *Newsweek* has called him "one of the Western world's most influential energy thinkers"; and *Car* magazine ranked him the 22nd most powerful person in the global automotive industry. With Hunter Lovins, he has shared the 1999 Lindbergh Award, the 1983 Right Livelihood Award (often called the Alternative Nobel Prize), and a 1982 Mitchell Prize. Hunter Lovins received Loyola's Alumni Award for Outstanding Service in 1975. The Lovinses amicably divorced in 1999, but they continue their strong collaboration as co–chief executive officers of RMI and as frequent coauthors.

Amory Lovins spends his free time playing and composing for the piano, climbing mountains, tying knots, taking landscape photographs, reading poetry, and studying languages. Hunter Lovins and her partner, Robbie Noiles, buy, sell, and train horses through their own Nighthawk Horse Company. They play polocrosse, which is like lacrosse but played on horseback, and their team placed second in the 1997 national championship. She helps out on neighboring ranches and rides rodeo, competing in barrel racing. She also works as a fire/rescue/emergency medical technician with her local fire department.

BIBLIOGRAPHY

Kupfer, David, "Amory Lovins," *The Progressive*, 1995; "Rocky Mountain Institute," http://www.rmi.org; "The Sage of Old Snowmass," *The Economist*, 1997; Warshall, Peter, "Lock-in," *Whole Earth*, 1998.

Lyons, Oren

(1930–)
Onondaga Chief

Faithkeeper of the Onondaga Council of Chiefs, Oren Lyons is also an artist, Hall of Fame lacrosse player, college professor, and internationally recognized leader of environmentalists and indigenous rights activists. Lyons has written and spoken widely about links between ecological crises and suppression of indigenous peoples. He has appeared at numerous international conferences, testified before Congress, and addressed the United Nations General Assembly. He edited an important study of American Indians and democracy, *Exiled in the Land of the Free* (1992), and contributed essays to several other collections, including Bill Willers's *Learning to Listen to the Land* (1991) and Christopher Vecsey's and Robert Venables's *American Indian Environments* (1980). He is professor of American studies at the State University of New York at Buffalo.

Oren Lyons was born in 1930 on Onondaga Indian land in upstate New York, with the name Jo-Ag-Quis-Ho (Bright Sun Makes a Path in the Snow). Lyons was the oldest of eight children. He dropped out of school in the eighth grade and after two years in the army, was earning his living by painting portraits of boxers for local bars. Lyons was also well-known for his abilities at lacrosse, a game invented by Lyons's Iroquois ancestors. The coach of the Syracuse University lacrosse team invited the head of the College of Fine Arts to view Lyons's painting of Jack Dempsey, and despite Lyons's lack of formal education, he was admitted to the university. He was an All-American lacrosse player, and during his senior year, with National Football League Hall

of Fame running back Jim Brown on the team, Syracuse went undefeated. (Lyons was elected to the Lacrosse National Hall of Fame in 1993.) He graduated from the College of Fine Arts of Syracuse University in 1958 and was awarded the Orange Key as an outstanding scholar-athlete.

Lyons moved to New York City to pursue a career as a commercial artist. He eventually landed a job with Norcross Greeting Cards, where he rose to become art and planning director, a job that entailed supervising 200 artists. His own painting also gained respect, and Lyons exhibited widely. In 1967, the clan mother of the Turtle Clan chose Lyons to be a faithkeeper, and he returned home to Onondaga. The Onondaga are one tribe in the Iroquois Confederation, and as a tribal chief, Lyons has long fought for recognition of Iroquois sovereignty. He has worked to exempt the nation from federal taxes, helped put forth claims for reparations for stolen lands, and served in the Traditional Circle of Tribal Elders, a council of leaders of Indian nations of North America. He has been particularly active in United Nations work around indigenous people's rights and helped establish the UN Working Group on Indigenous Populations in 1982. In 1990 he was a negotiator between the Mohawk Indians and Canadian government authorities in the standoff at Oka, Quebec, and helped the confrontation end peacefully, with both sides agreeing to look for long-term solutions to Mohawk demands.

The long view is central to Lyons's work and philosophy, shaped fundamentally by Onondaga beliefs. A consistent theme in his work on environmental issues is the need to care for future generations, to make decisions for "the seventh generation to come." He also stresses the belief that humans are connected to and equal with all other things, so that, for instance, how we treat other species will determine human fate as well. Thus Lyons compares Native Americans to the passenger pigeon and the wolf and suggests that Euramericans need to understand that the way the land is treated will affect their own survival.

Lyons has worked as both an activist and a spiritual leader and argues that the separation between the two is artificial and destructive. He has been active in the effort to demonstrate the influence of the Iroquois Confederacy on the Constitution of the United States and emphasizes the spiritual and political connections between the United States and Iroquois nations. Lyons has discussed his ideas in a wide range of publications and events and is a powerful speaker. In 1988 he addressed the first Global Forum of Spiritual and Parliamentary Leaders, held in England. The conference was attended by many influential leaders, including the Dalai Lama and Mother Teresa, and Lyons's speech brought the crowd to its feet. He helped convince the organizers to focus their 1990 meeting in Moscow entirely on environmental issues. In 1992 he led a delegation of the Iroquois Confederacy to the United Nations Conference on Environment and Development in Rio de Janeiro. Lyons's has been an influential voice in discussions of economic development, pointing out that "sustainable development" has often meant destruction of life and land for indigenous peoples. Lyons has worked to protect the Onondaga from environmental threats; in 1999 the Onondagas petitioned the Environmental Protection Agency for an "environmental justice investigation" of a plan to develop a gravel mine at the headwaters of Onondaga Creek.

Lyons's recent projects include a proposal to return stolen land, including the entire city of Syracuse, New York, to the Onondaga. In February 2000, he spoke at a national leadership summit, calling for presidential candidates to integrate spiritual and political concerns in an authentic way, rather than with simple statements of religious piety. He called on presidential hopeful ALBERT GORE, JR. to honor his environmental ideals with real action. Gore owns stock in Occidental Petroleum, which currently threatens the U'Wa people of South America. The corporation wants to pursue oil exploration on U'Wa land, and the Colombian military is conducting a campaign to force the U'Wa to allow exploitation of land they hold sacred. The case integrates several of Lyons's principal concerns: indigenous rights, environmental destruction, and the need for balance with natural and spiritual laws.

Lyons has received numerous awards and honors for his work in these areas. In 1993, he was awarded the Audubon Medal by the National Audubon Society. In 1995 he was given the Elder and Wiser Award by the Rosa Parks Institute for Human Rights and Self-Development. Oren Lyons is director of the Native American Studies Program at the State University of New York at Buffalo.

BIBLIOGRAPHY

Cornell, George, "Blight Knows No Border," *Los Angeles Times*, 1990; Moyers, Bill, *Oren Lyons the Faithkeeper* (PBS documentary), 1991; Schneider, Paul, "Respect for the Earth," *Audubon*, 1994; Tucker, Toba, *Haudenosaunee: Portraits of the Onondaga Nation*, 1999.

MacKaye, Benton

(March 6, 1879–December 11, 1976)
Regional Planner

Regional planner Benton MacKaye, a cofounder of the Wilderness Society (TWS), was the primary force behind the Appalachian Trail, built during the 1920s and 1930s to link a 2,000-mile corridor of mountain wilderness stretching from Maine to Georgia. He was a leader among regional planners, issuing perennial reminders in his writings and group discussions that planners should work toward improved habitability of the earth, which MacKaye, following his mentor Patrick Geddes, called "geotechnics."

Benton MacKaye was born on March 6, 1879, in Stamford, Connecticut, one of five children whose father, Steele MacKaye, was a well-known actor, director, and playwright. The family moved frequently throughout the northeastern United States but established a fixed summer home in the village of Shirley Center, Massachusetts, in the late 1880s. Steele MacKaye organized summer pageants, writing original plays and directing local actors for performances in the Shirley Center town hall. As a child, MacKaye was exposed to the best of urban centers: while living in Washington, D.C., he visited the Smithsonian Institution frequently and met personally such renowned explorers as Adm. Robert Peary and Major JOHN WESLEY POWELL. He also benefited from rural Shirley Center: he spent his teenage summers on mini-"expeditions" modeled after those of Peary and Powell, assiduously mapping forests, charting the course of the Squannacook River, and observing birds within a four-mile radius of his home. These childhood experiences were later to form the base for MacKaye's belief that for a healthy civilization, humans must have access to three types of environs: urban, rural, and "primeval" or wilderness.

MacKaye studied forestry at Harvard University, earning his B.A. in 1900 and his A.M. in 1905. He worked for GIFFORD PINCHOT's U.S. Forest Service for the first 13 years of his career, surveying forested areas in New England and the South to be incorporated into the National Forest system. Thanks to one of his reports, 30,000 acres of land in the White Mountains became the first parcel in the White Mountain National Forest. During the remainder of his career, he also served as

Benton MacKaye (Wilderness Society)

forester or planner for several other government agencies, including the Department of Labor, the Geological Survey, the Bureau of Indian Affairs, and the Tennessee Valley Authority. He also taught at the Harvard Forest School for many years. But it was as a member of independent organizations and intellectual groups that MacKaye distinguished himself.

During the early part of the 1900s, MacKaye developed a plan for what would blossom as the Appalachian Trail. The article he wrote for the *Journal of the American Institute of Architects* in 1921, "An Appalachian Trail: A Project in Regional Planning," proposed a 2,000-mile trail along the mountainous wilderness belt between Maine and Georgia. MacKaye suggested that the trail be constructed and maintained by local or regional groups and that the government protect the area in its wild state. The idea caught on almost immediately, and branches of the new Appalachian Trail Club formed to build their segments of the trail. By 1937, the trail was completed, and in 1938 the federal government moved to protect it.

MacKaye worked with a group of regional planners during the 1920s who called themselves the Regional Planning Association of America (RPAA). They criticized what they felt were the overdeveloped urban centers of the time, whose thirst and hunger were like tentacles reaching out and depleting nearby rural areas. Inspired by the Garden Cities movement of England, they proposed industrialized yet park-filled cities of a moderate size that would exemplify geotechnics or improved habitability. A 1925 issue of *Survey Graphic* magazine featured the ideas of RPAA members and

left its indelible print on the fields of regional planning and urban design. Fifty-one years later, RPAA member Lewis Mumford called the articles an accurate diagnosis of "the disorders of random metropolitan congestion and suburban scattering."

Despite being far ahead of their time in that respect, in another aspect the group neglected to foresee a problem with their all-out promotion of the automobile. The private car was seen by the RPAA school as a useful vehicle that could quickly carry urban inhabitants to rural and primeval settings. MacKaye proposed, in a 1931 article for *Harpers*, the "townless highway," which was basically the prototype for the modern freeway. What none of the RPAA planners foresaw, wrote Mumford in his 1976 tribute to MacKaye, was that the automobile would "wreck our far more efficient railroad system and in many areas completely wipe out virtually all public transportation, with all its useful auxiliary services—telegraph offices, taxis, baggage and regular freight deliveries."

In 1935, MacKaye, along with seven other proponents of wilderness protection, cofounded the Wilderness Society. All of the cofounders, who included Robert Marshall, Harvey Broome, Harold Anderson, Aldo Leopold, Ernest Oberholtzer, Bernard Frank, and Robert Sterling Yard, concurred that wilderness was an important antidote to increasingly inhumane and uninhabitable large urban areas. MacKaye served as president of the Wilderness Society from 1945 to 1950 and then as honorary president from 1950 until his death. His long life afforded him the enjoyment of seeing TWS grow from a tiny, underfunded fledgling organization to one of the most influential conservation lobbies in the United States.

MacKaye spent his many sabbaticals between government jobs and writing projects in Shirley Center. It was there that he spent his final years, under the affectionate care of his friends and especially his next-door neighbor, Lucy Johnson. He died at the age of 97, on December 11, 1976, in Lucy Johnson's Shirley Center home.

BIBLIOGRAPHY

Bryant, Paul T., ed., *From Geography to Geotechnics*, 1968; MacKaye, Benton, *Expedition Nine: A Return to a Region*, 1969; Mumford, Lewis, Stuart Chase, George Marshall, Paul H. Oehser, Frederick Gutheim, Harley B. Holden, Paul T. Bryant, Robert M. Howes, and C.J.S. Durham, "Benton MacKaye: A Tribute," *Living Wilderness*, 1976; Oehser, Paul H., "On Benton MacKaye's Centenary," *Living Wilderness*, 1979.

Mader, Ron

(November 8, 1963–)
Journalist, Web Publisher

Ron Mader is an environmental journalist specializing in ecotourism and sustainable development in Latin America as well as the Mexico–United States borderlands. His expansive "Planeta.com: Eco Travels in Latin America" (http://www.planeta.com/), which he established in 1995, is probably the most complete Internet source for this type of information.

Ronald Earl Mader was born in Fort Wayne, Indiana, on November 8, 1963. He became interested in travel and the environment as a child thanks to his parents—both schoolteachers—who took Ron and his older siblings, John Carl and Cheryl Barbara, on cross-country treks, visiting national parks and heritage sites in the United States. In Fort Wayne, Mader was an active member of the Fort Wayne Astronomical Society and a radio broadcaster, producing a book review program ("Page 35") for radio station WBNI. Mader studied telecommunications and film studies at Indiana University–Bloomington, graduating with a B.A. in 1986.

After graduation, Mader relocated to Los Angeles, California, where he attempted to sell his own movie scripts. Finding real life—and the stories he was hearing from Central America and Mexico—more interesting than his fictional work, he decided to refashion his career by improving international journalism and news coverage from the Americas. In 1988 Mader moved to Austin, Texas, and began studies at the University of Texas at Austin, where he earned a master's degree from the Institute of Latin American Studies in 1990. As Mader recounted in a 1999 interview, "I knew how to communicate but I didn't know what to say. Refocusing my career in Texas provided me with the impetus to make an impact in how U.S. citizens could improve their understanding of events and issues in Latin America."

Mader worked first as a reporter and editor for the *Mexico City News* (1992–1993) and then forged a successful freelance career, focusing on the U.S.-Mexico border, Mexico, and ecotourism in Latin America in general. He has written stories about economic and environmental problems and their potential solutions for such publications as *Transitions Abroad, Texas Environmental News, Honduras This Week, Mexico City News*, and *South American Explorer.* Mader is a member of two professional organizations: Mexican Writer's Alliance and the Society of Environmental Journalists.

Mader is best known in the tourism profession for the quarterly journal he published between 1994 and 1999, *El Planeta Platica*, now known by its down-sized title, *Planeta.* The journal included indepth information written by experts worldwide on sustainable development (economic development that can continue for generations and that empowers local communities), ecotourism (tourism that provides material assistance to conservation projects, empowers local communities, and is capable of sustaining itself financially), and environmental conservation throughout Latin America, one of the world's hot spots for biodiversity.

Once Mader realized the potential of on-line publishing for reaching interested readers in a timely and economical way, he founded a web site to archive back issues and provide links to other sources of information on similar topics. The "Planeta.com: Eco Travels in Latin America" web site (http://www.planeta.com/) is a rich, continuously updated source of data for students, researchers, travelers, businesspeople looking for like-minded contacts, and others. Mader developed this web site to promote what he calls "decentralized communications," putting environmentalists and those interested in the environment in contact with one another. Too often, information clearinghouses serve as filters and gateways rather than distributors of information, he laments. Besides being an archive, Planeta.com is an active forum for discussion. It has received recognition and awards from the *Dallas Morning News*, the *Miami Herald, PC Computing*, and most recently the Mexican government, which awarded Mader the Lente de Plata (Silver Lens) award for the web site's exhaustive coverage of Mexican ecotourism and environmental travel options.

In addition to his on-line work, Mader has written two ecotourism-oriented guidebooks, *Mexico: Adventures in Nature* (1998) and *Honduras: Adventures in Nature* (coauthored with James D. Gollin, 1998). These books, which emphasize the natural history–oriented destinations within each country, offer information about the history, culture, economy, and ecology of the respective nation and instruct visitors on how to keep their impact as positive as possible.

Mader currently lives in Mexico City and works as a correspondent for *Texas Environmental News* and *Transitions Abroad* magazine, in addition to his work hosting the Planeta.com web site. He is active on the lecture/conference circuit, continuously working to encourage dialogue on how ecotourism can contribute to sustainable development.

BIBLIOGRAPHY

"Eco Travels in Latin America," http://www.planeta.com/; Mader, Ron, "Bypassing the Power Structure (in Mexico)," *Forbes*, 1997; Mader, Ron, "Latin America's New Ecotourism," *Honduras This Week*, 1999.

Mander, Jerry

(May 1, 1936–)
Writer, Advertising Executive

Jerry Mander is a renowned advocate for ecological and social causes, including native rights. He is well known for his marketing work—he spent 15 years as a successful marketer, ultimately serving only nonprofit social groups—and has been described by the *Wall Street Journal* as "the Ralph Nader of advertising." Mander is the author of *Four Arguments for the Elimination of Television* (1978) and *In the Absence of the Sacred: The Failure of Technology and the Survival of the Indian Nations* (1991), among others. Mander continues to fight both the encroachment of technology in society and economic and industrial globalization.

Jerry Mander was born on May 1, 1936, in New York City to Jewish immigrants from eastern Europe. Although he was urged by his parents to take over his father's business in the garment industry, Mander found himself drawn to the flash and glamour of the marketing field, leading him to obtain a B.S. in economics from Wharton School of Business at the University of Pennsylvania in 1958. In 1959, he concluded his studies with an M.S. in economics from Columbia Graduate Business School.

During 1961 and 1962, Mander worked as assistant director of the San Francisco International Film Festival. In 1962, he formed Jerry Mander & Associates, a public relations firm of which he was president until 1965, when he joined a celebrated advertising company that became Freeman, Mander, and Gossage. As president and partner of the firm, Mander spearheaded its many successful campaigns for environmental and social advocacy groups, including the Sierra Club. Freeman, Mander, and Gossage authored such successful campaigns that Robert Glatzer of *New Advertising* credited the ads with "starting the whole ecology boom." The advertisements were characterized by "coupons" that could be torn out and mailed to political leaders. The most notable campaigns were instrumental in keeping dams out of the Grand Canyon and establishing, among other preserves, Redwood National Park.

Jerry Mander (Courtesy of Anne Ronan Picture Library)

In 1967, Mander wrote *The Great International Paper Airplane Book* with George Dippel and Howard Gossage. A simple account of the first Scientific American International Paper Airplane Contest, the book includes pseudoadvertisements written by Mander. The accusations that Mander and company were subtly undermining U.S. production of supersonic transport aircraft (SSTs) may have risen from the fact that an ad for the contest with cynical references to the SST appears on a page facing a real Lockheed ad advocating SST production. A U.S. SST was never built.

Conflicted interests became more apparent with the success of Mander's social campaigns, and large corporations such as auto manufacturers dissolved their accounts with the firm. The absurdity of the tenets and goals of marketing pressed hard on Mander, and the firm dissolved in the summer of 1972. That same year, he worked to create Public Interest Communications (PIC), the country's first nonprofit advertising agency. PIC catered exclusively to community, environmental, and social action groups and launched successful, albeit unprofitable, campaigns. In 1974 Mander took a leave from advertising to address his growing concern with the problems of marketing, and especially with television.

In 1978 Mander completed *Four Arguments for the Elimination of Television.* This potent critique of a widespread and readily accepted medium asserts that television is intrinsically dangerous. Mander's marketing experience gave him a deft understanding of how corporations use the power of television as a homogenizing force. He argues that television invades individual and mass consciousness and is so viciously harmful to personal sanity and health that it must be abolished.

For the next decade Mander worked on a book to follow *Four Arguments*, while continuing to advocate and work for many ecological and social action groups. Since 1980, he has been senior fellow at the Public Media Center, a group that has campaigned for Sierra Club, Greenpeace, Planned Parenthood, Friends of the Earth, and others. Mander continued to be influenced by his increasing ties with activist groups and especially by the struggles of Native American peoples. In 1991 his book, *In the Absence of the Sacred: The Failure of Technology and the Survival of the Indian Nations*, was published. The book grapples with two related issues, the flaws of technological society and native people's struggles against its encroachment. The critique of technological society revolves around a more widespread application of the attitude that Mander presented in *Four Arguments*. In this book, technology and the mechanization of society are accused of provoking depression, alienation, and social atrocities in modern society. The book discusses the subjugation of native populations around the world and the assault waged against them by technological society.

Mander subsequently cofounded the International Forum on Globalization, an organization of activists and writers worldwide. In 1996, Mander coedited (with British ecology movement founder Edward Goldsmith) *The Case against the Global Economy: And for a Turn toward the Local.* This volume presents 43 essays that decry economic and industrial globalization; contributors describe how international agreements such as the North American Free Trade Agreement and the General Agreement on Tariffs and Trade have led to stepped-up plundering of natural resources and disregard for human rights, especially in poor countries.

Mander continues to critique society and the expansion of the techno-corporate machine. He is program director of the Foundation for Deep Ecology and lives in San Francisco, California. He contributes articles to periodicals including *Co-Evolution Quarterly, City Magazine, San Francisco Chronicle,* and others.

BIBLIOGRAPHY

Mander, Jerry, "The Dark Side of Globalization: What the Media Are Missing," *The Nation,* 1996; Mander, Jerry, "Internet: The Illusion of Empowerment," *Whole Earth Review,* 1998; Mander, Jerry, *In the Absence of the Sacred: The Failure of Technology and the Survival of the Indian Nations,* 1991; Mander, Jerry, and Edward Goldsmith, eds., *The Case against The Global Economy: And for A Turn toward The Local,* 1996.

Manning, Richard

(February 7, 1951–)
Journalist, Author

Montana-based author Richard Manning has written about many important environmental issues pertaining to the western landscape and follows his convictions beyond his writing and into his daily life. While working as a reporter, he became one of the first journalists in the Northwest to undertake a major investigative report on the logging business and wrote a series critical of the timber industry that ended up costing him his job. He and his wife designed and built an environmentally sensitive house in western Montana using place-conscious methods and materials such as timber framing, an on-demand water heater, earth-sheltering, and a composting toilet. He chronicles the entire building process in *A Good House: Building a Life on the Land* (1993); he has also written books about his timber industry investigations, the land use history of native prairies, and the threat of mining to a Montana river.

Richard Dale Manning was born on February 7, 1951, in Flint, Michigan. He lived there until the age of five and then lived briefly in Tennessee and Ohio before returning at the age of ten to Alpena, Michigan, the hometown of his parents. Manning did well in school, earning a scholarship at the University of Michigan, where he enrolled in 1969. He originally intended to become a doctor but became involved in antiwar activism and ended up studying political science. When the draft ended in 1973, he left school, 12 credit hours short of a degree.

From 1974 to 1976, Manning worked as news director of a radio station in Alpena and then worked for two years as a reporter with the *Alpena News.* He moved west in 1978, and for the next six years he held reporting and editorial positions at three papers in southern Idaho. Then, drawn to the quiet mountains of Montana, Manning and his wife and young son moved to Missoula, where he took the position of political reporter with the *Missoulian*, one of the largest papers in the state. He reported on state politics and county government for about three years, until early 1988, when he took a new assignment as the environmental reporter.

Manning began working on a series of stories about the two dominant timber corporations logging in western Montana forests, Plum Creek International and Champion. Realizing that no one outside of these corporations knew how hard they were logging their land, Manning went to great lengths to uncover the truth of the situation. The two timber companies, reacting to economic and political trends in the 1980s, had abandoned sustained yield harvesting and begun massive clear-cutting. The ponderosa pine and western larch in Montana's northern Rockies were being liquidated at breakneck speed, destroying the natural systems in the forests and streams and threatening the long-term health of the local economy. When questioned by Manning, officials at Champion admitted that in less than a decade they had already cut lands that they had said would take 30 years to harvest.

Manning submitted his series, which exposed these findings, in late May, expecting after a week of editing that they would be in print. But the series, the most intense effort of Manning's reporting career, met with balking from *Missoulian* management, who wanted to avoid controversy. Finally, nearly six months after he wrote it, the series went

to print. During the week of its publication, Manning's marriage of 17 years ended in divorce. The following year was an unsettling time for him: Branded at work as a troublemaker, he found it hard to get straightforward responses from his superiors. His editor eventually confronted him, saying he was becoming too passionate and that he would be taken off the environmental beat. Having joined the newspaper profession because he believed its duty was to foment debate and "raise hell," Manning could no longer endure what he called corporate journalism, and rather than accept a different position, he quit. After Manning left the paper, his timber series won a C. B. Blethen Award for investigative journalism, which the *Missoulian* accepted. Without a job for the first time since he was 13, Manning embarked on a new freelance writing career. He started by recounting the story of his timber industry investigations and their impact on his life in his book *Last Stand: Logging, Journalism, and the Case for Humility*, which was published in 1991.

By early 1991, Manning had remarried and sought to establish his life with his new wife, Tracy, by building a house. Manning chronicled the story of designing and building their environmentally conscious home in *A Good House: Building a Life on the Land* (1993). He writes of his realization that building anything is an imposition on the land, though he and his wife planned every detail so as to minimize negative impact. They built on a south-facing slope, which allowed them to use passive solar heating and "earth-sheltering," with three sides of the house semiburied and insulated by the hill. Manning designed the house using timber framing—a system of supporting the house with large upright posts and beams, which is less wasteful than using smaller planks like two-by-fours. To conserve water, they installed a composting toilet, which uses less than a pint to flush, as opposed to five gallons per flush used by conventional toilets. This also allows the drain water from the rest of the house to be filtered and reused on the garden. In all, the house cost $50,000 to build and measured 1,200 square feet, much less than the median home size in the United States, which was 1,905 square feet in the late 1990s. In *A Good House* Manning argues that conservation cannot coexist with this trend toward expansion and that frugality must be built into daily life.

Manning's third book, published in 1995, studies one of the most disrespected biomes in the country—the native prairies of the Great Plains. *Grassland: The History, Biology, Politics, and Promise of the American Prairie* explores the history of this country's grasslands, from settlement to the slaughter of bison, to the farming and overgrazing that continue to deplete the soil. Manning advocates a massive restoration, including reintroducing bison as a more sustainable alternative to cattle, tearing out barbed-wire fences, and adapting agriculture to more natural methods. His next project was *One Round River: The Curse of Gold and the Fight for the Big Blackfoot* (1997), a book about the threat of a cyanide heap–leach gold mine on Montana's Blackfoot River, known to many as the river featured in Norman Maclean's *A River Runs Through It* (1976). Manning's book exposes the absurdity of mining the deposit, which is one of the lowest-grade ores ever considered for mining on such a massive scale: It would leach as little as one ounce of gold from 60 tons of ore.

The mine would leave a hole in the earth more than a mile in diameter and would require pumping and rerouting 15.8 million gallons of water each day, diverting water from creeks, springs, wetlands, and ponds in a 100-square-mile area. Manning takes the title for the book from an essay by ALDO LEOPOLD, who encouraged viewing a river as a cycle, not a straight line that can wash away pollution forever.

Manning published two books in 2000: *Food's Frontier*, an account of sustainable agriculture in the developing world, and *Inside Passage*, the case for a natural economy in the coastal temperate rain forests. He is also the founder of two Internet-based bioregional news services that compile environmental articles from regional newspapers; "Tidepool" covers the Pacific Northwest coastal forests, and "Headwaters" serves the Rocky Mountain region. Manning's work has won much recognition. In 1989 he received the Audubon Society Journalism Award, in 1993 he won the R. J. Margolis Award for environmental reporting, and his work has won three C. B. Blethen Awards for investigative journalism. He and his wife, the executive director of the Clark Fork Coalition, live in their home near Lolo, Montana.

BIBLIOGRAPHY

Connors, Philip, "One Round River: The Curse of Gold and the Fight for the Big Blackfoot," *The Nation*, 1998; "Headwaters," http://headwatersnews.org; Manning, Richard, *A Good House: Building a Life on the Land*, 1993; Manning, Richard, *Last Stand: Logging, Journalism, and the Case for Humility*, 1991; Schildgen, Bob, "Grassland: The History, Biology, Politics, and Promise of the American Prairie," *Sierra*, 1996; "Tidepool," http://tidepool.org.

Marsh, George Perkins

(March 15, 1801–July 23, 1882)
Scholar

George Perkins Marsh was a man of extraordinarily broad intellect, renowned in such fields as linguistics, philology, and etymology. During his long life, among other activities, he invented glass instruments, mined a marble quarry, served as Vermont's railroad commissioner, helped write the *Oxford English Dictionary*, bred sheep, and served for two terms in the U.S. Congress. But history esteems him most highly for his seminal contributions to the environmental sciences. Marsh was an avid student of the land, especially how human activity altered it. His book *Man and Nature, or Physical Geography as Modified by Human Action* (1864) made a deep impact on thinking about land use when he wrote it, and even today is considered an important primer.

George Perkins Marsh was born on March 15, 1801, in Woodstock, Vermont. By the age of five he had memorized an encyclopedia; then he went on to study Greek and Latin. At the age of eight, his eyes gave out, and he began to spend more time outside. He befriended nature, discovering a fantastic world of nonhuman beings. His father assiduously taught him the names of all the local trees and natural landmarks. Marsh developed powerful oral comprehension skills: because he was not able to read himself for several years, he had to learn from hearing others read to him. When he turned 15, Marsh entered Dartmouth College, where he immediately distinguished himself as the most brilliant student there. Graduating at the age of 19 in 1820, Marsh continued on to become a lawyer in 1825. But the legal field did not entirely satisfy his wide-reaching curiosity, and Marsh dabbled in a host of additional fields. He learned 20 languages, writing a book about Icelandic grammar. He collaborated with the authors of the *Oxford English Dictionary* and produced two volumes on the history and etymology of the English language.

Marsh was elected to the U.S. Congress in 1840. What most captured his interest during his two terms in Washington, D.C., was the debate about how to spend the enormous bequest to the government by James Smithson. Smithson had specified that he wanted the money spent on increasing and disseminating knowledge throughout the United States, and Marsh proposed that it be used to found a research institution and museum. The Smithsonian Institution owes its existence to Marsh's proposal, and Marsh also collected many of its early biological specimens.

Marsh first ventured into environmental sciences when he studied the decline of freshwater fish in Vermont's streams and rivers in 1857. Contamination by factories and mills and disruption of water flow by dams were obvious problems. But Marsh delved further, and highlighted other, more subtle issues. He saw that deforestation and careless cultivation denuded the land and that when it rained, floods were more likely to occur. The floods had several damaging consequences: they filled the waterways with sediment, which altered their courses and covered spawning grounds with silt. Without forests, there was less precipitation, and a reduced stream flow elevated the temperature of the water. This complicated explanation is reflective of modern ecological analysis, but it was innovative and totally unique in Marsh's time.

Marsh proposed a solution for Vermont's problem as well. After his experience in government, he had little faith that policy changes would improve the situation. Instead, he urged private landowners to protect their own forests and farmland. If good, accessible scientific knowledge spread throughout the populace, and individual enterprise prevailed, he thought this would be the most effective response.

Marsh had been sent as minister first to Greece from 1849 to 1854 and then, by President Lincoln, to Italy in 1861. There, in addition to performing his ministerial duties, he wrote *Man and Nature*. This book contained a sort of environmental history of certain areas of the world, recounting how human activity had altered each one. The Sahara Desert, for example, was created when humans' livestock deforested northern Africa; overgrazing was also the culprit in creating the arid Provence and Dauphiné regions of France. In the United States, this was not ancient history; Marsh's home state of Vermont had the highest rate of deforestation and soil erosion. In addition, Marsh traced the causes of natural disasters and was able to conclude that in each case, the behind-the-scenes culprit was human activity—usually deforestation. Marsh's conclusion was that human activity was inherently destructive to nature. His moral to the story: the earth was not given to humans to consume and waste, but rather to use with care.

Upon its release in 1864, the book was greeted with acclaim both in the

George Perkins Marsh (Library of Congress)

United States and in Europe, where Marsh was still living. European foresters heeded the book's recommendations, but Americans still did not seem fully ready to begin conserving their vast resources. In fact, some land speculators even turned around Marsh's description of the deforestation-desertification relationship, claiming to potential buyers that the West would become more humid and lush if new landowners planted enough trees!

Marsh died in Vallambrosa, Italy, on July 23, 1882. His ideas experienced a revival in the mid-1900s, with an international conference, "Man's Role in Changing the Face of the Earth," at Princeton University in 1955 and the reissue of *Man and Nature* by Harvard University Press in 1965.

BIBLIOGRAPHY

Lowenthal, David, *George Perkins Marsh: Versatile Vermonter*, 1958; Marsh, George Perkins, *Man and Nature*, David Lowenthal, ed., 1965; Strong, Douglas, *Dreamers & Defenders: American Conservationists*, 1988.

Marshall, Robert

(June 2, 1901–November 11, 1939)
Forester, Cofounder of the Wilderness Society

Intrepid mountaineer and professional forester Robert Marshall fought tirelessly during his short life for conservation of wilderness in its wildest state. During his two years as recreation director for the U.S. Forest Service, he restricted roads and development from some 14 million acres of national forest land; as the Bureau of Indian Affairs forestry director, he created 16 wilderness areas on Indian reservations. The Wilderness Society (TWS), which he cofounded, still leads the pack of conservation organizations on lobbying for new protected areas and environmental legislation. Marshall's deep personal love for wild places was complemented by a firm commitment to social justice. Marshall believed that too much time in the city would wear down the human spirit and that everyone—even society's poorest—deserved time in the wilderness to replenish.

Robert Marshall was born on June 2, 1901, into a liberal, privileged New York family. His father was a well-known constitutional lawyer who frequently came to the defense of oppressed minorities and lobbied lawmakers to preserve Adirondack forests. The Marshall family spent summers at their cabin on Lower Saranac Lake in the Adirondacks, and it was there that young Robert grew to love mountains and mountaineering. The family's private guide led Robert and his brothers up all 46 Adirondack peaks over 4,000 feet. Robert quickly overtook his guide, hiking 30 to 40 miles and climbing as many as 16 mountains per day.

In an attempt to combine his love for mountains with a profession, Marshall studied forestry at Syracuse University, earning a bachelor's degree in 1924. He then went on for a master's degree in forestry at Harvard University (1925). His thesis experiment at the Harvard Forest in Petersham, Massachusetts, taught him that selective cutting (cutting only certain trees) rather than clear-cutting (cutting all trees in a given area) led to healthier and more diverse secondary growth. Immediately after graduation, Marshall went to work for the U.S. Forest Service in the Montana Rockies, where he stayed for three years, until 1928. Marshall had the opportunity to meet a diverse group of westerners during these years, including poor, unemployed men who would sign on to fight fires in national forests when needed. Hearing their hard luck stories was real-world proof to Marshall of the cruelty of the modern industrial world's cutthroat capitalist exploitation.

When Marshall returned to the East Coast to continue his education, he studied plant physiology at Johns Hopkins and became active in the school's Liberal Club. Marshall was one of the first who combined his liberal politics with conservationism. He joined GIFFORD PINCHOT and other veteran foresters in signing "A Letter to Foresters," which criticized the forest industry's destructive clear-cutting practices and proposed a greater degree of control of the forest industry. It urged the creation of more public forests, a cause that Marshall would continue working toward during the rest of his life.

Marshall took a break from Washington in 1930–1931 and traveled to the Arctic village of Wiseman, Alaska. After studying the inhabitants of Wiseman, Marshall concluded that people were happier and more fulfilled in such a frontier environment than in any urban setting. His description of the colorful individuals and their lifestyle became *Arctic Village* (1933). For a book published during the Depression, when the general public could not afford to buy books, *Arctic Village* was considered a best-seller. True to his social commitment, Marshall split the royalties with the inhabitants of Wiseman.

During the 1930s, Marshall worked at the U.S. Forest Service and at the Bureau of Indian Affairs (BIA) forest service, always pushing for greater protection of wilderness and more public ownership of forests. His management style included much fieldwork, with at least half of his time spent roaming wild places in question. While at the BIA, Marshall felt that he was able to have vastly more useful interactions with Native Americans when he was hiking around their reservations than as another white official talking at them in a meeting hall. Marshall worked toward preserving wilderness on Indian

Robert Marshall (The Wilderness Society)

reservations and usually argued against plans for new roads cutting through them. A precursor to ecotourism advocates, Marshall felt that Indian-guided wilderness trips would contribute more to reservation economies than would one-shot clear-cutting or mining projects.

When in 1937 Marshall transferred to the U.S. Forest Service and became its recreation director, he worked hard to establish more public forests and keep existing national forest lands roadless and undeveloped. Although he advocated grand-scale recreation, he felt that the country's national forests should be free of motorized transportation and concessions such as hotels or lodges and stores, in order to preserve wilderness in a state as wild as possible and to assure accessibility for all economic classes. Whenever civil rights issues arose, he was quick to respond; Marshall fought against segregated campsites in the South and resorts that discriminated against minorities. In 1935 Marshall's strong beliefs and effective work earned him an accusation by New York congressman Hamilton Fish of being a communist. This was followed up in 1938 and 1939 with his inclusion on a list of government employees contributing to communism. Marshall was self-confident enough to brush off these attacks, but several of his colleagues lost their jobs and had their careers ruined by the House Un-American Affairs Committee.

Along with BENTON MACKAYE, Harvey Broome, Harold Anderson, ALDO LEOPOLD, Ernest Oberholtzer, Bernard Frank, and ROBERT STERLING YARD, Marshall was a co-founder in 1935 of the Wilderness Society (TWS). During its first years it was a humble operation funded almost in entirety by Marshall, but TWS has since grown to be one of the most active and influential conservation organizations in the nation.

Marshall died in the prime of his career, on November 11, 1939, at the age of 38, during an overnight train ride from Washington, D.C., to New York. His $1.5 million fortune was left in three equal parts to social advocacy organizations, civil liberties organizations, and TWS. Two years after his death, the Bob Marshall Wilderness Complex in northwest Montana was created and named for him. With the addition of the contiguous Great Bear and Scapegoat wilderness areas in the 1970s, the 1.5-million-acre Bob Marshall Wilderness Complex became the largest roadless area in the United States.

BIBLIOGRAPHY

Fox, Stephen, *John Muir and His Legacy: The American Conservation Movement*, 1981; Glover, James M., *A Wilderness Original: The Life of Bob Marshall*, 1986; Gottlieb, Robert, *Forcing the Spring: The Transformation of the American Environmental Movement*, 1993; Marshall, Robert, *Alaska Wilderness*, 1970; Marshall, Robert, *The People's Forests*, 1933.

Marston, Betsy, and Ed Marston

(July 6, 1940– ; April 25, 1940–)
Editor of *High Country News;* Publisher of *High Country News*

Ed and Betsy Marston are the publisher and editor, respectively, of *High Country News*, a highly respected biweekly newspaper specializing in western environmental issues.

Elizabeth (Betsy) Avice Pilat was born on July 6, 1940, in New York City to Oliver and Alice Avice Pilat. She was brought up in a literary household; her father was a New York City newspaperman and wrote ten books, mostly political biographies. She attended the University of Delaware, earning a B.A. in English in 1962, and the Graduate School of Journalism at Columbia University, where she earned an M.S. in 1963.

Edwin Herman Marston was born on April 25, 1940, in Brooklyn, New York, and grew up in Queens, New York. He says that his earliest memories are of wanting to be a journalist, long before he knew what a journalist was. But as he grew older, he decided he needed a more dependable income than he thought journalism could provide, so he went into the sciences. He graduated from City College of New York in 1962 with a B.S. in physics and earned his Ph.D. in physics in 1968 from the State University of New York at Stony Brook.

The Marstons met in 1961, at City College of New York, where Betsy was spending a year away from the University of Delaware and where both wrote for the student newspaper. They were married in 1966. Betsy became the first woman anchor on a New York television news program, for public television station WNET, winning an Emmy award for her three-part documentary on singer Paul Robeson in 1974. Ed was an assistant professor of physics at Queens College from 1968 to 1971 and an associate professor at Ramapo College in Mahwah, New Jersey, from 1971 to 1974. In 1976, he wrote the textbook *The Dynamic Environment: Water, Transportation, and Energy*, based on a physics course that he taught that used physics to describe how urban areas worked—how water flowed into a city, how oil flowed through the trans-Alaska pipeline, how elevators limited the height of skyscrapers.

By 1974 the couple had two small children, Wendy and David, and were tired of spending so little time with them because they worked such long hours. They decided to take a year off and retreat to their Colorado mountain cabin near the small town of Paonia. But after a month of their vacation, they realized they were "failures at leisure," and so they began working: Ed wrote environmental impact statements for local land management agencies, and Betsy made candles, mostly for fun. Two months later, after

realizing that they needed more intellectual stimulation to be happy, they started a local weekly newspaper, *The North Fork Times*. Despite the suspicion of local people about the radical politics of these urban dropouts, the paper quickly rose to become the county paper of record. The Marstons published it for six years before selling it in December 1980. They then took off enough time to build a passive solar home and then founded another paper, the biweekly *Western Colorado Report*, which focused on how the boom in energy development was affecting the small towns of western Colorado. When oil prices plummeted in 1983, and the boom went bust, the paper's readership dropped precipitously.

Just as they were wondering what to do with the ailing *Western Colorado Report*, the Marstons heard that *High Country News* (HCN), which was published out of Lander, Wyoming, needed a new staff. HCN had been founded in 1970 by wildlife biologist, teacher, and fifth-generation Wyoming rancher Tom Bell, who was outraged by the destruction being wreaked on his beloved western lands by ranchers, miners, loggers, politicians, and government bureaucrats. Bell edited the 16-page tabloid for four years, then passed it on to a series of professional journalists who shared his environmental passion and gave the paper a national reputation for reliable, in-depth reporting. Its few but devoted subscribers were so loyal that on the several occasions when the paper was threatened with bankruptcy, they sent unsolicited donations amounting to thousands of dollars to keep it afloat. When all three staffers quit the paper in 1983, the board of the nonprofit paper chose the Marstons to take the reins. An intern hauled the archives, the photo file, the mailing list, and an addressing machine from Lander to a former church in downtown Paonia, and the Marstons merged their *Western Colorado Report* subscription list with that of HCN and started their crash course in western politics and land and resource management.

The paper's network of freelancers continued to send in high-quality stories; the first scoop the Marstons published was a 1984 story by Tom Wolf about how the Glen Canyon Dam had almost collapsed when, during the 1983 flood year, the Colorado River ate away at the dam's core and spillway channels. The Bureau of Reclamation later acknowledged the veracity of the story and released photos that showed how close the dam was to breaking apart. Every two years the paper published a four-issue series that delved deep into a problem. The 1986 series, "Western Water Made Simple," examined the Columbia, Missouri, and Colorado Rivers and their various development projects. It received a prestigious George Polk award for environmental reporting and was published in book form by Island Press in 1987, as was a 1988 series on the resurgence of extractive industries, entitled *Reopening the Western Frontier.*

Even though the Marstons themselves are liberal environmentalists, and the paper is known for its environmental advocacy, the Marstons do recognize that in the West, where most nonurban residents live directly or indirectly off the land, strong local communities are as important as protected public lands. The Marstons feel that condominium development and rural sprawl pose threats, much like the decades-long impacts created by subsidized extractive industries.

HCN features include views from the many sides of every debate, and for this reason the paper's readership has expanded to include not only environmentalists but also high-level decision makers, reporters looking for story ideas, and western miners, loggers, and ranchers.

When they first took on HCN, the Marstons believed that the paper's number of subscribers would never rise to more than a few thousand. The ten states covered by HCN—Washington, Oregon, Idaho, Montana, Wyoming, Utah, Colorado, Nevada, Arizona, and New Mexico—are now the fastest-growing region in the United States. By 2000, there were 21,000 subscribers to the paper, and the paper is generally regarded as an important regional institution. Columns and essays by HCN writers are offered to other newspapers through the Writers on the Range syndicated column service, currently subscribed to by some 50 newspapers throughout the west. And Betsy Marston hosts a weekly radio show, "Radio High Country News," produced at Paonia's community station KVNF and broadcast on ten stations in Colorado, Utah, and New Mexico. The paper also operates a web site (www.hcn.org) with an easily accessible archive of seven of the paper's 30 years, the radio show on real audio, the current issue, and an online newsletter. The paper became a nonprofit foundation in its first home in Wyoming; it is the primary project of the High Country Foundation. Sixty percent of the paper's operating costs are from subscriptions, and 30 percent more comes from contributions; advertising, which takes up only one to one and a half pages of each issue, brings in less than 5 percent of the paper's total operating budget. Since the Marstons began producing the paper, close to 200 interns have come to the paper's home in Paonia to learn the western environmental beat. Since 1998, interns have been paid, and they stay for four months. Many of these interns now freelance for HCN or other news media throughout the country or are engaged in other environmentally oriented activities.

Ed Marston has written the text for *Colorado: 1870–2000*, which features photographs taken of Colorado landmarks by two great photographers, W. H. Jackson, who lived in the nineteenth century, and John Fielder, who works in the present. Marston works with a number of local and national organizations, including the Delta-Montrose rural electric cooperative, for which he is currently president; Friends of the Earth, where he was a board member from 1987 to 1993; and the Regional Advisory Council of the Rocky Mountain Office of Environmental Defense. He was a John S. Knight Journalism Fellow at Stanford University for the academic year 1990–1991.

The Marstons live on a mesa just outside Paonia, Colorado.

BIBLIOGRAPHY

Burkhead, Rebecca, "Watching the West, The Duo Who Keep High Country News Aloft," *Columbia Journalism Review*, 1993; "High Country News," http://www.hcn.org; Hill, David, "High Country News: Small Paper, Strong Voice," *Washington Journalism Review*, 1989; Jackson, William Henry, John Fielder, and Ed Marston, *Colorado: 1870–2000*, 1999; Marston, Betsy, "The Little Paper that Could," *High Country News*, 1995; Marston, Ed, ed., *Reopening the Western Frontier*, 1989; Marston, Ed, ed., *Western Water Made Simple*, 1987.

Martínez, Dennis

(August 10, 1941–)
Restoration Ecologist, Ethnoecologist

Dennis Martínez has been involved with restoration ecology since 1969—before the field of restoration ecology had even formally come into existence. Martínez is of O'odham, Chicano, and Anglo heritage and has been adopted by the Pretty Weasel Family of the Whistling Water Clan, Absorooka Tribe at Crow Agency, Montana. He melds western science, native cultural practices, and traditional ecological knowledge in his work, recognizing that all are necessary for successful ecological restoration. He specializes in North American temperate forest, desert mountain, and tropical dry forest ecosystems. Martínez assists traditional communities in Canada, Mexico, and the United States (including Hawai'i) with the ecological restoration of their tribal land bases and ancestral lands and the renewal of their traditional relationships with the land.

Dennis Martínez was born on August 10, 1941, on a ranch near Selma, California, in the San Joaquin Valley. He was raised by his grandparents and grew up hunting and fishing in the Sierra Nevada mountains. Even as a child, he noticed negative changes in the San Joaquin Valley caused by the area's increased population and urban development. Before he left the area for college, the salmon disappeared from the San Joaquin River, deer populations crashed, groundwater sources dried up, and wetlands with their native waterfowl were lost. Martínez studied at the University of California at Berkeley, receiving his B.A. in history and philosophy of science (Darwinian evolution) in 1976.

Martínez began doing landscaping work in the San Francisco Bay area in 1969. After meeting a Pomo Indian elder and learning from him about traditional uses of native plants, Martínez became very interested in incorporating these plants into his landscaping projects. In 1975, David Amme, Don Cook, and later, David Kaplow, along with Martínez launched a nonprofit group, Design Associates Working with Nature (DAWN). The first major restoration contractor on the West Coast, DAWN pioneered the use of little-known native species in its restoration projects. DAWN worked in public parks, U.S. military lands, and plots owned by such conservation groups as The Nature Conservancy and the Audubon Society, in an attempt to replicate the ecosystems that existed before European colonists changed the California landscape.

In the decades since then, Martínez has worked as a contractor or consultant on scores of ecological restoration projects. He works primarily with indigenous people to restore the landscape that their ancestors had inhabited without depleting for thousands of years. Traditional indigenous cultures of North America influenced the ecological structure, composition, and function of their habitat by

selective harvesting of useful plants and animals and by regular light burnings. What European colonists considered "virgin wilderness" was in many cases actually more like a carefully tended plant and animal reserve. Martínez has come to recognize that ecological restoration and the restoration of traditional caregiving relationships with the land go hand-in-hand, and he has worked with many traditional communities on both of these aspects. From 1989 to 1990, he worked with the Ya-ka-ama ("Our Land") Indian Development and Education Corporation's project as nursery manager and tribal liaison, servicing more than 20 rancherias and reservations in five California counties. From 1987 to 1992, he helped design a restoration plan for 7,000 acres of ancestral lands for the Sinkyone Intertribal Wilderness Council, which had been clear-cut by Georgia Pacific lumber company. The Takelma Intertribal Project, which he currently cocoordinates with Takelma/Siletz elder Agnes Pilgrim, is a collaborative effort among the local indigenous community, the U.S. Forest Service, and scientists from the Pacific Northwest Forest Experimental Station and Oregon State University. The project strives to restore the precontact cultural landscape of oak/pine savanna, maintain it with regular light burns, and preserve and restore culturally significant plants and animals as cultural and natural resources. The local indigenous community celebrated its First Rites Salmon Homecoming Thanksgiving ceremony in May 1994 for the first time in 150 years and has continued to celebrate it annually. The Hawai'ian island reserve of Kaho'olawe, which is sacred to native Hawai'ians and in ancient times was an important departure and arrival point for boats to other islands of Polynesia, was ecologically devastated by the goats that Captain Cook left there and then later was bombed by the U.S. Navy for 50 years. Martínez has consulted for the Kaho'olawe Island Reserve Commission, a state office responsible for the restoration of the dry tropical forest ecosystem, focusing on the cultural and spiritual aspects of ecological restoration. Another of his consulting projects is with the Mountain Maidu of the northern Sierra Nevada in California, who have a contract with the Pluma National Forest, focusing on cultural landscape restoration.

Martínez emphasizes that restoration work can take many years before it can be considered completed. For instance, the restoration of tallgrass prairie at the University of Wisconsin's arboretum, the Curtis Prairie, after 60 years boasts only half of the species that grew in the precontact tallgrass prairie. An attempt begun in 1850 to restore the full species composition of a meadow in England is just now reaching completion.

Martínez has been a member of the Society for Ecological Restoration (SER) since 1989 and has served two terms on its International Board of Directors. He has been cochair of its Science and Policy working group and serves on the Peer Training working group and the International Awards committee. With Chuck Striplen (Mutsum Ohlone) he codirects the Indigenous People's Restoration Network work group. He has succeeded in integrating the issue of cultural survival for Native peoples into the programs and annual meetings of SER. He reminds his restorationist colleagues that national laws governing land management and protection of indigenous cultural resources are inadequate. Among the problems that traditional communities face, which he addressed in a presentation at SER's 1999 annual conference, is the fact that the National Forest Management Act does not "recognize American Indian natural resource management practices as having a role in forest planning. Traditional elders are not consulted. . . . There is no First Amendment constitutional protection for sacred sites lacking obvious artifacts or structures. . . . Landscape change is ignored in the delineation of tribal cultural resource areas." Martínez has defended burial and village sites sacred to indigenous peoples from development and has acted as a liaison between Indian people and government land agencies during negotiations about the protection of cultural resources. Through the Indigenous People's Restoration Network, Martínez organized a 1996 meeting for the United Nations Environmental Program (UNEP) on Human Values of Biodiversity, with 23 Native presenters from North American at the San Xavier Tohono O'odham Reservation in Arizona. The proceedings from this meeting have been collected in the 2000 book *Cultural and Spiritual Values of Biodiversity*, edited by Darrell Posey. Martínez has also served on several Forest Service interdisciplinary teams, including the Applegate Partnership's Ecological Assessment Needs Team.

In addition to his work with SER, Martínez is a member of the Traditional Knowledge Council of the American Indian Science and Engineering Society and of the Karuk Tribal Team for Cultural Resource Protection. With Judith Vergun, he is cochair of the Oregon State University Pacific NorthWest Traditional Ecological Knowledge Project. He serves as member of the advisory council, board, or steering committee of the following organizations: Baca Institute of Ethnobotany, the Institute for Sustainable Forestry, the Cultural Conservancy, Black Mesa Permaculture Project, the Indigenous Permaculture Center, the Kaho'olawe Island Reserve Commission, Earth Legacy, Sombra Buena Organic Forest Products (based in Honduras), and Native American Food Systems Project. He is a member of the Alliance of Forest Workers and Harvesters. SER recognized Martínez in 1997 with its John Rieger Service Award for his substantial contribution to ecological restoration. Martínez speaks widely, at tribal and nontribal conferences, and has published articles in a number of journals and books, including a Forest Service publication on special forest products and an Environmental Protection Agency publication called "This Place Called Home: Tools for Sustainable Communities."

Martínez resides in an intentional community in the Klamath Mountains near Glendale, Oregon, called Mountain Grove Center for New Education. The community sustains itself in part through selective thinning of high-quality hardwood timber and small-diameter softwoods, milling them locally to add value to the commercial product. The community's aquatic ecosystems have been restored to the point that native coho salmon have begun to return after a long absence. Martínez earns part of his income from working in the woods as a vegetation surveyor, seed collector, restoration contractor, nursery grower of native plants, and restoration thinner and logger.

BIBLIOGRAPHY

"Dennis Martínez: Member of Earth Legacy Advisory Board," http://www.earthlegacy.com/about/dmartinez.html; Martínez, Dennis, "Salmon Homecoming," *Intricate Homeland*, 2000; Martínez, Dennis, and Jesse Ford, *Ecological Applications* (Special Invited Feature on Traditional Ecological Knowledge), 2000; Posey, Darrell, *Cultural and Spiritual Values of Biodiversity*, 2000; "Society for Ecological Restoration," http://ser.org.

Mather, Stephen

(July 4, 1867–January 22, 1930)
National Park Service Director

Stephen Mather was the first director of the burgeoning U.S. National Park Service from 1915 to 1929 and is remembered for improving, enlarging, and consolidating the country's national parks. With a team including assistant HORACE ALBRIGHT and publicist ROBERT STERLING YARD, Mather convinced Congress to create a National Park Service, transformed inaccessible and unknown national parks into attractions that the public began to visit in droves, reformed the chaotic concessions system, and doubled the acreage conserved as national parks and monuments. A millionaire, Mather never hesitated to donate his own money

Stephen Mather (Bettmann/Corbis)

if not enough federal or private money was allotted for a particular national parks cause.

Stephen Tyng Mather was born on July 4, 1867, in San Francisco, California. His mother, English-born Bertha Jemima Walker, had followed her husband, Joseph Mather, when he traveled west for his Gold Rush accounting job. She never liked California and left her husband and children for the East Coast when Mather was six years old. Mather stayed in California, spending free time on horseback and camping trips in the state's vast wilderness, until he graduated with a degree in journalism from the University of California at Berkeley in 1887. After graduation, Mather traveled east to see his mother and began working as a journalist at the New York *Sun*. He spent five years writing for the *Sun* before marrying Jane Thacker Floy in 1893 and entering the borax mining business with his father.

Borax, a water-softening chemical found in quantity in California and Nevada, was little known and little used until Mather got into the business. In order to create a market for borax, Mather, as the Pacific Coast Borax Company's publicity manager, paid anyone who had a letter about one of the many wonders of borax published in a small-town newspaper or women's magazine. This campaign resulted in 800 published letters in 250 publications in 33 states.

Borax became a household word, and Mather, eventually a co-owner of the company, became a millionaire.

Like many wealthy Americans at the dawn of the twentieth century, Mather and his family visited Europe and marveled at the luxury and comfort of visitors' lodgings and services in Europe's mountain parks. Disappointed with the quality of facilities in national parks in the United States, Mather fired off a disgruntled letter to Secretary of the Interior Franklin Lane. Secretary Lane was just then wondering whom he could hire to improve the parks on the shoestring budget his department had for the job. Lane invited Mather to take the job and improve the parks as he saw fit. Mather felt ready to move out of industry and into conservation; it has been said that he felt it was time to give something back to the earth since up until that point he had been taking so much from it. So in 1915 he accepted the invitation and became assistant secretary of the interior in charge of national parks.

Mather approached his new job with the same entrepreneurial spirit that had earned him his millions from borax. He hired an able assistant, Horace Albright, who was to accompany him during his entire tenure with the parks and would succeed him after his resignation in 1929. Mather and Albright made a checklist and got to work. They immediately identified the need for greater public interest in the national parks. Mather and his publicity assistant, Robert Sterling Yard, placed over 1,000 magazine articles extolling the national parks and had photo-filled brochures about the parks sent to the country's wealthiest citizens and power elite, including all members of Congress. Within a year, most members of Congress knew much more about the parks, and many had visited them personally, some on trips for influential people that were organized by Mather. In 1916, Congress passed the National Park Service Act, which created the National Park Service as a bureau of the U.S. Department of the Interior. Under this act, all parks would fall under the jurisdiction of the National Park Service, which would facilitate administration and coordination.

Another major task on the checklist was to ease travel and lodging in the parks so that more people would be able to visit. Once World War I began and U.S. tourism in Europe ceased, more Americans wanted to visit their own national parks. At that point, arriving at most existing parks was a daunting task. The automobile was still a modern contraption, so roads to and within the parks were either nonexistent or in terrible condition. Rail connections to the parks were also scarce, and once passengers got off in the towns nearest the parks, it was hard to find transportation and guides to the parks. Mather campaigned successfully to award concessions to only one company per park or area of park, a system that resulted in more orderly, reliable transportation and lodgings but earned him enemies among those who lost their right to work in the park.

Some approaches by Mather to popularize the national parks and please visitors included tactics that today would be prohibited, such as building a tunnel through a giant sequoia tree, public feeding of the bears in Yellowstone, and heaving bonfires off Yosemite Falls. Current critics of the Mather legacy complain that he built too many roads through areas that should have been left more remote. Mather did have strict ideas of what was acceptable, however. When a Glacier National Park concessionaire did not dismantle a sawmill on time after building a new lodge, Mather lost patience. He invited his daughter and a large party of visitors to observe as he personally dynamited the sawmill.

Mather's personal charisma was a key ingredient to his success in raising interest and funds for the parks. He charmed potential patrons on the dozens of exploratory pack trips he organized for them. Congressmen and wealthy conservationists were impressed both by the scenery to which he escorted them and by his energy and his vision. He always invited the wives of these individuals, knowing that once the men returned to their jobs they would forget about their trip but that the women would remember and remind them to follow up on their commitments.

Mather was also a skillful personnel manager, identifying talented parks employees, encouraging them to seek posts where their gifts would be of use, and often offering generous gifts or loans to help them reach their potential.

In addition to his work on behalf of national parks and monuments, Mather encouraged the growth of the state park movement. His vision included a state park for every 100 miles of state highway. This would allow cross-country mo-torists to camp in a different beautiful place every night and would also serve as a pressure valve for the National Park Service. Senators and congressmen could be pushy about having their particular local attraction declared a national park, and Mather felt that many of these attractions did not deserve the stature of national park but should be conserved under some designation.

After 14 whirlwind years of nonstop work on behalf of the national parks, criss-crossing the country repeatedly and at breakneck speed, Mather suffered a stroke in 1928. He resigned from his post as director of the National Park Service in 1929 and died of a second stroke in Brookline, Massachusetts, on January 25, 1930. He has been memorialized with a Mount Mather in Alaska, the Mather Memorial Highway in Mount Rainier National Park, and the Mather Memorial Arboretum at the University of California.

BIBLIOGRAPHY

Albright, Horace M., *The Birth of the National Park System: The Founding Years, 1913–1933*, 1985; Fox, Stephen, *The American Conservation Movement: John Muir and His Legacy*, 1986; Hartzog, George B., *Battling for the National Park Service*, 1988; Mackintosh, Barry, *The National Parks: Shaping the System*, 1991; Shankland, Robert, *Steve Mather of the National Parks*, 1970; Wirth, Conrad L., *Parks, Politics and the People*, 1980.

Matthiessen, Peter

(May 22, 1927–)
Writer

Peter Matthiessen has forged a subgenre of outdoor adventure writing by covering unexplored territories, wildlife, and native peoples as far away from contemporary modern life and its accoutrements as he could find them. Author of more than 20 books, his fictional and nonfictional accounts of the Amazon, East Africa, the Miskito Coast, and the Himalayas are found, respectively, in his most famous works: *At Play in the Fields of the Lord* (1965), *The Tree Where Man Was Born* (1972), *Far Tortuga* (1975), and *The Snow Leopard* (1978), which won a National Book Award. While Matthiessen describes himself as "a generalist," his eye and ear for details in the natural world have helped him produce lasting work in American literature and have made him a standard bearer for creative nature writers.

With ancestry dating back to seventeenth-century Danish whaling captain Matthies the Fortunate, Peter Matthiessen was born the fortunate son of Elizabeth Carey and Erard A. Matthiessen on May 22, 1927, in New York City. From his earliest days, in a house overlooking the Hudson River, or in his family's Fifth Avenue apartment across from Central Park, or visiting his family's vacation home on Fishers Island off Long Island, Matthiessen had privileged access to untamed woods, the sea, and the city. His life as a man, in addition to his life as a writer, is marked by his rejection of the amenities of affluence and his quest for an unmediated experience with nature.

As a boy, Matthiessen and his younger brother collected snakes, and they earned their sea legs early on deep-sea fishing trips off Montauk, which instilled him with a life-long love of the ocean. After a tour in the United States Navy at Pearl Harbor (1945 to 1947), Matthiessen attended the Sorbonne at the University of Paris (1948 to 1949) and received a B.A. in English from Yale University in 1950. While he was still in college, his fiction writing was already gaining attention at the *Atlantic Monthly* and Farrar, Straus publishers. He married Patricia Southgate in 1951 and moved to France, where, along with writers William Styron, George Plimpton, and Ben Bradley, he cofounded the *Paris Review* in 1953. Parisian notoriety aside, Matthiessen's early novels did not earn him much acclaim or money, so he returned to Long Island to work as a commercial fisherman to support his family and his writing career. He later relates how idyllic it was for him to return to the ocean after several years of urban high life. But during this period, his wanderlust overcame him, and his marriage disintegrated. He set out to visit every wildlife reserve in the United States for a book on the state of the country's endangered wildlife. This project showed Matthieson how his writing could serve both as a warning about species on the brink of extinction and an homage to species already gone (the great auk, the passenger pigeon, the near extinction of the bison). The descriptive power of his writing was nationally recognized in the resulting book, *Wildlife in America* (1959).

Wildlife in America also put Matthiessen "on the map" in another sense. A growing rate of extinction further motivated his quest to commune with most endangered species before they disappeared and to seek out, as he told *Buzzworm* in 1993, the wild "people and places that still have their own integrity." He seized opportunities that took him far from the familiarity of his Long Island home. The *New Yorker* magazine underwrote five months of trekking in South America between Tierra del Fuego and Lima, Peru, for articles later yielding the book, *The Cloud Forest: A Chronicle of the South American Wilderness* (1961). But as would be the case with other journalistic projects, he would reserve the best writing for his novels. Matthiessen has compared his nonfiction writing to cabinetmaking, that is, the crafting of something utilitarian, while he says his novels are the work of artistry, like fine sculpture. *At Play in the Fields of the Lord* (1965) is set in the Amazon and involves the fate of the fictitious Niaruna tribe after the intervention of Christian missionaries. The book reflects Matthiessen's distrust for Western civilization's effects on the preindustrial people left in the world. The research that Matthiessen did for *New Yorker* pieces about turtling off the Miskito Coast of Honduras resulted in his greatest novel, *Far Tortuga* (1975), which some critics have acclaimed as second to *Moby Dick* among U.S. sea novels.

With the Harvard-Peabody expedition in 1961, Matthieson moved to a remote corner of New Guinea to live with the aboriginal Kurelu people, believed to be the last Stone Age culture. Matthiessen's account of this experience, *Under the Mountain Wall: A Chronicle of Two Seasons in the Stone Age* (1962), decries the inevitable loss of culture to the industrial juggernaut of the twentieth century.

Peter Matthiessen (Courtesy of Anne Ronan Picture Library)

Later in the 1960s he went to live with the East African Hazda, a tribe whose graceful integration into what otherwise could be described as a harsh, uninviting environment fascinated Matthiessen enormously. The Hazda mythologize that man climbed down from the baobab tree, and it was this folktale that inspired Matthiessen to write *The Tree Where Man Was Born* (1972). Of his fascination for "the small people," the Eskimos, sherpas, and pygmies he has written about, he told *New York Times Magazine* in 1990, "I just love being with people who know the wilderness and wild things so well. They're not the least bit sentimental about it, the way conservationists and naturalists are." His interest in the non-Westernized indigenous peoples is also part of his life-long study of the ease with which certain tribes coexist with nature.

Matthiessen continued to write books about animals after the publication of

Wildlife in America. These included *Shorebirds of North America* (1967), reprinted as *The Wind Birds* in 1973; *Sand River* (1981), which focused on the large game left in the East African Selous Game Reserve; and his award-winning nonfiction work, *The Snow Leopard* (1978), written after trekking the Crystal Mountain in northwestern Nepal. *The Snow Leopard* is similar to his previous books in that he kept a copious journal and wrote as he traveled through a remote region of the world in search of endangered species and cultures, but he did not see the snow leopard even once. Instead, he had a chance encounter with a high-altitude solitary monk, Buddhist teacher Shey Gompa, at the Crystal Monastery. In this book, Matthiessen's philosophical discourse achieves greater heights, charting his own spiritual journey.

Matthiessen has been equally adventurous internally. After the death of his second wife, Deborah Love, from cancer in 1972, Matthiessen began sitting zazen, and after years of practice and study, he was ordained a sensei (teacher of Zen). Appropriately, his Zen Buddhist name is Muryo, meaning "Without Boundaries." When he is not on the road, he lives in Sagaponack, Long Island, New York, with his wife since 1980, Maria Eckhart, an editor at *Conde Naste Traveler.* Matthiessen is a member of the American Academy and Institute of Arts and Letters; the New York Zoological Society, for which he served as trustee from 1965 to 1978; and the American Academy of Arts and Sciences. He was also elected to the Global 500 Honour Roll, United Nations Environment Program.

BIBLIOGRAPHY

Dowie, William, *Peter Matthiessen,* 1991; Houy, Deborah, "A Moment with Peter Matthiessen," *Buzzworm,* 1993; Iyer, Pico, "The Laureate of the Wild," *Time,* 1993; Matthiessen, Peter, *The Peter Matthiessen Reader: Non-fiction 1959–1991,* ed. McKay Jenkins, 1999; Trip, Gabriel, "The Nature of Peter Matthiessen," *New York Times Magazine,* 1990.

McCloskey, Michael

(April 26, 1934–)
Executive Director and Chair of the Board of the Sierra Club

Michael McCloskey was executive director of the Sierra Club from 1969 to 1985 and chairman of the board of directors of the Sierra Club from 1985 to 1999. He was instrumental in bringing about the organization's transformation from a 17,000-member excursion-focused club to one of the largest, most influential conservation and advocacy groups in the United States.

John Michael McCloskey was born on April 26, 1934, to John Clement and Agnes Margaret Studer McCloskey in Eugene, Oregon. He spent his youth learning to hike and climb in the Three Sisters Primitive Area of western Oregon, which was virtually an extension of his backyard. In his book *The History of the Sierra Club*, Michael Cohen writes that McCloskey "had learned the true spiritual value of an old-growth forest while reveling in the forests, being swallowed up in the green wilderness." McCloskey went away to college at Harvard University, secure in his knowledge that the wild places he had grown up in would never disappear. He earned a bachelor's degree in 1956 and then spent several years in the army before returning to Oregon in the late 1950s, only to find that a large amount of his favorite forests and wilderness areas simply did not exist anymore.

McCloskey had returned to Oregon to study law at the University of Oregon and while he was completing a law degree, he jumped headlong into the local conservation movement. In his final year of law school, he found that environmental issues in the Pacific Northwest had grown to the point that the outdoor groups were in desperate need of staff assistance to link together all of the various activists and campaigns. Regional conservation groups began to raise money to hire a staff person, and the Sierra Club, based in San Francisco (having no staff outside of San Francisco at the time), agreed to match whatever funds were raised. The money was raised, and in 1961 McCloskey began to represent numerous conservation groups in the Northwest, including the Sierra Club. He kept this job for four years, spending more than half of his time on the road, working to organize isolated conservation groups, encouraging them, and providing successful strategic models for them to emulate.

In 1965, McCloskey moved to San Francisco to begin work as assistant to the president of the Sierra Club, Will Siri.

McCloskey had already proven that he was overqualified for such a position, and it was not long before he was assigned to create a new conservation department whose sole purpose would be to focus on organizing the Sierra Club's conservation activities. McCloskey served as conservation director from 1966 to 1969, during which time the Sierra Club was fighting for the preservation of the Grand Canyon and for the creation of a Redwood National Park, as well as for a new national park in Washington's Northern Cascades.

McCloskey became executive director of the Sierra Club in 1969 and served in this position until 1985. Then he became chairman of the Sierra Club in Washington, D.C., a position he occupied until his retirement in 1999. During McCloskey's tenure as executive director, the Sierra Club grew dramatically. Both the club's membership and its net worth increased fivefold. When McCloskey started with the organization in 1961, the Sierra Club had a membership of 17,000, most of whom joined because they enjoyed going on the outings sponsored by the club. By 1982, the Sierra Club had grown to include nearly 300,000 members and had become much more organized and results-oriented. In an interview in a 1982 *Sierra* magazine, McCloskey talks about the Sierra Club's overarching philosophy and how the club has been able to maintain a common philosophy considering the growth it has undergone and the diversity of members it has attracted. There is, he believes, a philosophy that binds Sierra Club members together. What it boils down to is a "respect for the needs of future generations, of other creatures and of the processes that make life on the planet possible." It was this respect that allowed the Sierra Club to address such issues as pollution, energy, population growth, urban planning, economics, and transportation under McCloskey's leadership.

A 1996 memo of McCloskey's received a significant amount of attention in the United States. The memo, entitled "The Limits of Collaboration," was designed to "spur discussion" among the Sierra Club's board of directors, but it ended up being published in *High Country News* and *Harper's Magazine*. In the memo, McCloskey questions the true effectiveness of collaborative decision making in resource management, which at that time was coming into vogue as a management strategy. He argues that while collaborative, community-based management does empower the communities doing the managing, it also disenfranchises the significant urban constituency of large environmental organizations such as the Sierra Club. McCloskey also argues against the effectiveness of the consensus decision-making aspect of collaboration, stating that "only ideas of the lowest common denominator survive." "The public and the environment," in his opinion, "deserve better." This opinion of his is in direct opposition to the opinion of many environmentalists who believe that consensus-based decision making plays an important part in environmental policy decisions.

McCloskey retired from the Sierra Club in 1999. He remains on the faculty of the University of Michigan's School of Natural Resources and the Environment, where he teaches courses on the management of advocacy organizations and on international environmental policy. He has been a faculty member since 1988. Since 1965, McCloskey has been married to Maxine Mugg Johnson McCloskey. An important conservationist in her own right, Maxine McCloskey has worked for the protection of marine mammals. Michael McCloskey has contributed to numerous books and periodicals and has received several important awards for his conservation work. He was presented with the John Muir Award in 1979, was named to the UN Environmental Program Global 500 Honor Roll in 1992, was given a Lifetime Achievement Award by the Wilderness Foundation in 1992, and received an Honor award from the Natural Resources Council of America in 1999.

BIBLIOGRAPHY

Cohen, Michael P., *The History of The Sierra Club 1892–1970*, 1988; Glendin, Frances, "A Talk with Mike McCloskey, Executive Director of the Sierra Club," *Sierra*, 1982; McCloskey, Michael, "The Limits of Collaboration," *Harper's Magazine*, 1996.

McDonough, William

(February 20, 1951–)
Architect, Designer

William McDonough is a prize-winning architect whose design of the headquarters of the Environmental Defense Fund (EDF) in 1985 helped launch the "green" building movement. As the so-called Green Dean of the University of Virginia's School of Architecture, he founded the Institute for Sustainable Development to foster innovative design and promote environmentally sustainable business practices. Designer of products ranging from athletic shoes to recyclable carpeting, McDonough has expanded the idea of environmental design from buildings to products to manufacturing processes.

William McDonough was born on February 20, 1951, in Tokyo, Japan, the son of Bari and James E. McDonough, a U.S. executive with Seagram's. He was raised in Hong Kong. After receiving his undergraduate degree from Dartmouth College in 1973, he worked in Jordan for a year as a planning consultant. McDonough returned to the United States to attend Yale, but after receiving his architecture degree in 1976, he once again went overseas, this time to pursue an interest in energy-efficient building. He is recognized as having designed one of the first solar-heated homes in Ireland. His early exposure to the architecture of Asia and the Middle East had a lasting influence on McDonough. Impressed by native dwellings that responded to the landscape and climate, such as the energy-efficient Bedouin tent, he aimed for a similar harmony between building and environment. Among his professional influences, McDonough cites Joseph Paxton, Louis Sullivan, Frank Lloyd Wright, and Mies van der Rohe. He also admits to being inspired by idea people from a broad cross section of disciplines, from BUCKMINSTER FULLER to organic gardening entrepreneur ROBERT RODALE.

McDonough joined Davis, Brody and Associates in New York in 1977; he represented the firm in a joint venture, working in the office of Jacquelin Robertson. In 1981 he founded the firm now called William McDonough + Partners in New York. He developed a design approach based on the premise that an architect should consider not only the needs of the client but also the needs of the surrounding landscape and of the people who will use the building. The firm's guiding principles—best summed up as above all, do minimum harm—formed the foundation of McDonough's philosophy, fully expressed in his "Hannover Principles" almost twenty years later.

One of McDonough's early projects was designing the headquarters of the Environmental Defense Fund in New York in 1985. Concerned about the "sick building" syndrome that was widespread in tightly sealed buildings built in the energy-conscious 1970s, the EDF specified that its headquarters be designed for optimal indoor air quality. The building featured what would become McDonough's basic building blocks: minimal use of toxic materials and maximum use of natural light. McDonough more than met his goal of improving indoor air quality; the building provided a fresh air exchange six times the national standard. The EDF building is generally recognized as one of the innovative designs that sparked the "green architecture" movement.

McDonough was often frustrated in his attempts to use safe building materials because of the dearth of suitable products such as nontoxic paints and adhesives. He realized that "green architecture" could not develop without a similar "greening" of industry. Materials, products, even manufacturing and building processes needed to be created with environmental considerations in mind—in short, industry would have to be reengineered from the bottom up. In 1995, McDonough formed a sister company, McDonough Braungart Design Chemistry, to address this need. With his partner, German chemist Michael Braungart, McDonough has designed many nontoxic, recyclable, or biodegradable products, including cosmetics, toys, food containers, athletic shoes, upholstery fabric, and carpeting. He has also served as a consultant to major companies, including Ciba Geigy, Unilever, Monsanto, Ford Motor Company, and IBM, working with them to improve their products and manufacturing processes.

In 1992, McDonough participated in the UN Conference on Environment and Development (also known as the Earth Summit) in Rio de Janeiro, Brazil, where he represented the American Institute of Architects (AIA). McDonough moved his company from New York to Charlottesville, Virginia, in 1994 when he accepted a position at the University of Virginia. In 1996, McDonough founded the university's Institute for Sustainable Design. The institute advocates innovative, ecologically sound design and promotes the development of environmentally sensitive business practices in the global marketplace. McDonough, the dean of the School of Architecture, became known as the Green Dean.

The *Business Week* Architectural Record Award for the best corporate headquarters was awarded to McDonough in 1998 for his design of Gap Inc.'s new headquarters in San Bruno, California. In addition to his customary use of low toxicity materials and abundant natural lighting, McDonough also used design features—such as a sound buffer of soil, native grasses, and wildflowers on the roof—to create a healthy working environment for employees. The headquarters showcases McDonough's ability to design buildings that are not only environmentally friendly, but also people friendly.

Other McDonough + Partners projects of the 1990s included Nike's European headquarters, Coffee Creek Center (a pedestrian-oriented development) in Chesterton, Indiana, and the Joseph Lewis Center for Environmental Studies at Oberlin College in Oberlin, Ohio. McDonough designed the Oberlin center as a teaching tool for students that demonstrates practical solutions to environmental problems, such as using an artificial wetlands to filter waste water. The center received both a *Chicago Atheneum* American Architecture Award and an AIA Committee on Architecture for Education Honor Award in 1999.

During the years 1993 to 1996, McDonough served as adviser to the President's Council on Sustainable Development. In 1999, he received the only Presidential Award for Sustainable Design ever awarded to an individual. In addition to his professional awards, McDonough has been recognized in many popular forums. In 1995, he was selected as one of 100 Visionaries Who Could

Change Your Life by *Utne Reader Magazine*. Four years later, McDonough was named one of *Time* magazine's Heroes for the Planet.

McDonough, known as an eloquent and impassioned speaker, has been praised as a visionary and criticized as a utopian. McDonough's philosophy of sustainability is outlined in a set of guidelines he wrote for the city of Hannover, Germany, host of EXPO 2000. The nine principles range from the ineffable "Respect relationships between spirit and matter" to the practical "Create safe objects of long-term value." McDonough believes that the twenty-first century will bring a new industrial revolution in which business, not governmental regulation, will be the driving force for environmental progress.

McDonough stepped down from his position of dean at the University of Virginia School of Architecture in 1999, but he continues to teach in the architecture and business schools. Reflecting its increased focus on promoting environmentally sound business practices, the institute was renamed the Institute for Sustainable Design and Commerce and moved from the School of Architecture to the Darden School of Business in the summer of 2000. McDonough serves as creative director of the institute and holds the position of alumni research chair at the Darden School of Business. McDonough serves on the board of Second Nature, a nonprofit organization in Boston that helps higher education institutions incorporate environmental sustainability into their curriculum.

McDonough lives in Charlottesville, Virginia, with his wife, Michelle, and children, Drew and Ava.

BIBLIOGRAPHY

Braungart, Michael, and William McDonough, "The Next Industrial Revolution," *Atlantic Monthly*, 1998; "Institute for Sustainable Design and Commerce," www.virginia.edu-sustain/; "The Hannover Principles: Design for Sustainability," http://www.mcdonough.com/principles.pdf; Klimint, Stephen, "Green Giant," *Planning*, 1999; "McDonough + Partners: Architects and Planners," www.mcdonough.com; Rosenblatt, Roger, "The Man Who Wants Buildings to Love Kids," *Time*, 1999.

McDowell, Mary

(November 30, 1854–October 14, 1936)
Settlement Worker, Garbage Activist

Known affectionately as Chicago's "Garbage Lady," Mary McDowell was a leader in the settlement movement in Chicago and was instrumental in cleaning up the city's garbage and sewers. She led the University of Chicago Settlement, located in "Packingtown," one of Chicago's poorest neighborhoods. She was active in many of the Progressive movement's issues, including union organizing, literacy campaigns, justice for African Americans, and fair hous-

Mary McDowell (R) stands with Julia Lathrop (L), a social reformer who fought for the betterment of conditions in mental institutions and childcare facilities, and Jane Addams (M). (Corbis)

ing, and was eventually appointed commissioner of public welfare. For her work with the least powerful of Chicago's citizens, Mary McDowell was dubbed the "Angel of the Stockyards."

Mary McDowell was born on November 30, 1854, in Cincinnati, Ohio, to Jane and Malcolm McDowell. When she was seven, the Union army called her father into service, and McDowell developed a keen identification with President Lincoln and the cause of abolition. One biographer suggests her lifelong interest in the rights of African Americans stemmed from her experiences during the Civil War. When McDowell's father returned from the war, he moved the family to Chicago. Malcolm McDowell started a successful steel foundry, and the McDowells became prominent Chicago citizens. McDowell had her first experience with social work during the great Chicago fire of 1871. She helped people evacuate the city and then organized and distributed relief supplies. Rutherford B. Hayes, then the governor of Ohio, was an old friend of the McDowell family, and he delivered his state's aid directly to the McDowell home. In the early 1880s the McDowells moved to Evanston, Illinois, where Mary met Frances Willard, leader of the Woman's Christian Temperance Union. McDowell became a close friend of Willard and worked as a state organizer for the Young Woman's Christian Temperance Union.

In 1890, McDowell went to work with JANE ADDAMS at Hull House in Chicago. The settlement movement aimed to combat urban poverty on a neighborhood level. Settlement houses were centers where people could seek education, childcare, help in gaining employment, and housing aid. The workers were mostly middle-class Progressive reformers and in many ways were the first professional social workers in the United States. Though McDowell only taught kindergarten at Hull House for a few months, it was to prove a turning point in her life. In 1894, the new Department of Sociology at the University of Chicago decided to found a new settlement, and Addams recommended McDowell for the position of head resident. McDowell accepted the position, moved into the University of Chicago Settlement, and remained there for the rest of her life. The settlement was "back of the yards," in the neighborhood nearest the meat packing yards. The situation in Packingtown was dismal: makeshift, overcrowded housing; massive unemployment; and utter lack of sanitation facilities. McDowell's arrival in the community was greeted with suspicion. There had been a massive, violent meat packers' strike a few months earlier, and many workers thought McDowell was a spy, sent by plant owners to identify labor leaders. McDowell eventually earned the trust of her neighbors through hard work and effective advocacy. She founded the area's first kindergarten, organized a successful neighborhood cleanup campaign, and fought for and won public bathing facilities. She was given the nickname "Fighting Mary" for her labor organizing. McDowell helped organize one of the first women's unions, the Women's Trade Union League of Chicago, and actively supported the workers during the packing strike of 1904. McDowell published articles in the local press in support of the strikers and helped raise money; she is credited with helping maintain the peace. The plant owners brought in African American replacement workers, and McDowell intervened to stop the strikers from carrying out reprisals, which would have taken the form of lynch mobs.

McDowell also led the fight for a safer, healthier environment. Packingtown was bordered on one side by the meat plants, on another by the city's vast, open garbage pits, and on the third side by an arm of the Chicago River called Bubbly Creek because of its high content of carbonic acid gas. The combination of these factors and the overcrowded, impoverished living conditions created serious health hazards. McDowell organized for years to get the city to clean up and cover the river and did make headway, though modern waste treatment facilities were not built during her lifetime.

McDowell was more successful with garbage. The municipal garbage dump bordering Packingtown was an exhausted clay mine, owned by a powerful alderman who charged private haulers a hefty dumping fee. The alderman used his influence to block any efforts to improve sanitation at the dump. McDowell's first success was to get the city to extend garbage collection services to her neighborhood. But this failed to alter the pollution of the pits themselves. McDowell began a huge press campaign to generate public support and political pressure. In 1911, a supporter financed McDowell's trip to Europe, where she researched the disposal methods of large cities in England, France, Holland, and Germany, cities with better sanitation programs than Chicago. She gathered information that could be useful in a public relations campaign and on her return began speaking to numerous groups around the city. Mass meetings were organized throughout Chicago, women's clubs began organizing garbage committees, and the city council began to take notice. In 1913, the first modern garbage reduction plant was built, in which greases were extracted from the garbage and resold for a number of commercial purposes. The open dumps were closed, ending a significant threat to the health of Packingtown.

McDowell continued her work in the settlement throughout her life and saw a number of achievements. She was instrumental in generating the publicity that pushed Pres. THEODORE ROOSEVELT and the U.S. Congress to enact inspection regulations at the packing plants. She helped create networks of parks and community gardens. In 1923 she was appointed commissioner of public welfare by Mayor William Dever, in which capacity she set up a Bureau of Employment and lobbied for a permanent Housing Commission. McDowell's work earned her the respect of her neighbors, Chicago's political machine, and the profession of social work. She died on October 14, 1936. The University of Chicago Settlement was renamed the Mary McDowell Settlement in her honor.

BIBLIOGRAPHY

Davis, Allen, *Spearheads for Reform*, 1967; Taylor, Lea, "The Social Settlement and Civic Responsibility—The Life Work of Mary McDowell and Graham Taylor," *Social Service Review*, 1954; Wade, Louise, "Mary E. McDowell," http://www.kentlaw.edu/ilhs/mcdowell.html; Wilson, Howard, *Mary McDowell: Neighbor*, 1928.

McHarg, Ian

(November 20, 1920–)
Landscape Architect, Environmental Urban Planner

Ian McHarg is a landscape architect and environmental urban planner. His famous book *Design with Nature*, released in 1969, calls for ecologically sensitive urban planning and for the recognition that humans can and must create cities that exist in harmony with nature. McHarg is responsible for bringing ecological awareness to urban planning and landscape architecture, and he is responsible for elevating both of these fields to national prominence, in both the political and social spheres.

Son of John Lennox and Harriet Bain McHarg, Ian Lennox McHarg was born on November 20, 1920, in Clydebank, Scotland. He spent his childhood living in between examples of two very different types of human occupation of land. The large and smoking city of Glasgow was only ten miles to the east of his home. The blast furnace flames lighting the eastern horizon at night made a lasting impression on him. To the west were countryside and the Firth of Clyde, with its estuary opening into the Atlantic Ocean. In the introduction to *Design with Nature* McHarg writes, "During all of my childhood and youth there were two clear paths from my home, the one penetrating further and further to the city and ending in Glasgow, the other moving deeper into the countryside to the final wilderness of the Western Highlands and the Islands." According to McHarg, the career he eventually pursued was a response to the simple choice laid out for him in his childhood: beautiful countryside or city? He chose both. He discovered landscape architecture at the age of 16 and saw in it the opportunity to provide for those who lived in cities the positive experiences he had found in nature.

In 1939, McHarg entered the British Army, eventually serving as an officer with the Second Independent Parachute Brigade Group in Italy for two years beginning in 1943. After World War II, McHarg spent four years at Harvard University in Cambridge, Massachusetts, studying landscape architecture. He earned his bachelor's degree in 1949, a master's in landscape architecture in 1950, and a master's in city planning in 1951. In 1947, McHarg married Pauline Crena de Longh. They would have two sons together. In 1950, McHarg took his new family back to Scotland. He was "determined to practice my faith upon that environment of drudgery that is Clydeside." He worked as a planner for the Scotland Department of Health from 1950 to 1954. During this time, he discovered that he had contracted pulmonary tuberculosis. He spent the first six months of his convalescence in a hospital on the outskirts of Edinburgh, a hopeless and horrible experience for him. After six months, he transferred to a Swiss sanatorium for British parachutists, where he recovered. For the next half year, he hiked the Swiss countryside, an experience that impressed upon him the healing powers of the natural world, not just for the spirit, but for the flesh as well.

In 1954, McHarg was invited to join the faculty at the University of Pennsylvania. He accepted. At the University of Pennsylvania, he established the Department of Landscape Architecture and Regional Planning. McHarg remained at the University of Pennsylvania full-time for the next 32 years, serving as department chair from 1960 until 1986, when he retired. Today he is professor emeritus.

In the late 1960s, Russell Train, then head of the Conservation Foundation (he would later become head of the Environmental Protection Agency during the Nixon administration), approached McHarg and requested that he write a book on the ecological aspects of urban planning. McHarg accepted and spent one year of his life producing *Design with Nature*, which was published in 1969. This book was extremely well received and reviewed. It was reprinted four times in its first year alone and has become the book of authority for the design and planning professions, as well as for students of ecology. When *Design with Nature* was published, there was no legislation requiring an ecological understanding of areas to be developed (the role environmental impact statements fulfill today). As McHarg wrote in his autobiography, *A Quest for Life*, published in 1996, "The book contributed to the efflorescence of environmental legislation. Moreover, it has increasing relevance for countries only now confronting the crisis of the environment and the need for ecological planning."

In *Design with Nature*, McHarg pioneered layered mapping techniques that were the precursor to the geographic information systems (GIS) used in much of modern mapping. McHarg created a system whereby all of the different physical attributes of a site—soil, drainage, vegetation, groundwater—are recorded on separate, transparent maps that are laid over one another to determine which areas are most suitable for development.

His book explains that the dynamic processes of nature should be considered in the design process and that "changes to parts of the system affect the entire system." *Design with Nature*, which has been described as "almost a book of poetry," is still required reading in many universities' departments of landscape architecture. It was nominated for a National Book Award in 1971.

In 1963, McHarg helped to create the firm Wallace-McHarg, Roberts and Todd, architects, landscape architects, and planners in Philadelphia. The firm grew to include offices in Los Angeles, Miami, Denver, and San Francisco. Through this firm, McHarg was involved in such projects as a 1962 plan for four contiguous Maryland river valleys, in which it was proposed, for the first time, that development rights be transferred in order to preserve a landscape. He also helped to design the 600-acre Pardisan Environmental Park outside of Tehran, Iran, in 1976. Fee disputes over this project led to McHarg's leaving the firm.

McHarg has maintained a busy schedule in his retirement. He was a visiting professor at the University of California, Berkeley, in 1986 and 1987 and at Pennsylvania State University and Harvard in 1994. In 1996, he published his autobiography, *A Quest for Life*. And in 1998 he released *To Heal the Earth: Selected Writings of Ian L. McHarg*, a compilation of some of his most important writings. In 1977, three years after the death of his first wife, McHarg married Carol Anne Smyser, with whom he has two sons. He and his family live together on a 35-acre plot of land 40 miles away from the University of Pennsylvania. McHarg has received numerous awards for his accomplishments, including the Bradford

Williams Medal of American Society of Landscape Architects, 1968; an outstanding achievement award from Harvard University, 1992; the Thomas Jefferson Medal from the University of Virginia, 1995; and the Pioneer Award from the American Institution of Certified Planners, 1997.

BIBLIOGRAPHY

Holden, Constance, "Ian McHarg: Champion for Design with Nature," *Landscape Architecture*, 1977; Knack, Ruth, "Utopia!" *Planning*, 1997; Landecker, Heidi, "In Search of an Arbiter," *Landscape Architecture*, 1990; McHarg, Ian L., *Design with Nature*, 1969; McHarg, Ian L., *A Quest for Life*, 1996.

McKibben, Bill

(December 8, 1960–   )
Writer

Bill McKibben came to prominence as an environmental writer and naturalist with the 1989 publication of his groundbreaking book, *The End of Nature*. His overriding concerns have been rampant consumerism and the effects of human encroachment on nature, which are altering the global ecosystem. McKibben's books follow a progression from apocalyptic to homiletic: In *The End of Nature*, he sets the argument, warning of the disastrous results of proliferating greenhouse gases; then, in his next book, *The Age of Missing Information*, he addresses the causes of human indifference to the environment. His recent books, while maintaining his oratorical style and the theme of materialism and its consequences, offer advice for change and are increasingly personal expressions of hope for the future of the natural world.

William Ernest McKibben was born in Palo Alto, California, on December 8, 1960, and reared in Lexington, Massachusetts. His parents, Gordon McKibben and Margaret Hayes McKibben, were both journalists for the *Boston Globe*. Throughout his childhood, young McKibben envisioned himself as "a newspaperman" and at Harvard University worked on the student newspaper, *The Crimson*.

McKibben graduated with a B.A. in 1982 and went to work for the *New Yorker*, whose editor, William Shawn, had admired McKibben's college journalism. At the *New Yorker*, he wrote hundreds of articles, primarily "Talk of the Town" stories, as well as longer, general-interest features and humorous fiction. In 1987, he quit the *New Yorker* to become a freelance magazine writer and to work on his first book. He continues to contribute articles to a wide range of publications, including the *Atlantic*, the *New York Review of Books*, the *New York Times*, *Natural History*, *Outside*, *Rolling Stone*, *Esquire*, *Audubon*, and many others.

The End of Nature was published in 1989 and became an almost instant bestseller. Dire, alarming, and provocative, it was a source of great controversy and gave McKibben a reputation as a prophet of ecological doom. He warned of imminent global disaster, basing his ideas on scientific evidence about the greenhouse effect. With clarity, precision, and universally praised eloquence, he outlined the implications to the environment of the industrial world's proliferation of methane and carbon dioxide, which prevent heat from escaping the atmosphere, causing the earth's temperatures to increase, the polar ice caps to melt, oceans to rise, and droughts to occur.

Decades of civilization, McKibben wrote, have created devastating pollution and are dominating the forces of nature. "The 'thermal equilibrium'—the heat storage—of the oceans may be saving us at the moment. But if so it is only a sort of chemical budget deficit. Sooner or later our loans will be called in. . . . We have done this to ourselves, by driving our cars, cutting down our forests, turning on our air conditioners," he wrote. He took to task not only the consumer society but government agencies and environmentalist groups, which couch their arguments in terms of nature's economic benefits and make compromises to achieve what is politically possible in the short term.

Noting that our ecological problems are "huge and growing," McKibben called for radical solutions, notably ending the use of machines that burn oil and coal. He eschewed traditional answers, such as planting more trees to absorb carbon monoxide, describing them as too little too late and even misguided, when the solutions must come by changing our consumerist way of life.

How humans live, particularly those few Americans who use most of the earth's resources, is an ongoing thread through McKibben's writings. In 1992,

with *The Age of Missing Information*, he examined the causes of human indifference to the environment, the twin culprits of conspicuous consumption and television, "our main information source."

To make his case, McKibben taped 103 television channels during a 24-hour time span, then watched every hour of every channel. He found that the endless blast of information from sports, weather, and news channels; shows about people and comedy; MTV and its imitators; and so on are not only empty of content but exacerbate the boredom we already fear. He concluded that Americans are "grounded" by "intravenous entertainment," ranging from an average eight hours a day of television watching to computers and "the phone in the airplane-toilet. . . . our minds are jazzed . . . we fear boredom . . . we are hooked on infodrug . . . [and] any break in the action seems unnatural, a vacuum." We are therefore distracted from nature and any imperative toward understanding and solving our environmental problems.

He followed his television blitzkrieg with a week alone in the mountains. The "three rarest commodities" of contemporary life are "solitude, silence and darkness," he wrote. Quiet contemplation in the midst of nature allows one to become observant, refreshed, and content and to hear "natural broadcasts . . . sense the presence of the divine . . . that has marked human beings in every culture as far back as anthropologists can go."

In 1995, McKibben published *Hope Human and Wild: True Stories of Living Lightly on the Earth*, an attempt to "convince myself and others that it is not completely pie-in-the-sky to imagine there could be other ways to conduct ourselves. . . . The point of the book is to counter despair," he told Michael Coffey of *Publishers Weekly* in 1995.

In *Hope Human and Wild*, McKibben illustrates the nuts-and-bolts, everyday specifics of three locales: Curitiba, a mountain city in Brazil; Kerala, a community in India; and his own home in the Adirondack Mountains of New York, where he found enormous environmental and social progress thanks to human imagination and prudent government.

McKibben tackled issues of overpopulation in 1998 with *Maybe One: A Personal and Environmental Argument for Single Child Families*. In this attempt to persuade Americans to consider reducing strain on the earth's resources by having just one child, McKibben spoke frankly of how he and his wife agonized over the issue and finally decided to have just one. In the book's final section, he describes at length his vasectomy.

"When we think of overpopulation, we usually think first of the developing world . . . [b]ut . . . we fool ourselves when we think of population as a brown problem," he wrote. He listed statistics showing, for example, that one "American uses seventy times the energy of a Bangladeshi, fifty times that of a Malagasy, twenty times that of a Costa Rican. . . . During the next decade, India and China will each add to the planet about ten times as many people as will the United States . . . [but] the 57.5 million northerners . . . added during this decade will add more greenhouse gases to the atmosphere than the roughly 900 million southerners."

Between publication of his major books, McKibben has produced several smaller ones, based on his spiritual viewpoint, including *The Comforting Whirlwind: God, Job and the Scale of Creation* (1994), in which he challenged current thinking about the environmental crisis through a reading of the book of Job. McKibben teaches Sunday School and is a lay minister at the Methodist church of his rural community. McKibben's new edition of HENRY DAVID THOREAU's *Walden* emerged in 1997. Whereas most recent scholarship has focused on Thoreau as forest ecologist, McKibben's annotations and introduction to the 1854 edition place Thoreau back in his role as spiritual and cultural seer. In 1998, *Hundred Dollar Holiday: The Case for a More Joyful Christmas* was published. Originally begun as a project for his church, it offers tips on simplifying the frenzied materialist holiday against "those relentless commercial forces" and returning it to a time of fellowship with each other and the natural world.

McKibben lives in Johnsburg, New York, with his wife, writer Sue Halpern, and their daughter, Sophie.

BIBLIOGRAPHY

Coffey, Michael, "Bill McKibben: Environmental Hope in Conservative Times," *Publishers Weekly*, 1995; McKibben, Bill, *The Age of Missing Information*, 1992; McKibben Bill, *The Comforting Whirlwind: God, Job, and the Scale of Creation*, 1994; McKibben, Bill, *The End of Nature*, 1989; McKibben, Bill, *Hope, Human and Wild: True Stories of Living Lightly on the Earth*, 1995; McKibben, Bill, *Hundred Dollar Holiday: The Case for a More Joyful Christmas*, 1998.

McPhee, John

(March 8, 1931–)
Writer

John McPhee is a reporter and a writer of creative nonfiction whose focus on the interactions and relationships between people and the natural world has been instructive and revealing. He is not an editorialist, and his environmentally focused books do not preach any kind of message. Rather, his works such as *Encounters with the Archdruid* (1972), *Coming into the Country* (1977), and *The Control of Nature* (1989) simply present the characters of the people and of the land they live in and allow readers to draw their own conclusions.

John Angus McPhee was born on March 8, 1931, in Princeton, New Jersey. He was the youngest child born to Harry Roemer McPhee and Mary Ziegler McPhee. His father, a physician, was a specialist in sports medicine and was the regular physician for Princeton University's athletic teams. McPhee grew up watching Princeton's various athletic teams practice and became, despite his small stature, an accomplished athlete himself. Throughout his childhood, McPhee spent his summer months at Keewaydin Camp in Vermont, a place he would later refer to as the most important educational institution he attended. He attended Princeton High School, where he was active in sports and where his skills as a writer began to take shape. He remembers his English teacher, Olive McKee, as having a considerable influence on him.

McPhee graduated from Princeton High School in 1948, at the age of 17. He had already been accepted into Princeton University, but his mother, believing that he needed another year to mature, sent him to Deerfield Academy in Massachusetts. At Deerfield, McPhee studied under the guidance of the headmaster, Frank L. Boyden, who many years later was the subject of McPhee's book, *The Headmaster* (1966). McPhee entered Princeton University in 1949. At Princeton, he studied English and appeared as the regular teenage panelist on the radio and television quiz show, *Twenty Questions*. He graduated from Princeton with a B.A. in English in 1953. He satisfied his senior thesis requirement by submitting an unpublished novel, *Skimmer Burns*. In doing so, he set a precedent for future students in the Princeton English Department. McPhee spent his next year in England, doing postgraduate work at Cambridge University. He studied literature and in his spare time toured with a basketball team that played against teams from schools such as Oxford and the London School of Economics.

Returning to the United States in 1954, McPhee settled for a short time in New York City in an attempt to establish himself as a freelance writer. Biographer Michael Pearson quotes McPhee about this period of his life: "I didn't rule out anything as a younger writer. I tried everything." McPhee wrote speeches and articles for W. R. Grace and Company in New York City. He wrote short stories and sold several television scripts to a popular show of the time, "Robert Montgomery Presents." In 1957, McPhee became a reporter for *Time*. For five years, he wrote the magazine's "Show Business" column and also produced nine cover stories on such celebrities as Joan Baez and Barbara Streisand. He continued selling articles to various periodicals, and in 1963, the *New Yorker* purchased "Basketball and Beefeaters," an account of his experiences playing basketball in England. McPhee sold another article to the *New Yorker* two years later. Entitled "A Sense of Where You Are," the article was about Princeton basketball star and general All-American hero, Bill Bradley. The article was published in book form in 1965 and was praised by critics for its clarity and its capacity to inspire without preaching.

Soon after the publication of *A Sense of Where You Are*, McPhee accepted a position as a staff writer with the *New Yorker*, which he still occupies today. The *New Yorker* pays employee benefits but does not provide a salary. McPhee is paid only when articles of his are accepted for publication. He can, however, sell any of his articles that are rejected by the *New Yorker* to other periodicals. Since 1974, McPhee has also been the Ferris Professor of Journalism at Princeton University.

McPhee has written more than 30 books on a wide variety of subjects, many of them addressing, in one form or another, the natural world and the people who live in and around it. In *Encounters with the Archdruid* (1972), he tags along with environmentalist DAVID BROWER in a series of three confrontations/discussions with an exploration geologist, a dam builder, and a developer. In these encounters, McPhee draws no conclusions of his own but presents the "characters" with all of their complexities and allows readers to choose sides for themselves. Another book of his that focuses on the natural world, *Coming into the Country*

John McPhee (Courtesy of Anne Ronan Picture Library)

(1977), is perhaps his most popular environmentally focused work. Its plot line focuses on a search for a new Alaskan capital to replace Juneau. According, again, to Michael Pearson, this book asks questions about the type of self-confidence and skill that exists among the human dwellers of this daunting landscape. It asks, "Will this survive, especially if the wild country is lost or even tamed into more civilized parks?" As in *Encounters with the Archdruid*, McPhee does not answer this question for himself. He allows his subjects to decide. And the answers are never simple or clear, if only because McPhee provides so many different points of view.

In another book, *The Control of Nature* (1989), McPhee examines the relationships between humans and the natural world in three different parts of the world:

the Mississippi River, an Icelandic volcano, and the San Gabriel Mountains of southern California. He portrays three scenarios in which humans attempt to force uncooperative nature to comply with their demands. Along the Mississippi, humans attempt to prevent the totally natural shift of the river out of its current bed and on to a different course. In Iceland, people fight the eruption of a volcano with fire hoses in an attempt to save the commercially valuable harbor on the island of Heimaey. And in southern California, suburbanites, against all common sense, occupy hillsides prone to massive mud slides. McPhee has also produced a series of four books about American geology. Called *Annals of the Former World*, the series is composed of *Basin and Range* (1981), *In Suspect Terrain* (1983), *Rising from the Plains* (1986), and *Assembling California* (1993). These books describe McPhee's tours of the United States with professional geologists and are considered by many the most ambitious of McPhee's work. They provide a perspective into the dramatically short period of time humans have occupied the planet by presenting humans as but a minor character in the vast history of earth.

While the majority of McPhee's books focus on the natural world, he has written on a variety of other subjects such as tennis, oranges, and Russian art. He has received many awards for his writings, including the Woodrow Wilson Award from Princeton University in 1982, the John Wesley Powell Award from the United States Geological Survey in 1988, and the John Burroughs Medal in 1990. McPhee lives in Princeton, New Jersey, with his wife, Yolanda Whitman, whom he married in 1972. He has four grown children from a previous marriage: Laura, Sarah, Jenny, and Martha, as well as four stepchildren: Cole, Andrew, Katherine, and Vanessa Harrop.

BIBLIOGRAPHY

Hamilton, Joan, "An Encounter with John McPhee," *Sierra*, 1990; Jones, Daniel, and John D. Jorgenson, eds., *Contemporary Authors*, 1999; Pearson, Michael, *John McPhee*, 1997.

Meadows, Donella H.

(1941–)
Founder and Director of the Sustainability Institute, Writer

A self-proclaimed grassroots worker, Donella H. Meadows arose as a voice for environmental consciousness in the early 1970s with the publication of *The Limits to Growth*, a slim yet controversial book that sold nine million copies and linked population and economic growth with the environment.

Since then her reputation as an advocate for sustainable systems has flourished. Founder and director of the Sustainability Institute, the prolific Meadows also publishes pamphlets and writes a range of magazine and earth-friendly articles—the most visible being her syndicated column, "The Global Citizen." In addition to her writing, Meadows teaches environmental studies at Dartmouth College and lives on a small farm where she practices sustainable agriculture.

Often labeled professionally as a systems analyst and international coordinator of management systems, the multifaceted yet independent and private Donella "Dana" Hager Meadows defies any such convenient categorizing and prefers not to publicize her personal history. Instead she chooses to rechannel any attention that comes her way toward the ideas she considers vital both to the welfare of the environment and its resources and to human communities and the human spirit. What is open knowledge about Meadows's personal life is that she was born in 1941 in the United States, received her Ph.D. in biophysics from Harvard University in 1968, and was married to Dennis L. Meadows, who today is a professor of systems management and director of the Institute for Policy and Social Research at the University of New Hampshire.

With the help of Dennis Meadows and Jorgen Randers, currently a policy analyst and president emeritus of the Norwegian School of Management, "Dana" Meadows was the lead writer of the 1972 book *The Limits to Growth* and the 1992 update *Beyond the Limits: Confronting Global Collapse, Envisioning a Sustainable Future*. While working together at the Massachusetts Institute of Technology (MIT) in the late 1960s and early 1970s, the three researchers developed computer models, based on the exponential concept of doubling, that demonstrate a range of future scenarios for our global economy. *The Limits to Growth* presented 12 computer graphs, from the most negative case predicting an exhaustion of nonrenewable resources and hence the deterioration of human life, to the most optimistic case promoting a European sustainable standard of living. Twenty years later in *Beyond the Limits*, the authors again used computer models to restate their concerns, with graphs showing a growing world population stripping food supplies and natural resources by 2100. In 1985—between work on *The Limits to Growth* and *Beyond the Limits*—Donella Meadows, along with photographer Jenny Robinson, published *The Electronic Oracle: Computer Models and Social Decisions*, a critical look at computer modeling as a method for social, economic, and political analysis. Then in 1991 Meadows authored *The Global Citizen*, a collection of personal stories, research experiences, and environmental insights. This book gave its title, tone, and format to her weekly opinion column—a column that the Context Institute's on-line publication *In Context* calls "one of the most reliable places we know for holistic and humane commentary on world affairs." Meadows's articles also appear in *Amicus Journal, Earth Island Journal*, and *Organic Gardening* and in *Whole Earth*, where she is a contributing editor.

Regardless of the format, Dana Meadows tenaciously works to counter the "religion" of consumption and growth in hopes of turning the tide toward moderation and sustainability. Indeed, as a writer she might be compared with RACHEL CARSON for a natural prose style that combines readability with scientific discourse to convey a futuristic warning. Similarly, Meadows uses her prose with its snappy journalistic style as a venue to temper scientific theories with a positive vision. Indeed, while she is compelled by the predictability of her computer model calculations that forecast a dubious future of depleted earth resources, she also believes we all have the responsibility to make informed choices. Consequently in recent years she has allowed spirituality to influence her mindset, which had previously been focused solely by scientific method. For Meadows, therefore, sustainability has come to take on an encompassing meaning—a meaning that she clarified at a roundtable discussion of environmentalists conducted by *The Amicus Journal* as the "protection of the environment, social justice, and all of the other issues . . . [getting off] growth and productivity and quantitativeness and reductionism, into wholeness and wellness and living with limits."

Despite her persistent fight for sustainability, in an interview with *Social Justice* Meadows has admitted that she is "disheartened to hear the political debate just talk about growth, growth, growth, growth and to hear nobody stand up and challenge them." Meadows's seemingly indefatigable desire for action led to the development of the Sustainability Institute—"a think-do tank"—which sponsors people, programs, and initiatives targeted toward raising awareness about the consequences of our lifestyle decisions. A Pew Scholar in Conservation and Environment and a MacArthur Fellow, Dana Meadows lives the life she professes, from farming organically to driving a gas-electric hybrid car. Her latest project is an environmental studies textbook written, naturally, from a systems perspective.

BIBLIOGRAPHY

"Environmental Futures," *The Amicus Journal*, 1995; Meadows, Donella H., "The Global Citizen," http://iisd1.iisd.ca/pcdf/meadows/default.htm; "Rio's Success," *In Context #33*, http://www.context.org/ICLIB/backi.htm; Walljasper, Jay, "Rethinking the Left," *Social Justice*, 1996.

Meany, Edmond

(December 28, 1862–April 22, 1935)
Historian, President of the Washington Mountaineers

Edmond Meany became the president of the Washington Mountaineers in 1907, one year after its inception, and held that position until his death in 1935. Until his death, Meany was the backbone and inspiration of the burgeoning outdoors club that still combines outdoors expeditions with conservationist activism. After the Sierra Club, no outdoors group has had a greater impact on wilderness recreation in the American West than the Mountaineers.

Edmond Stephen Meany was born on December 28, 1862, in East Saginaw, Michigan. His family moved to Washington while he was still a teenager. During the summer of 1880, Meany's father drowned while working on a steamer on the Skagit River. Meany, not yet 18, was left as the primary supporter of his mother, sister, and baby brother. In the years that followed, Meany operated a small dairy, pasturing cows on grass that grew along the sides of Seattle's unpaved streets, delivered newspapers, kept books for a grocer, and worked as a janitor both at his church and at the only bank in town.

Showing promise and determination, Meany entered the University of Washington and graduated in 1885 at the top of his class. He became a reporter at the *Seattle Press* and was promoted to city editor. In 1889–1890, the *Seattle Press* sponsored the Press Exploratory Expedition into Washington's Olympic Range; the explorers thought enough of Meany to name a mountain after him. He would later get the chance to climb his namesake peak on the first Mountaineers outing.

In 1891, Meany was elected to the state legislature and was instrumental in securing land in northern Seattle for the current University of Washington campus. The decision to move the university was controversial, as the original campus was conveniently located in downtown Seattle. Despite its panoramic views of the Cascades to the east, the Olympics to the west, and Mount Rainier to the south, the new campus was some distance out of the city, at the very end of a streetcar line. After serving for two sessions in the state legislature, Meany became the registrar at the university before becoming a history lecturer. In 1897, he was appointed as head of the history department, and in 1906, he became the managing editor of the *Washington Historical Quarterly;* he held both positions for the rest of his life.

Though not included among the charter members of the Mountaineers (founded in 1906), Meany joined just after the charter rolls closed. The Mountaineers offered an ideal combination of his major interests: aboriginal legends, Northwest history, and outdoor recreation. Meany quickly became a popular member of the club, speaking at monthly meetings and sharing his considerable knowledge of native lore. He succeeded the club's first president, Henry Landes, in 1907 and helped outline the club's mission, which remains the same almost a century later:

> To explore and study the mountains, forests and water courses of the Northwest and beyond; to gather into permanent form the history and traditions of these regions and explorations; to preserve by example, teaching and the encouragement of protective legislation or otherwise the beauty of the natural environment; to make expeditions and provide educational opportunities in fulfillment of the above purposes; to encourage a spirit of good fellowship among all lovers of outdoor life.

Meany took his role as president very seriously and set very high standards of conduct for club activities. He established standards of mountaineering ethics that prioritized group safety and minimized the impact of environmental degradation caused by large groups in mountainous and wilderness areas. These mountaineering ethics made the Mountaineers' concern for the outdoors among the foremost among outdoors groups in the nation. During Meany's leadership, the Mountaineers formed its own legislative committee to concentrate specifically on preservation issues in both the national and state arenas.

Meany was also an affable and amiable leader. He was a frequent participant in campfire entertainment on outings, telling stories of the early explorers in the Northwest, or from native mythology, or reciting his own poetry. Meany's pleasant attention to all members, old or new, made him a beloved figure the club could rally behind during its formative years. Originally just an outdoors exploration club, the Mountaineers, under Meany, expanded into rock climbing, skiing, snowshoeing, and even theater performances. Nevertheless, conservation remained a primary mission for Meany. Writing in the Mountaineers' 1910 annual, Meany postulated: "This is a new country. It abounds in a fabulous wealth of scenic beauty. It is possible to so conserve parts of that wealth that it may be enjoyed by countless generations through the centuries to come."

Meany was also a devout Christian. He always led the Sunday morning worship service on club outings. Meany bought land at the eastern end of the Stampede Tunnel, adjacent to the Northern Pacific Railroad tracks, and donated it to the club for the purpose of building a ski lodge. The Meany Ski Hut was built on this land, the sole proviso from its benefactor being that the hut be closed on Easter Sundays for religious observation.

Meany died on April 22, 1935, of a stroke, just before he was to give a lecture at the University of Washington. He was survived by his wife, Sarah Elizabeth Ward of Seattle, whom he married in 1889, and two of their four children.

BIBLIOGRAPHY

Frykman, George, *Seattle's Historian and Promoter: The Life of Edmond Stephen Meany*, 1998; Gowen, Herbert H., "Meany, the Road Maker," *Washington Historical Quarterly*, 1935; Kjeldsen, Jim, *The Mountaineers: A History*, 1998; McMahon, Edward, "Professor Meany as I Knew Him," *Washington Historical Quarterly*, 1935; Sieg, Lee Paul, "Edmond S. Meany: The Value of a Man," *Washington Historical Quarterly*, 1935; Todd, Ronald, "A Selected Bibliography of the Writings of Edmond Stephen Meany," *Washington Historical Quarterly*, 1935; Warren, James, "History of Edmond Meany," *Seattle Post-Intelligencer*, 1984.

Merchant, Carolyn

(July 12, 1936–)
Professor, Author, Environmental Historian

As a professor of environmental history and the author of several classics of environmental theory, Carolyn Merchant has broadened society's awareness of how humans and their natural environment impact each other. In her pathbreaking book, *The Death of Nature: Women, Ecology, and the Scientific Revolution* (1980), Merchant elucidated the emerging concept of "ecofeminism" and helped bring it into wide usage in environmental discourse. By delineating age-old connections between women and nature and exposing the forces that have oppressed them both, she offered a gender-based perspective as a new context for interpretation of environmental values and interactions. In *The Death of Nature* and other books, Merchant also provides historical analyses of the expansion of power over nature by social institutions, and in doing so, she puts into perspective current arguments over the importance of environmentalism and suggests guidelines for a future sustainable partnership with the natural world.

Carolyn Merchant was born on July 12, 1936, in Rochester, New York, the daughter of George and Elizabeth (Barnes) Merchant. She attended Vassar College, graduating with a bachelor's degree in 1958. She earned a master's degree four years later in 1962, from the University of Wisconsin at Madison, where she also earned her Ph.D. in 1967. Beginning in 1969, Merchant taught as a lecturer at the University of San Francisco, becoming an assistant professor of history of science in 1974, and staying until 1978. The following year she became an assistant professor at the University of California at Berkeley (UCB), and one year later she was promoted to associate professor of environmental history, philosophy, and ethics.

Intrigued by the interactions between humans and their natural environment over time, Merchant focused her research on American environmental and cultural history within the larger contexts of philosophy and the history of science. In examining the relationship of certain ideas to cultural trends, Merchant began looking into the women's and ecology movements and eventually recast her field of study to encompass the historical roots of both. The culmination of these scholarly investigations was *The Death of Nature: Women, Ecology, and the Sci-*

Carolyn Merchant (Courtesy of Anne Ronan Picture Library)

entific Revolution (1980). Merchant emphasizes the age-old association between women and nature and explores these historical connections in light of the exploitation that has left both women and the environment oppressed by culture and the economy. She undertakes a critical reassessment of the scientific revolution in the book and demonstrates how scientific progress has contributed to the domination of both women and nature. Though the term *ecofeminism* had been around since the mid-1970s, it was not until the publication of trail-blazing books such as *The Death of Nature* that the connection between feminist and ecological ideas became a popular framework for interpreting interactions between humans and the environment. New dialogues arose as people began to look at contemporary views of nature and society through a gender-based perspective, leading to new analyses of arguments about the causes of environmental degradation. For example, patriarchal parenting roles were examined and implicated in the way that some boys, as they grow up, come to want control over their environment, and people's differing outlooks on the natural world now took on gendered explanations. Merchant's book helped to establish ecofeminism as an intellectual movement and helped it develop into new approaches to action.

In a later book, *Radical Ecology: The Search for a Livable World* (1992), Merchant discusses another new concept—radical ecology—that she sees as an ethic that seeks fundamental changes in the way society and its institutions relate to nature and the environment. An offshoot of social ecology, which Merchant describes as the interactions among people and the various political and social customs that shape their views of nature and its resources, radical ecology urges new patterns of thought, challenges the political and economic order, and supports social movements in the restoration and protection of the environment. In *Radical Ecology*, Merchant discusses how other concepts such as deep ecology, ecofeminism, sustainable development, and green politics fit into the definition of radical ecology and shows how, by presenting alternative views, it raises public consciousness and exposes some of the social and scientific assumptions that underlie the mainstream environmental movement.

Her next book, *Earthcare: Women and the Environment*, was published in 1995. It is a collection of essays that brings together a comprehensive summary of 15 years of Merchant's work. The range of essays provides a good balance between the scholarly and the practical. Some deal with the various definitions of ecofeminism and how they depict the factors that curtail the advancement of women and the protection of the environment. Others give examples of how women have participated in the conservation movement through history and up to the present. She concludes the collection by extending the hope that humans can be equal partners with nature rather than trying to control it and that diversity as well as biodiversity should be valued.

Merchant's research into the interactions between nature and culture includes *Green Versus Gold: Sources in California's Environmental History* (1998), which she edited. It is an investigation of California's environmental history: from the American Indians who lived there, to the gold rush, logging, cattle ranching, water use, and urbanization. She brought together documents from California's ecological history along with other primary sources and interpretive essays to illuminate the state's ongoing struggle between environment and economy.

Merchant's work has helped put the contemporary disharmony over environmentalism into useful context and has provided guidelines for policy reassessment. Her research has been recognized and encouraged widely—in 1984 she was awarded a Fulbright senior scholarship in Sweden, in 1991 she was the ecofeminist scholar at Murdoch University in western Australia, and she has also received fellowships from the Center for Advanced Study in the Behavioral Sciences at Stanford University, the American Council of Learned Societies, and the Guggenheim Foundation. She has served on the executive committees of the American Society for Environmental History and the History of Science Society and is on the advisory boards of *Environmental History Review, Environmental Ethics*, and the Association for the Study of Literature and the Environment. She continues to teach at UCB and lives in Berkeley, California.

BIBLIOGRAPHY

Gottlieb, Robert, *Forcing the Spring*, 1993; Kaufman, Polly Welts, "Earthcare: Women and the Environment," *Pacific Historical Review*, 1997; Merchant, Carolyn, *The Death of Nature: Women, Ecology, and the Scientific Revolution*, 1980; Merchant, Carolyn, *Earthcare: Women and the Environment*, 1995; Merchant, Carolyn, *Radical Ecology: The Search for a Livable World*, 1992; Merchant, Carolyn, ed., *Green Versus Gold: Sources in California's Environmental History*, 1998; Mills, Stephanie, "The Death of Nature: Women, Ecology and the Scientific Revolution," *Whole Earth Review*, 1996.

Mills, Enos

(April 22, 1870–September 21, 1922)
Naturalist, Conservationist

Mountaineer and naturalist Enos Mills is remembered as the "Father of Rocky Mountain National Park." From age 14 he roamed the Colorado mountains and guided visitors in forests below Longs Peak. A chance meeting with JOHN MUIR in 1889 led Mills to a career as a nature writer and conservationist. He authored 16 books, including *Wild Life on the Rockies* (1909) and *Your National Parks* (1917), which along with his public speaking tours taught a postfrontier generation to revere wild places. He lobbied for almost a decade to have the federal government establish a national park to protect the Rocky Mountains. His original vision called for a park along the crest of the Rockies from Wyoming to New Mexico. In 1915, Congress protected a small fraction of that with the establishment of Rocky Mountain National Park, which now includes 400 square miles of land near Estes Park in northern Colorado.

Born outside Pleasanton, Kansas, on April 22, 1870, Enos Abijah Mills grew up in a Quaker farm family. His parents, Enos and Ann (Lamb) Mills, were originally from Indiana but had traveled west, mining in the Rockies before homesteading in Kansas. A sickly child, Mills often missed school but read widely on his own. At age 14 he was sent to visit relatives in Colorado with hopes of improving his health. Mills stayed with his uncle, the Reverend Elkanah Lamb, at his ranch at the base of 14,255-feet-high Longs Peak. The Lamb family guided people up the mountain and led Mills on his first ascent.

Mountain air and a better diet improved Mills's health immediately. Within a year he had built a small cabin homestead across the valley from the Lamb Ranch. Taking seasonal jobs around the West, he traveled as far away as Butte, Montana, where he started as a toolboy in the copper mines. On a trip to San Francisco in 1889, Mills was walking along the beach in Golden Gate Park when by chance he met the famous mountain man and wilderness writer John Muir. Muir encouraged Mills to share his strong interest in nature by writing for magazines. Mills began to document his treks, keeping journals and carrying an Eastman Kodak pocket camera. Soon he was publishing quick-paced, inspirational essays in national magazines such as *Harper's* and *Saturday Evening Post*. He would write 16 books and take more than 15,000 photographs during his lifetime.

In 1902 Mills bought the Lamb Ranch from his uncle. Changing its name to the Longs Peak Inn, he used it to develop his Trail School, a guide service offering nature walks popular with visitors of all ages. He soon expanded the inn's facilities, designing furniture and buildings to fit the surroundings while adding modern amenities such as steam heat and telephones. Around 1906 he became the official Colorado snow observer, measuring wind speeds and snow depths for the state Department of Agriculture. At the same time he took a more active stance in the national environmental conservation movement, joining others in the fight to preserve Niagara Falls in 1906.

The growing threat to nature in other areas of the country, and the conservationist response to it, alerted Mills to the urgency of protecting the best parts of the Rocky Mountains. Mills envisioned a huge park preserve that would stretch along the mountain crest from Wyoming to New Mexico. Persuading legislators to create even a smaller version of this park turned into a seven-year fight. Mills wrote hundreds of letters and lectured around the country, emphasizing the importance of protecting natural areas from settlement. He learned valuable lessons in politics while lobbying with Muir against the damming of California's Hetch Hetchy Valley. Although they lost the fight, the struggle tested Mills's activism and gave him experience in Washington. He also won a minor battle. To balance the Hetch Hetchy verdict, the Wilson administration gave the country something Muir, Mills, and other preservationists had been requesting for several years: a separate National Park Service within the Department of the Interior, whose sole role was to create and protect national parks. This was a great improvement over the previous arrangement, in which many separate entities shared the administration of the parks, and there had been no effective coordination.

Mills's sometimes ferocious and always untiring activism finally bore fruit. In 1915 an Act of Congress signed by Pres. Woodrow Wilson set aside 358.5 square miles of land, near the heart of Colorado, to be preserved and protected as a national park. The *Denver Post* praised Mills's "single-handed" efforts and gave him sole credit for the park's creation.

Mills's travels had yielded another success as well. At a lecture in Cleveland, Ohio, a young woman named Esther Burnell was swayed by Mills's words and in time decided to visit the Longs Peak Inn. After completing college, in 1916 she homesteaded near Estes Park and started working for the curmudgeonly

Mills as his secretary. She married him in 1918. Their daughter, Enda, was born during a snowstorm the next year.

Mills believed passionately that nature was a necessary adjunct to civilization, teaching that "wilderness is the safety zone of the world." The park that exists because of his vision now draws more than three million visitors a year. On September 21, 1922, Mills died suddenly at age 52 of an unknown illness, at his home in Colorado. His homestead cabin, now on the National Register of Historic Places, is maintained by his descendants as an informal museum dedicated to Mills's life and work.

BIBLIOGRAPHY

Drummond, Alexander A., *Enos Mills: Citizen of Nature*, 1995; "Enos Mills Cabin," http://www.home.earthlink.net/~enosmillscbn/index.htm; McKibben, Bill, "Hero of the Wilderness," *New York Review of Books*, 1989; Mills, Enos, *Radiant Days: Writings by Enos Mills*, 1994; Mills, Enos, John Fielder, and T. A. Barron, *Rocky Mountain National Park: A 100 Year Perspective*, 1995.

Mills, Stephanie

(September 11, 1948–)
Bioregionalist, Writer

Stephanie Mills's well-known commencement address at Mills College in 1969, in which she dramatically extolled the virtues of population control, catapulted her into the limelight literally overnight, and she has been active as an organizer, editor, and author ever since. Her current focus is on bioregionalism, a philosophy that is based on the idea that humans should become familiar with the natural processes of the land upon which they live, and having achieved an appreciation for the land, act appropriately towards it.

Stephanie Mills was born on September 11, 1948, in Berkeley, California, to Robert and Edith Mills. She spent her childhood in the city she refers to in her memoir, *Whatever Happened to Ecology?* as "Anywhere, USA." The rest of us refer to it as Phoenix, Arizona. Due to the suburban nature of the Phoenix area, Mills does not feel as if she grew up in the desert. But in spite of the generic immediate scenery of her childhood, the desert landscape of Arizona instilled in her an appreciation for vast open spaces, the preciousness of water, and the desert's native plants and animals.

Mills attended Mills College in Oakland, California, where she witnessed first hand the fervor and passion of the antiwar movement. For a variety of reasons, however, she did not participate. She saw one of the shortcomings of the student movement as its not being comprehensive enough. The "war against the planet," the degradation of the environment, was being ignored, and Mills was an ambivalent, apolitical participator in the generation-wide effort of the 1960s to (as she puts it), "Fix It and Realize the

Stephanie Mills (Courtesy of Anne Ronan Picture Library)

Ideal." In college, Mills had an environmental epiphany while reading *The Place No One Knew: Glen Canyon of the Colorado*, which was published by the Sierra Club in 1963 and contains photographs by Eliot Porter of pre-dam Glen Canyon. This book, coupled with the marijuana she was using at the time she read it, connected her to the threats facing planet earth and inspired a desire to preserve all remaining wilderness intact. As student speaker at her 1969 commencement, Mills spoke about the problem of overpopulation, stating that she was "terribly saddened that the most humane thing for me to do is to have no children at all." Her speech received quite a bit of attention in the national media and catapulted her to celebrity status.

After her graduation, Mills began a career of activism, writing and editing in the San Francisco Bay area. She worked as a Planned Parenthood campus organizer, as editor in chief of *Earth Times*, and as a conference coordinator for her alma mater, Mills College, during the years between 1969 and 1974. She then went to Georgia for a year and worked as a writer for a birth control center. Returning to San Francisco in 1975, Mills took a position as director of outings program with DAVID BROWER's Friends of the Earth. She remembers the staff as being outgoing and volatile. It was, she says, "maddeningly unbureaucratic and refreshingly nonprofessional. It really couldn't afford to be otherwise." She moved up to director of membership development, a fundraising position that she herself claims she was not very good at. She kept the job because she liked the company, and her "big break" came when she was offered the editorship of Friends of the Earth's main publication, *Not Man Apart*. She edited the journal for the years of 1977 and 1978, in the process coming to a realization that the hierarchical structure of Friends of the Earth and other similar organizations rendered them ineffective, incapable of producing a fundamental change in culture. At this point she became curious about the potential of bioregional organizations.

Bioregionalism focuses on natural history in an attempt to address that which is wild and native and to help humans achieve a "sense of place." It teaches people to really know where they live, to recognize native plants and animals, to understand local weather patterns, and to have an appreciation for such overlooked features as soil, wildflowers, fire, and moon cycles. In an article for *Sierra*, Mills writes, "Only by alighting—and staying put—do we stand a chance of finding out who we are, where we are, and what we are going to do about it."

After leaving Friends of the Earth, Mills helped to plan a conference called "Technology: Over the Invisible Line?" that took place in 1979. Immediately after the conference, Mills took a position with the journal *Coevolution Quarterly* as an assistant editor, and in 1981, she was invited to guest-edit a special issue on bioregions. After this issue's success, she became a full editor of the magazine, which has since been renamed *Whole Earth.* She remained in this position for two years, until 1982, when she became editor in chief and research director for California Tomorrow, a nonprofit organization dedicated to helping to build a society in which diversity of race, culture, and language are embraced as being our greatest strengths. Then, from 1983 to 1984 she acted as director of development for World College West in San Rafael, California.

Since 1984, Mills has been a freelance writer and lecturer, publishing articles in such periodicals as *Sierra* and *Whole Earth Review.* She has written two books, *Whatever Happened to Ecology?* (1989) and *In Service of the Wild* (1995), which is in two parts. The first part describes the ecological restoration of her farm in Michigan's Upper Peninsula, and the second tells five other land restoration stories, among them ALDO LEOPOLD and his "shack" and the Nature Conservancy's restoration of savanna and prairie in northern Illinois. She also contributed to and edited *In Praise of Nature* (1990), a collection of reviews and excerpts from many of the best books about nature and humans in nature. *In Praise of Nature* is built around five essays, "Earth," "Air," "Fire," "Water," and "Spirit" and contains work from such authors as HERMAN DALY and PAUL EHRLICH. Mills has continued to be active in various environmental and activist organizations. She was vice president of Earth First! Foundation for three years, starting in 1986. She has also been active on the Northern Michigan Environmental Action Council and with the Great Lake Bioregional Congress and the Oryana Natural Foods Cooperative. Mills serves on advisory boards for the Earth Island Institute, the Center for Sustainable Development and Alternative World Futures, and the Northwoods Wilderness Recovery. In 1987, she received an award from Friends of the United Nations Environment Program. She is currently in the process of editing *Neoluddite Papers.*

Mills lives in Maple City, Michigan. She is divorced and has no children, just as she promised in her commencement address.

BIBLIOGRAPHY

Baker, Will, "Whatever Happened to Ecology?," *Whole Earth Review*, 1991; Mills, Stephanie, "The Journey Home," *Sierra*, 1997; Mills, Stephanie, *Whatever Happened to Ecology?* 1989.

Mitchell, George J.

(August 20, 1933–)
U.S. Attorney, U. S. Senator from Maine, Diplomat

A former senator from Maine, George Mitchell has often led the fight to pass key environmental legislation throughout his lengthy career in politics. A consensus builder on the Committee on Environment and Public Works for 14 years, he championed the first major acid rain bill, Superfund toxic waste cleanup, and the reauthorization of the Clean Air Act, among other important pieces of legislation. Appointed Senate majority leader in 1988, he authored the book *World On Fire* (1990), which focused on the greenhouse effect. After his retirement in 1995, Mitchell assumed the role of special adviser to the president on Irish economic issues. In 1998 he chaired the Northern Ireland peace talks, where his presence was crucial to the historic peace agreement.

George John Mitchell was born on August 20, 1933, in Waterville, Maine, to working-class parents George and Mary (Saad) Mitchell. Mitchell spent his youth in Waterville. After receiving his bachelor's degree from Bowdoin College in 1954, he entered the military and served as an officer in the United States Army Counterintelligence Corps until 1956. He married Sally Heath in 1959. They had one daughter, Andrea. In 1960 he earned a law degree from Georgetown University. An attorney in government and private practice for much of the 1960s and 1970s, he served as a trial lawyer in the Antitrust Division of the Justice Department for two years.

He then made a life-changing move, taking a position in Washington, D.C., as executive assistant to Maine's senior congressman, Sen. EDMUND MUSKIE, who became Mitchell's most important political mentor. Muskie was a known conservationist and political powerhouse; Mitchell never forgot his influence. Though he returned to private law practice in Maine in 1965, he remained active in the Maine Democratic Party. In 1968 and 1972, Mitchell served as deputy director for Muskie's vice presidential and presidential campaigns. Mitchell made an unsuccessful run for governor of Maine in 1974, but his political advancement was not slowed. In 1977, he was appointed U.S. attorney for the state of Maine, after Muskie had recommended him to Pres. JIMMY CARTER. Both Muskie and Carter were pleased with Mitchell's performance, and in 1979, the president appointed Mitchell to a newly created U.S. district court judgeship in Bangor, Maine.

Mitchell's star continued to rise. In 1980, when Muskie was named U.S. secretary of state, he recommended that Mitchell fill the remaining two years of his congressional term. Gov. Joseph Brennan appointed Mitchell to the post, beginning his 14-year career in the Senate. Senator Muskie had been working on legislation pertaining to windfall profits tax, waste treatment costs, veterans' education, and approval of the Strategic Arms Limitation Talks (SALT) II treaty. Senator Mitchell was thrust into committee work, eventually serving on the Environment and Public Works, Finance, Veterans Affairs, Governmental Affairs, and Senate Democratic Steering Committees.

Among Mitchell's first tasks was working with Sen. William Cohen to achieve the passage of the Maine Indian Lands Claim settlement of 1980.

In 1982, the state of Maine faced a new threat. Growing concern about waste materials from nuclear power plants led to numerous studies on long-term storage options. One federal report suggested that parts of Maine could act as radioactive waste repositories. Citing the state's faultline geology, Mitchell successfully fought this plan. His constituents were pleased. That same year, Mitchell won his first Senate race with 61 percent of the vote.

Other major legislation championed by Mitchell included the Clean Water Act, passed in 1987. Mitchell worked to override Pres. Ronald Reagan's veto of the act, criticizing the president for his failure to keep his promises about funding clean water programs. The Clean Water Act was a highpoint in Mitchell's congressional record. Three years later, he worked to expand the landmark legislation. A 1990 amendment provided financial resources to small communities for upgrading waste treatment facilities and dealing with runoff or nonpoint-source pollution. After being reelected by a landslide 81 percent of the vote in 1988, Mitchell authored the nation's first oil spill prevention and cleanup bill, the Oil Pollution Act of 1990. The act preserved the right of states to regulate standards for oil transport more strictly than federal law.

Mitchell was instrumental in the 1990 passage of the Clean Air Act, which he brought to the floor and pushed until it was signed into law. The original Clean Air Act, passed in 1970, had aimed at obvious sources of pollution; the 1990 law was aimed at invisible pollutants and the secondary effects of cleanup technologies, such as acid rain caused by high-stack chimneys. While preparing for the Clean Air Act fight, Mitchell put his research on the environment to good use by writing a book about pollution's effects on the atmosphere. *World On Fire*, published in 1990, described the new "greenhouse effect" and suggested ways to slow the process of global warming.

Mitchell had also inherited Senator Muskie's files on the Superfund legislation, which established a Hazardous Substance Response Trust Fund from fees on oil and chemical industries. The fund paid for certain losses resulting from releases of hazardous chemicals. The new Comprehensive Environmental Response Compensation and Liability Act also defined "hazardous substances" more clearly than had been previously outlined in the Clean Air Act, the Clean Water Act (1972), the Solid Waste Disposal Act (1965), and the Toxic Substances Control Act (1976). Mitchell's work on this and other legislation earned him the Wilderness Society's prestigious Ansel Adams Award in 1994. Among other achievements, he counted successes in the areas of childcare, affordable housing, civil rights, campaign finance reform, and universal health care.

Mitchell also accumulated political bonuses. In late 1984 he was appointed chairman of the Democratic Senatorial Campaign Committee and led the Democrats in gaining 11 seats and majority control of the Senate. He was then appointed to the Select Committee on the Iran-Contra Affair in 1987, from which he reminded Oliver North that "God does not take sides in American politics." His performance during the Iran-Contra hearings helped him win the post of Senate majority leader in 1988, succeeding Robert Byrd.

In March 1994, George Mitchell announced his decision to retire from the Senate at the end of his term, after six years of being voted the "most respected member" of that organization. A month later, when Justice Harry A. Blackmun retired from the U.S. Supreme Court, Mitchell quickly became Pres. Bill Clinton's top choice to replace the exiting justice. Though Mitchell turned down the nomination, he found other ways to serve the president. He remained active in international politics, becoming special adviser to the president and the secretary of state for economic initiatives in Ireland in 1995. He compiled the Mitchell Report, released on January 24, 1996, which called for a phasing-out of guerilla weapons in Northern Ireland in addition to elections prior to the opening of peace talks. After serving as moderator for more than 22 months, Mitchell's mission was eventually completed on April 12, 1998, with the signing of a multilateral peace agreement, approved by public referendum. His efforts won him a nomination for the 1998 Nobel Peace Prize.

Mitchell now lives in Washington, D.C., with his second wife, Heather MacLachlan, and their son, Andrew. He is a lawyer in private practice. He has remained in the spotlight, most recently as chairman of the Ethics Committee of the U.S. Olympic Committee. In 1999, he returned to Bowdoin College to found the Senator George J. Mitchell Scholarship Research Institute, which has provided more than $1.6 million in funds for worthy Maine students. Bowdoin College also serves as repository for the senator's papers.

BIBLIOGRAPHY

Blumenthal, Stanley, "The Wisdom of George Mitchell," *New Yorker*, 1994; "George J. Mitchell Papers," http://library.bowdoin.edu; Mitchell, George J., *Making Peace*, 1999; Mitchell, George J., *World on Fire*, 1990; "The Mitchell Institute," http://www.mitchellinstitute.org.

Mittermeier, Russell

(November 8, 1949–)
Zoologist, President of Conservation International

Russell Mittermeier is an internationally recognized expert on primates and reptiles. Since the mid-1970s he has been a proponent of conservation worldwide. As president of Conservation International (CI), a nongovernmental organization that facilitates conservation in the most biodiverse regions of the world, Mittermeier acts as an influential advocate of biodiversity preservation.

Russell Alan Mittermeier was born on November 8, 1949, in Bronx, New York. His mother stimulated his interest in the natural world by taking him to the Bronx Zoo and the American Museum of Natural History and allowed him to assemble a collection of pet reptiles and amphibians. By the age of six, Mittermeier had decided to become a jungle explorer.

Mittermeier studied at Dartmouth College and was given the opportunity to spend his senior year in Central America, with a fellowship to study monkeys. He was a member of Phi Beta Kappa and graduated *summa cum laude* from Dartmouth in 1971. Continuing on to Harvard University, Mittermeier earned an M.A. (1973) and a Ph.D. (1977), both in biological anthropology. His doctoral research took place in Surinam, one of the most biodiverse rain forests in the world. Mittermeier developed a great familiarity with and affection for Surinam's flora and fauna, as well as for its Maroon people, rain forest–dwelling descendants of escaped slaves who, according to a *Time* profile of Mittermeier, say that the forest goes "to the heart of our society."

Mittermeier's extensive travels in Surinam, Madagascar, and other poor tropical countries have convinced him that poverty is the greatest culprit in environmental destruction. The land that poor people must farm is often unsuitable for agriculture, and cultivation results in erosion of soil and sometimes, on steep slopes, devastating landslides. When the best agricultural land is claimed by wealthier owners, poor people must colonize forested areas to grow their food. And when economies are designed so that small-scale subsistence farming does not even produce enough food for families, farmers who live near forests resort to poaching or selling lumber. Mittermeier has worked with several international conservation organizations that collaborate with local governments and nonprofit organizations to promote environmental conservation. Since 1976, he has served the conservation movement in a variety of capacities, ranging from conservation associate at the New York Zoological Society (1976–1977), to Species Survival Commission member of the World Conservation Union (since 1981), to chairperson of the World Bank Task Force of Biological Diversity (1988). Other organizations he has worked with include the World Wildlife Fund–U.S., the World Health Organization, and conservationist organizations in Peru and Brazil.

Since 1989, Mittermeier has been president of Conservation International, which focuses its efforts on areas of the world that harbor the most biodiversity. CI concentrates on two types of biodiverse regions: Global Biodiversity Hotspots, the 25 areas of the world that are most biodiverse but are threatened by human activity, and the as yet untouched biodiverse wilderness. Mittermeier is a proponent of a "megadiversity country" protection plan, whereby certain highly biodiverse countries would be declared world protection zones.

Mittermeier's familiarity with Surinam has allowed CI to help that country's government set aside an enormous area called the Nature Reserve, which composes one-tenth of Surinam's entire area. CI has raised money for a trust fund that the government can use for the area's protection, and it works with the Maroon population on limiting human activities within the reserve.

In addition to his work with CI, Mittermeier continues his work in academia. He has been an adjunct professor at the State

University of New York at Stony Brook since 1977, is a scientific fellow of the New York Zoological Society, and a member of the Linnean Society of London. He has written five books and over 200 scholarly papers. Married to Christina Goettsch, he resides in Washington, D.C.

BIBLIOGRAPHY

"Conservation International," http://www.conservation.org; Mittermeier, Russell, *Lemurs of Madagascar*, 1994; Mittermeier, Russell, *Primate Conservation in the Tropical Rain Forest*, 1987; Rosenblatt, Roger, "Into the Woods," *Time*, 1998.

Montague, Peter

(November 6, 1938–)
Director of the Environmental Research Foundation, Editor of *Rachel's Environmental and Health Weekly*

Peter Montague has been providing solid, understandable, scientific information about the effects of toxic substances on human health since 1980, when he founded the Environmental Research Foundation (ERF). He edits a weekly newsletter entitled *Rachel's Environmental and Health Weekly*, which covers technical issues that are either ignored, censored, or just poorly treated by the mass media. It is sent for free by e-mail to more than 15,000 subscribers, with the goal, according to the ERF mission statement, of "helping people find the information they need to fight for environmental justice in their own communities."

Peter Montague was born on November 6, 1938, in Westport, Connecticut, with what he calls a "justice gene" that he says has always prevented him from just sitting back and watching bullies picking on other people, men abusing women, humans taking advantage of nonhumans. His earliest concern about environmental problems came during the late 1950s, after reading a newspaper article about radioactive fallout coating the northeast-

ern United States, fogging film at the Eastman Kodak plant in Rochester, New York. He tried to research radioactivity, but the two books he found at his local library were impossible for him to understand. Rather than abandoning the topic, he taught himself the science he needed to understand it and began to follow the issue in the news. During the early 1960s, Montague was impressed that biologist BARRY COMMONER, through his scientific research on radioactive fallout and his ability to clearly communicate its danger, was able to convince Pres. Kennedy to sign an above-ground atomic test ban treaty with the Soviet Union in 1963.

Montague earned a B.A. in journalism at the University of the Americas in Mexico City in 1962 and then obtained an M.A. in English at Indiana University, Bloomington, in 1967 and a Ph.D. in American studies in 1971 from the University of New Mexico (UNM). In 1971, he cofounded the Southwest Research and Information Center (SRIC), a small public interest research organization based in Albuquerque, New Mexico, that focuses on local and regional environmental problems. From 1974 to 1979, he and Katherine Montague edited its monthly publication, *The Workbook*, a rich source of information about environmental, social, and consumer problems "aimed at helping people gain access to vital information that can help them assert control over their own lives," according to its mission statement.

During the environmental awakening following the first Earth Day in 1970, university administrators asked Montague—by then well known locally for his environmental activism—to teach a course on the environment. His environmental courses for the School of Architecture and Planning at UNM, which he taught until leaving New Mexico in 1979, transformed students into environmental investigators. They chose "really bad ideas" of government and industry, he recounted to David Case of TomPaine.com, and then would immerse themselves in research, emerging with enough information to challenge policy makers and experts. Their strategy was often to "clog the toilet," to focus on the waste disposal processes of these projects. This had particular relevance for nuclear power plants: without a safe way of disposing their radioactive waste, they had difficulty winning public support for continued operation.

While teaching at UNM, Montague bolstered his scientific knowledge through frequent conversations with a fellow activist and physicist friend, Charles Hyder. After years of daily, informal tutoring sessions, Montague had gained much knowledge about many scientific issues pertaining to the environment. He published many articles and several books during this period, including *Mercury*, cowritten with Katherine Montague and published by the Sierra Club in 1971.

In 1979, Montague moved to Princeton University in New Jersey, where he worked first as a research fellow for its School of Engineering/Applied Science Center for Energy and Environmental Studies and then as project administrator for the school's Hazardous Waste Research Program. These positions allowed him to research the generation and disposal of hazardous materials in New Jersey, including radioactive and hazardous waste. The university suffered from a conflict of interest, however, because it was the recipient of large donations from many of the corporations with the most serious records for contamination in the state. In December 1983, the university discontinued the Hazardous Waste Research Program and transferred Montague to work on computer support and network development.

Concurrently with his work for Princeton, Montague set up the Environmental Research Foundation, through which he hoped to provide community activists with important technical information about environmental and health matters that usually remained the domain of a small, elite community of "experts." ERF's first project was an on-line database, the Remote Access Chemical Hazards Electronic Library, which was called RACHEL (as in RACHEL CARSON). Montague edited RACHEL and, beginning in 1986, the weekly newsletter *Rachel's Hazardous Waste News*. These resources were offered to hazardous waste activists who needed solid scientific information that was written in a style accessible to nonscientists.

In December 1990, Montague left Princeton, and the following month he took a position with the Washington, D.C.–based Greenpeace USA toxics campaign as a senior research analyst. His role was to provide research support to ten community organizers and grassroots campaigners who worked with citizen activist groups fighting hazardous waste incinerators. He stayed with Greenpeace for one year before going to work full time for the Environmental Research Foundation in 1992, at its new Annapolis, Maryland–based headquarters.

In keeping with its mission to provide information, ERF has developed a web site (http://www.rachel.org) that offers an extensive conference calendar, links to thousands of environmental organizations throughout the world, and a searchable archive of ERF's hundreds of reports and newsletters, many available in Spanish. The newsletter continues to be a major focus of Montague, who writes or edits every weekly issue. Renamed *Rachel's Environmental and Health Weekly* to reflect a broader focus, it is now published electronically. Each issue, distributed for free to a subscriber list of more than 15,000, takes a specific issue pertaining to environment and public health and synthesizes important findings for readers who are not experts. Typical of the newsletter's informative, easily understood style is Issue 617, "Landfills Are Dangerous," published September 24, 1998, which describes a recently released New York study about increased rates of bladder cancer and leukemia in women living near landfills and explains the danger of landfill gas in the following way:

> Landfill gas consists of naturally-occurring methane and carbon dioxide, which form inside the landfill as the waste decomposes. As the gases form, pressure builds up inside a landfill, forcing the gases to move. Some of the gases escape through the surrounding soil or simply move upward into the atmosphere, where they drift away.

> Typically, landfill gases that escape from a landfill will carry along toxic chemicals such as paint thinner, solvents, pesticides and other hazardous volatile organic compounds (VOCs), many of them chlorinated.

> The New York state health department tested for VOCs escaping from 25 landfills and reported finding dry cleaning fluid (tetrachloroethylene, or PERC), trichloroethylene (TCE), toluene, 1,1,1-trichloroethane, benzene, vinyl chloride, xylene, ethylbenzene, methylene chloride, 1,2-dichloroethene, and chloroform in the escaping gases.[1]

[1]State of New York Department of Health, INVESTIGATION OF CANCER INCIDENCE AND RESIDENCE NEAR 38 LANDFILLS WITH SOIL GAS MIGRATION CONDITIONS, NEW YORK STATE, 1980–1989 (Atlanta, Ga: Agency for Toxic Substances and Disease Registry, June, 1998). Available from the National Technical Information Service in Springfield, Virginia [1–800- 553–6847]; request publication PB98–142144.

Issue 617 goes on to cite six other studies that found increased rates of bladder cancer and leukemia in people who lived near landfills, five studies showing low birth weight, and three showing increased rates of birth defects. All of the studies are cited in footnotes, and a reader wanting to learn more about TCE, for example, can search the archive and find the January 8, 1992, issue (267), "Popular Solvent, TCE, Seems to Cause Serious Birth Defects in Animals, Humans," and a dozen more newsletters that mention the chemical.

ERF's primary goal of empowering citizen activists with easily understood scientific knowledge furthers its complementary goal: that grassroots action be "the effective lever for change in our neighborhoods and that informed citizens are the essential backbone of a strong democracy and a healthy environment."

In addition to his writings for *Rachel's*, Montague has written and edited more than 130 papers, journal articles, book chapters, and books. He sits on the board of directors of the SRIC and that of Sustainable America, a coalition of economic development organizations. He was awarded the Joe A. Callaway Award for Civic Courage from the Shafeek Nader Trust for the Community Interest in December 1996. Montague resides in Annapolis, Maryland.

BIBLIOGRAPHY

"Environmental Research Foundation," http://www.rachel.org/; "Southwest Research and Information Center," http://www.sric.org; "TOMPAINE.com: "Peter Montague: A Common Sense Civic Hero," http://www.tompaine.com/features/2000/04/25/2.html.

Moses, Marion

(January 24, 1936–)
Physician, Founder of the Pesticide Education Center

Founder of the Pesticide Education Center, Marion Moses is a physician and advocate for health and justice for farmworkers. She began working with CÉSAR CHÁVEZ, DOLORES HUERTA, and the United Farm Workers (UFS) in the 1960s and went to medical school to gain knowledge that could be useful in their struggle. She is an acknowledged scientific expert on the health effects of pesticide exposure and an active part of the struggle against environmental racism.

Marion Moses was born on January 24, 1936, in Wheeling, West Virginia. She graduated from Georgetown University in 1957 with a B.S.N. and from Columbia University in 1960 with an M.A. in nursing education. In 1964 she was living in Berkeley, California, when she became interested in the plight of farmworkers. She joined Citizens for Farm Labor, a group working to publicize the workers' cause, and reactivated the University of California's Student Committee on Agricultural Labor. For five years she volunteered as a nurse with the United Farm Workers in Delano, California. This was a vital period in the organization's history, the height of the grape boycott and strikes, and Moses was a key participant in this struggle. The workers were striking for better wages and working conditions, including better health and safety regulations. At this time there was no requirement for sanitary facilities in the fields and no protection against pesticide exposure. Moses decided to pursue a medical degree in order to become a more effective advocate on these issues. She received an M.D. from Temple University in 1976 and completed her residency in internal medicine at the University of Colorado Medical Center in 1977. She completed a residency in occupational medicine at Mount Sinai Medical Center in New York City in 1980 and was board certified by the American College of Preventive Medicine and Public Health in 1980.

Moses returned to work with the farmworkers as the medical director of the National Farm Workers Health Group, a post she held from 1983 to 1986. In 1988 César Chávez held a widely publicized hunger strike to call attention to a new grape boycott, this time focused primarily on the issue of worker exposure to pesticides. Moses was one of the attending physicians during his fast. Her work with the UFW makes Moses's views on the effects of pesticide exposure particularly compelling. In the mainstream media, this issue is often described from the point of view of consumers, fearful of the trace residues left by pesticide use in the fields. Moses emphasizes that agricultural workers have far more to fear and are much more vulnerable than the average American consumer. Workers are subject to doses many times larger than those of consumers and experience exposure over a more prolonged period. Moses puts the issue of pesticide safety into an international context. Multinational corporations produce and use pesticides to produce crops as cheaply as possible, at the expense of the health of farmworkers. These corporations work hand in hand with the U.S. Department of Agriculture, which subsidizes research in chemical means of production, at the expense of research into organic methods. And farmworkers, the least powerful actors in the agricultural field, pay for these subsidies with higher cancer rates and neurodevelopmental problems among their children.

In 1988 Moses founded the Pesticide Education Center to increase awareness of the dangers of pesticides among workers and consumers. The center's mission is "to educate consumers to make more informed choices to protect themselves, their families, their pets, their neighbors, and the environment from toxic pesticides." The center collects the results of recent research into the health effects of pesticides and produces educational materials, including the video *Harvest of Sorrow* and Moses's 1995 book *Designer Poisons*. In 1990 Moses led a landmark study, funded with a $500,000 appropriation from Congress, to assess the danger pesticides pose to farmworkers. The study will follow the same group of workers for many years, in the hopes of gathering definitive epidemiological evidence about the effects of exposure. Increased risk of cancer and other disease is notoriously hard to prove. Each pesticide is chemically unique, and it is difficult to demonstrate an undeniable link between a particular pesticide and a particular instance of disease. Moses's work aims to establish this link, by thoroughly documenting every aspect of the workers' daily environment, over a prolonged period of study.

In 1991 Moses served on the National Advisory Committee of the First National People of Color Environmental Summit, which played a pivotal role in making the concerns of people of color heard in mainstream environmental organizations. Moses has also been active as an adviser to the Environmental Protection Agency (EPA), serving on a number of committees, including the Toxic Substances Advisory Committee and the National Advisory Committee of the Pesticide Farm Safety Center. In April 1999,

Moses led a walkout by environmental groups of another committee, the EPA's Tolerance Reassessment Advisory Committee. Moses, representing the Pesticide Education Center, joined Consumer's Union, Natural Resources Defense Council, Farmworker Justice Fund, and other national groups in resigning in protest of what they saw as the Clinton administration's capitulation to the chemical industry. Moses continues to direct the Pesticide Education Center in San Francisco, California.

BIBLIOGRAPHY

Newton, David, *Environmental Justice*, 1996; "Pesticide Education Center," http://www.igc.apc.org/pesticides/; Ruttenberg, Danya, "Interview with Dr. Marion Moses," *Sojourner*, 1999.

Moss, Doug

(November 13, 1952–)
Publisher

Doug Moss publishes *E The Environmental Magazine*, the nation's only independent, bimonthly magazine devoted to the environment.

Born in Norwalk, Connecticut, on November 13, 1952, Douglas Edward Moss grew up catching frogs and fishing in local ponds but traces his environmentalism to an event that occurred much later. While living in New Haven, Connecticut, after graduating with a degree in marketing from Babson College in 1974, Moss one day watched a television report about the clubbing of baby harp seals. Moss, a business forms salesman who had no history of political activity, was outraged. His first impulse was to call the television station and complain, but then he realized the television was only the messenger, so instead he joined a local antifur group. When he found that he enjoyed activism, he began running with a local community of left-of-center intellectuals and activists. Moss started to spend his free time on such activities as gathering signatures on petitions for BARRY COMMONER's 1980 candidacy for president and the successful Norwalk nuclear freeze referendum of 1982.

In 1979, Moss left the Burroughs Corporation, whose forms he had been selling since 1974, and started his own company, Douglas Forms. Most of his clients were magazines, and from his work producing renewal forms for them, he learned more about the business end of publishing. Moss and a few other members of the local chapter of Friends of Animals, which he had helped found, decided to publish an animal rights magazine. In late 1979 the first issue of *Animals' Agenda* appeared.

After nine years of publishing *Animals' Agenda*, Moss tired of the narrow range of issues covered by the animal rights' movement. He and his wife, Deborah Kamlani, decided to move on. It was the summer of 1988, which Moss now refers to as the "greenhouse summer" because it was so hot that scientists' predictions of global warming were beginning to seem more real to the American public. In a 1999 interview, Moss recalled that while he and Kamlani ate breakfast in a Westport, Connecticut, deli one hot morning, they read a *New York Times* article about the medical waste washing up on New Jersey beaches. Global warming, contaminated oceans, and other related issues triggered the idea to use their publishing skills in a new nonprofit venture. Moss and Kamlani decided to leave *Animals' Agenda* and put their publishing experience to use on a new, independent magazine that would focus on a broad range of environmental topics.

E The Environmental Magazine incubated during 1988 and 1989, and its first issue appeared on newsstands in December 1989 with a January/February 1990 cover date. It got an early boost from two events that mobilized the public on environmental issues: the now infamous April 1989 *Exxon Valdez* spill in Prince William Sound and the 20th anniversary of Earth Day on April 21, 1990. Unbeknownst to Moss, Kamlani, and cofounder Leslie Pardue, two more environmental magazines were being inaugurated at the same time, *Buzzworm* and *Garbage*. For several years, there was enough momentum in the new environmental movement that the market supported all three magazines. By the mid-1990s, however, *Buzzworm* and *Garbage* both had folded.

E serves as a bimonthly "clearinghouse" for environmental information, according to its mission statement, providing information that is both accessible for the general public and detailed enough to be of use to serious environmentalists. One unique and very useful feature of the magazine is that almost all of its articles end with contact addresses and phone numbers for reader follow-up. Its departments include news, commentary, and interviews with such environmental leaders as Clinton EPA head Carol Browner, environmental justice specialist Dr. Robert Bullard, and biotech expert Jeremy Rifkin. The magazine's Green Living section features consumer and health shorts, travel options, and recommendations for greener earning and spending.

Moss's business and marketing training have contributed to *E*'s survival. Knowing that publishing ventures suffer a notoriously low profit margin, Moss founded the nonprofit corporation Earth Action Network to solicit grant support for *E*. The first few years were slim for *E*, and Moss and Kamlani took out personal home equity and business loans to kick start the magazine. But thanks to a 1997–1998 debt resolution campaign, several foundations stepped in to help retire *E*'s debt. Moss calls this "capitalization catch-up"; he is finally getting the funding he really needed earlier for the magazine's development. *E* currently has a circulation of 56,000, sold primarily through subscriptions but also on newsstands and in bookstores across the country. *E* also peddles its editorials through several syndicates. The Los Angeles Times Syndicate, the New York Times Syndicate, Alternet, and several others frequently reprint *E* articles; many also are reprinted in books and textbooks. *E*'s materials are widely available electronically, including in all major library research databases.

Moss and Kamlani live in Westport, Connecticut, with their two sons, Tim and Jeff.

BIBLIOGRAPHY

Booker, Vonetta, "The Staying Power of *E*," *Fairfield County Weekly*, 1999; *"E Magazine,"* http://www.emagazine.com; Sledge, Anne, *"E* Magazine Remains True to Its Eclectic Roots," *Fairfield County Business Journal*, 1999.

Muir, John

(April 21, 1838–December 24, 1914)
Writer, Naturalist, First President of the Sierra Club

Bearded, long-haired John Muir was America's prototypical wild man, happiest when he was wandering wild lands. He is remembered for his passionate writings in defense of the wilderness, his untiring fight for their preservation, and his role as founding president of the Sierra Club.

Born on the North Sea coast in Dunbar, Scotland, on April 21, 1838, John Muir emigrated with his family to Portage, Wisconsin, in 1849, in pursuit of religious freedom. His father was a preacher for the evangelical Disciples of Christ branch of Christianity. The Muir children and their mother tended the farm. The father had a strict approach to child rearing: he prohibited his eight children from singing or reading anything besides the Bible, even from eating a balanced diet and keeping the house comfortably warm in winter. The Muir children were nurtured by their mother, however, who loved nature and art. All of the Muir children, especially John, became avid naturalists. Whereas their father subscribed to a religion based on fear, John and his siblings recognized God in every natural and beautiful place.

Muir borrowed books from neighbors to read by kerosene lamp in the dark of night; he also spent nights tinkering in his workshop, inventing clocks, barometers, hydrometers, and more fanciful apparatuses, such as a bed that dumped out the sleeper when it was time to wake up and a desk that rotated books around in a circle so that each could be studied the same amount of time. These inventions eventually provided Muir with a way to escape the oppressive family farm. Young John Muir traveled to Madison to exhibit his inventions at the 1860 Wisconsin state fair. There he became a star and decided to stay in Madison and attend the University of Wisconsin. Muir studied botany and geology until 1863, when he fled to Canada to avoid conscription for the Civil War. After a factory accident in Canada that left him temporarily blind, he reevaluated his life plans and decided to abandon his work with machinery and factories. In 1867, he began a 1,000-mile walk through the post–Civil War countryside, walking 25 miles a day, sleeping outside where he could, penniless most of the time, but feeling rich in freedom. Throughout his walk he collected botanical specimens, drawing and studying them. Eventually he boarded a steamer for California and from the docks in San Francisco he walked to Yosemite Valley.

The dramatic mountains and waterfalls and lush flower-filled valleys of

John Muir (Corbis)

Yosemite enchanted Muir. He stayed in the area for ten years, working first as a farmhand and shepherd, and later at a sawmill cutting fallen logs for tourist cabins. His home consisted of an 8-foot by 6-foot extension of the top floor of the sawmill, hanging over a river. It was packed with his botanical collections and his notes. During his free time, he hiked through Yosemite's wilderness, becoming more and more interested in the geological phenomena that formed it. Over time, Muir developed a theory that ran counter to the accepted explanation of how Yosemite had been formed—that earthquakes were responsible for the jagged mountainscape. Muir came to believe that a glacier had gradually carved out the valley, leaving the granite protrusions Half Dome and El Capitan. Muir described his ideas to the tourists he guided through Yosemite, and his theories traveled by word of mouth. Eventually he was persuaded to write them down in the first of his many articles for national magazines ("Yosemite Glaciers," published in *New York Daily Tribune*, 1871). Today, geologists accept his theory but make two amendments: that there were two glaciers rather than one and that water too was important in carving the valley.

While Muir was intensely interested in the natural history of Yosemite, he became also increasingly concerned about its preservation. During the late 1800s, loggers and ranchers were rapidly exploiting California's forests, clear-cutting huge stands of the state's unique ancient redwoods. Vast tracts of primary forest were in the hands of a small group of wealthy owners, who were eager to make as much money as quickly as possible by exploiting their land. Muir began to spend his winters writing articles promoting preservation of beautiful natural areas. He favored the creation of national parks that would set aside spectacular sites like Yosemite, protecting them forever from development. In his vision, they would be reserves to replenish the spirit of an American people who, Muir predicted, would need more contact with nature as urban areas grew more crowded and frenetic. He believed that when tourists visited natural areas, the beauty and peace of nature would transform them into preservationists. Through his writings, Muir succeeded in persuading politicians to establish numerous national parks, including Yosemite in 1890 and Mount Rainier in 1899.

When University of California professors Henry Senger and William Armes decided in 1892 to found the Sierra Club to work for conservation, they asked Muir to attend the first meeting. Although Muir was shy and feared social occasions, he believed so deeply in the cause that he did attend the meeting and accepted a nomination to serve as the Sierra Club's first president.

Muir became so well known through his writing and Sierra Club activism that Pres. THEODORE ROOSEVELT asked him to be his personal guide through Yosemite during a 1903 tour of the West. The two men spent three entire days hiking together, and Roosevelt left California convinced of Muir's preservationist vision.

Muir's final preservationist battle was against a dam that would flood the Hetch Hetchy meadows just outside of Yosemite. In his arguments for their preservation, he compared them to cathedrals and churches and claimed that "no holier temple has ever been consecrated by the heart of man" (*The Yosemite*, 1988). Despite a fierce and energetic battle, in which he collaborated with Sierra Club member WILLIAM COLBY and *Century* editor ROBERT UNDERWOOD JOHNSON, the fight to preserve Hetch Hetchy was lost in 1913, and the valley now lies under a reservoir that collects water for San Francisco.

Muir was married to Louisa Strenzel and for over 30 years lived on her family's farm in Martinez, California. They raised wine grapes and fruit trees and, thanks to Muir's hard work and shrewd business sense, quickly became quite wealthy. The couple had two daughters, Annie and Helen, both of whom, as adults, accompanied Muir on mountaineering expeditions. Louisa was a quiet yet strong presence in the household. She helped Muir with all of his writings and, throughout their marriage, always encouraged him to take time off from the farm and explore the natural wonders that inspired his awe. Louisa Muir died in 1905. On December 24, 1914, shortly after the Hetch Hetchy defeat, John Muir died of pneumonia.

BIBLIOGRAPHY

Cohen, Michael P., *The Pathless Way: John Muir and the American Wilderness*, 1984; Fox, Stephen, *The American Conservation Movement: John Muir and His Legacy*, 1986; Muir, John, *My First Summer in the Sierra*, 1988; Muir, John, *The Story of My Boyhood and Youth*, 1965; Muir, John, *A Thousand Mile Walk to the Gulf*, 1981; Nash, Roderick, *Wilderness and the American Mind*, Revised Edition, 1973; Wilkins, Thurman, *John Muir: Apostle of Nature*, 1995; Strong, Douglas H., *Dreamers & Defenders: American Conservationists*, 1988.

Mumford, Lewis

(October 19, 1895–January 26, 1990)
Social Philosopher, Urban Planner

Lewis Mumford's interests ranged from history to fine arts and literary criticism, but he is best known for his contributions to the fields of urban planning and architecture. The author of more than 30 books, Mumford believed that the modern obsession with technology obscured those human values that created great civilizations and that unplanned growth and mechanization contributed to the breakdown of biological, social, and personal well-being.

Lewis Charles Mumford was born on October 19, 1895, in Flushing, Queens, New York, the illegitimate son of a German Protestant, Elvina Baron Mumford, and Lewis Charles Mack, a Jewish businessman, whom the boy never met. Nor did he meet the man whose name he carried, John Mumford, an Englishman his mother had married briefly 12 years before his birth.

Mumford's mother kept a boardinghouse in Manhattan. "I was a child of the city. New York exerted a greater and more constant influence on me than did my family," he wrote in his autobiography, *Sketches from Life*. The teeming, diverse, energetic New York of his childhood was home to more than a million foreign-born immigrants. Every weekend, his grandfather took Mumford on long walks throughout the city, "saunters [that] furnished the esthetic background of my childhood," he wrote.

In 1909, Mumford entered Stuyvesant High School to prepare for an engineering career, but by 1912, upon enrolling in City College of New York, he had decided to become a writer. At 20, Mumford discovered the works of Patrick Geddes, who would be his greatest influence and who shaped the young man's future as a social philosopher and urban planner. Geddes was an early environmentalist, a Scottish botanist who used his scientific training to help revitalize and rehabilitate the choking, grimy slums of industrial Edinburgh. Following Geddes's pattern, Mumford made long, solitary explorations of New York, studying its streets, neighborhoods, buildings, bridges, and geology. Reading Geddes's University of London Extension Lectures, Mumford learned to take notes as he walked, recording city life and its supporting activities and, as if in the wilderness, noting where the city's humanistic ecosystems were disrupted. In 1917, Mumford and Geddes began corresponding; they did not meet until 1923.

Geddes's ideas about using the past to design future cities inspired Mumford, who came to believe that "the city is an age-old instrument of human culture, essential to its further development." History and antiquity would inform Mumford's theory of "organic planning," in which he cited cities of medieval Europe and other eras as models of situations where physical layout gave rise to cultural growth, artistic expression, and human contact. "When one considers the amount of space and fine building given to Pompeii's temples ... markets ... law courts ... public baths ... stadium ... theatre ... one realizes that American towns ... do not, except in very rare cases, have anything like this kind of civic equipment, even in makeshift form," he wrote in *Technics and Civilization* (1934), one of four volumes of his Renewal of Life Series. Mumford described organic planning as embracing the myriad processes by which a city evolves historically, each generation building on the accomplishments of the previous.

In 1920, after a year in the navy and a stint on the staff of the literary magazine *Fortnightly Dial*, Mumford went to London as acting editor of *Sociological Review*. This move was to "mold the rest of my life," he wrote, for there he met Victor Branford, a financier and Geddes collaborator. Branford's interpretations of science and art and the role of religion as the binding element of human communities would form a crux of Mumford's later thoughts.

Mumford left London in 1921 to marry Sophia Wittenberg. His first book, *The Story of Utopias*, published in 1922, emphasized the role of artists of all disciplines in the process of social transformation. His notion of artists' responsibility for the reconstruction of our inner and outer worlds was a step beyond Geddes's belief that a systematic sociology must be linked to the good life. Mumford wrote that artists could contribute to social reform by suggesting images of a more balanced, spiritually satisfying life, which could be integrated into the plans of regional surveyors. Like Geddes, Mumford advocated festivals and pageants, such as those of the Middle Ages, to celebrate and understand the city's diverse history. The *Story of Utopias* was followed in 1924 by *Sticks and Stones*, his first book about architecture and the architect's role as artist-reformer.

Meanwhile, Mumford joined three great American planners—Clarence Stein, Henry Wright, and BENTON MACKAYE—to form the Regional Planning Association of America (RPAA), which attempted to stop unplanned urban growth and restore human scale to cities. RPAA proposed small satellite cities—suburbs—separated from a parent city by open space. Mumford moved into an RPAA-planned community in Sunnyside, Queens, which featured clustered buildings and a large common garden. He helped plan Radburn, a community in Fairlawn, New Jersey, and served in various other urban development organizations, including the New York Housing and Regional Planning Commission. While Mumford was involved in the RPAA, he met CATHERINE BAUER, who would later become well known as a planner and advocate of improved housing for poor people.

The Golden Day, a discussion of the utilitarian, "timekeeper" culture brought to American shores by Protestant settlers, was published in 1926, the same year Mumford helped found and edit *The American Caravan*, a periodical dedicated to the works of emerging writers. He also produced a biography of Herman Melville.

Although he had never received a college degree, Mumford began a part-time visiting professorship at Dartmouth College in 1929 and soon became a popular lecturer and teacher, with appointments at Stanford University (where he helped design the humanities program), the University of Pennsylvania, and the Massachusetts Institute of Technology, among others. He was appointed to the New York City Board of Higher Education.

In 1931, Mumford joined the *New Yorker* magazine staff with his column "The Sky Line," which he wrote until 1963. He challenged readers to consider the costs to people of politically motivated urban projects and spoke out against industrialists, promises of "progress," and what he called "the myth of the machine." He attacked "bigger-is-better" proponents and led bitter campaigns against the policies—such as building a railroad through Washington Square—of then New York commissioner of public works Robert Moses. Through "The Sky Line," Mumford battled developers and skyscrapers, which he labeled "elegant monuments to nothingness." Most of his fights were lost, but thanks to Mumford's outspoken courage, such areas as preservation, social issues, and urban development could no longer be the sole, secret purview of planners and politicians but became public issues, matter for community dialogue.

After World War II, Mumford came out against proliferation of the atomic bomb. In the 1960s, he protested against U.S involvement in the war in Vietnam. But across time, he became increasingly despairing about the dehumanization and desocialization born of mechanization and industrialism: unemployment, isolation, invisibility of the individual, dissolution of families and communities, the future of children and the natural environment, the pollution of air and water.

Mumford won the National Book Award for *The City in History* in 1962, the Presidential Medal of Freedom in 1964, the National Medal for Literature in 1972, the French Prix Mondial del Duca in 1976, and many other awards, including the honorary Knight Commander of the British Empire in 1977. Although he never completed his college education, he was awarded an honorary LL.D. from the University of Edinburgh in 1965 and an honorary doctorate of architecture from the University of Rome in 1967.

His final book, *Sketches from Life* (1982), chronicled Mumford's early years into the 1930s and was nominated for an American Book Award. In 1986, Mumford was awarded the National Medal of Arts. He was overjoyed to be recognized as an artist. Yet, according to biographer Donald Miller, it was bittersweet. With old age, Miller wrote, Mumford "was frustrated, sometimes to the point of anger, that he would no longer . . . do any good in the world. 'Resignation would be easier' [he said], 'if the world at large were in a more hopeful state.'"

Mumford and his wife, Sophia, had two children: a son, Geddes, born in 1925, and a daughter, Alison, born in 1935. Geddes Mumford was killed in combat during World War II and was memorialized by his father in a 1947 biography titled *Green Memories*.

Lewis Mumford died on January 26, 1990, at the age of 95 at his home in Amenia, New York.

BIBLIOGRAPHY

Hughes, Thomas P., and Agatha C., eds., *Lewis Mumford: Public Intellectual*, 1990; Miller, Donald L., *Lewis Mumford: A Life*, 1989; Miller, Donald L., *The Lewis Mumford Reader*, 1986; Mumford, Lewis, *The City in History*, 1961; Mumford, Lewis, *The Culture of Cities*, 1938; Mumford, Lewis, *Sketches from Life: The Early Years*, 1982; Mumford, Lewis, *The Story of Utopias*, 1922.

Murie, Mardy, and Olaus Murie

(August 18, 1902– ; March 1, 1889–October 21, 1963)
Conservation Activist; Wildlife Ecologist, Naturalist, Cofounder of the Wilderness Society

Arguably the "first family" of Alaskan wilderness, Olaus and Mardy Murie were highly influential players in the interwar wilderness movement that culminated in the 1964 passage of the Wilderness Act. The focus of their own work was the preservation of Alaska's wild lands. As a member of the U.S. Biological Survey, Olaus Murie conducted pioneering wildlife studies, predominantly in Alaska and Wyoming. Once they married, Mardy frequently joined him on his wilderness expeditions. After the death of Olaus in 1963, Mardy continued the work they had begun together, joining the Governing Council of the Wilderness Society and working for the protection of wild Alaska.

Olaus Johan Murie was born in Moorhead, Nebraska, on March 1, 1889, the son of Norwegian immigrants. His father died when he was quite young, leaving his mother to care for him and his brother, Martin, and his half-brother, Adolph. Murie spent much of his childhood picking potatoes, working on truck farms, and delivering milk to help make ends meet at home. Growing up in natural surroundings led to his early interest in the environment and conservation. One of his early influences was ERNEST THOMPSON SETON,

Mardy and Olaus Murie (Courtesy of Anne Ronan Picture Library)

U.S. Biological Survey and hunted predators in the Olympic Peninsula in Washington during the winter of 1916–1917. Whereas his contemporary, ALDO LEOPOLD, was an avid predator hunter as a young man, Murie seemed to lack such enthusiasm. His record that winter was exceptionally poor: two bobcats, one cougar, and no wolves killed. His future wife, Mardy, would later say that he never believed strongly in predator control. His criticisms of the predator control policy, however, were mild at first. Writing to his superior from Alaska in 1923, Murie hinted that predators might not represent such a problem as was believed. "I have a theory," he wrote, "that a certain amount of preying on caribou by wolves is beneficial to the herd, that the best animals survive and the vigor of the herd is maintained." Such views were well ahead of their time and practically heresy in their day. Accordingly, Murie tempered his criticisms so as not to jeopardize his job. Over the years, however, Murie became an increasingly vocal opponent of predator control, believing that predators played a crucial role in maintaining balance in the ecosystem.

In 1920, Murie was sent to Alaska and northern Canada to conduct the first "life history" of the caribou in that region. Accompanied by his brother Adolph, Murie conducted studies on a variety of wildlife, including waterfowl, bears, and especially elk. While in Alaska in 1924, Murie met Margaret Elizabeth Thomas, soon to be his lifelong partner for his future wilderness studies.

Margaret Elizabeth Thomas—fondly called the Grandmother of the Conservation Movement—was born August 18, 1902, in Seattle, but she spent her childhood in Fairbanks, Alaska. In 1924, she

whose books in the local library Murie nearly wore out from avid reading. Following Seton's example, Murie sketched and painted the animals he observed on his wanderings. He studied biology at Fargo College in North Dakota and later zoology at Pacific University in Oregon, where he received a B.A. in 1912. Murie worked as a conservation officer for the Oregon State Game and Fish Department. Within two years, he was hired as a field mammal curator for a Carnegie Museum of Natural History expedition to Canada. It would be the first of his many trips to the northern wilds.

After serving as a balloonist and observer in World War I, Murie joined the was the first woman to graduate from the University of Alaska at Fairbanks. That same year, she married Olaus and joined him on many of his wilderness expeditions. Their honeymoon was a dogsled caribou research expedition that covered some 500 miles of Alaska's Brooks Range. Subsequent trips included their three children: Martin (named for Olaus's brother who died during the 1922 influenza epidemic), Joanne, and Donald. The Murie family's adventures are described in Mardy Murie's book, *Two in the Far North.*

The couple moved to Wyoming to study elk in 1926. In Jackson Hole, they built a cabin that Mardy lives in to this day. From this home base, Olaus Murie wrote the works that his fame rests on: about wilderness preservation and the delicate ecological balance of the wild and human intervention in it. In a 1927 study in Wyoming, Murie concluded that the decline in the elk population in the region was a result of a reduction in their natural predators. Because predators were not keeping elk populations in check, elk numbers ballooned until the animals were forced to stray from their normal feeding patterns in order to find food. Murie demonstrated that the bushes the elk had started to feed on were ripping their mouths (a disease called sore-mouth), causing fatal infections. Murie's findings and his appeals against the slaughter of predatory animals were instrumental (along with Aldo Leopold's later assessments) in the reversal of previous predator control policies.

Called "Mr. Wilderness" by the *Washington Post*, Murie, along with BENTON MACKAYE, ROBERT MARSHALL, and several others, cofounded the Wilderness Society in 1935. Murie served as a member of the society's council, beginning in 1937, and as a director from 1945 until his death. In that capacity, Murie was a leading member of the conservation movement's lobbying for the congressional protection of the nation's wilderness areas. He was awarded the Audubon Medal of the National Audubon Society in 1959 and the John Muir Award of the Sierra Club shortly before his death. Murie died on October 21, 1963, in Jackson Hole, just months before the passage of the Wilderness Act in 1964, the realization of his life's work.

Mardy Murie attended the signing of the Wilderness Act by Pres. Lyndon Johnson, in her late husband's stead. Working with her husband before his death and independently once she was a widow, Mardy Murie wrote letters and articles, traveled and lectured, and promoted wilderness preservation to the public and the government. After the Wilderness Act, she continued to push for wilderness preservation, especially in Alaska. She played a significant role in the 1980 passage of the Alaska National Interest Lands Conservation Act, the greatest land preservation act in U.S. history. Testifying for this act, Murie said: "I am testifying as an emotional woman and I would like to ask you, gentlemen, what's wrong with emotion? Beauty is a resource in and of itself. . . . I hope the United States of America is not so rich that she can afford to let these wildernesses pass by, or so poor she cannot afford to keep them."

Murie also served on the Council of the Wilderness Society, received an honorary doctorate from the University of Alaska and the prestigious Audubon Medal, and was named an honorary park ranger by the National Park Service. She was also on the founding board of the Teton Science School. Her most recent award, in 1998, was the Presidential Medal of Freedom, which Pres. Bill Clinton bestowed on her for her lifetime service to conservation.

On September 16, 1998, the Murie Center was formally opened with a gathering of some of the nation's top conservation leaders on Mardy Murie's front porch. In the time since, the Murie Center has launched a series of projects that continue the Murie legacy of land conservation and wilderness protection.

BIBLIOGRAPHY

Glover, James M. "Thinking like a Wolverine: The Ecological Evolution of Olaus Murie," *Environmental Review*, 1989; "Mardy Murie: Medal of Freedom Award Winner," http://www.wilderness.org/profiles/murie.htm; Murie, Adolph, *A Naturalist in Alaska*, 1961; Murie, Olaus, *The Elk of North America*, 1951; Murie, Olaus (finished by Mardy Murie), *Wapiti Wilderness*, 1966.

Muskie, Edmund

(March 28, 1914–March 26, 1996)
U.S. Senator from Maine

Edmund Muskie served as a member of the Maine state legislature from 1946 to 1951, as governor of Maine from 1955 to 1959, and as U.S. senator from Maine from 1959 to 1980. As senator, he sponsored the Clean Air Act of 1963, the Water Quality Act of 1965, and a 1967 act that authorized more than 400 million dollars for pollution control. He served as chairman of the Environment and Public Works Subcommittee on Environmental Pollution and was a supporter of the 1966 Model Cities Act.

The second child of Stephen and Josephine Muskie, Edmund Sixtus Muskie was born on March 28, 1914, in Rumford, Maine. His father had emigrated to the United States from Poland in 1903 to escape the oppressive environment of his home country. Muskie's mother was from a large Polish family in Buffalo, New York. In his book, *Journeys* (1972), Muskie remembers his childhood as being "as healthy and happy a childhood and family life as a boy could wish." He pursued the bulk of his interests out-of-doors. He fished, hunted, and played baseball and football, and in winter he skied. The natural environment of Maine had a lasting impact on him. "My journey towards a place in the environmental sun," he writes in *Journeys*, "began in my backyard, in the environment of the place I was born and raised."

While studying at Bates College in Lewiston, Maine, Muskie became interested in politics, finding that his sensibilities were more in alignment with the New Deal philosophy of the recently elected president, FRANKLIN D. ROOSEVELT, than with the traditionally Republican views of most residents of Maine. Muskie graduated *cum laude* from Bates College in 1936. He received a scholarship to Cornell University Law School and received his LL.B. degree in 1939. He was admitted to the Massachusetts bar in 1939 and to the Maine bar in 1940.

In 1940, Muskie moved to the small town of Waterville, Maine, and practiced law for a short period of time before enlisting in the naval reserve in 1942. He served as an engineering officer on destroyer escorts in both the Pacific and Atlantic theaters of operation in World War II and was released to inactive duty in 1945. He returned to Waterville to revive his law practice and to become active in the Maine Democratic Party. In 1946, Muskie was elected to the Maine House of Representatives. He was elected to three consecutive terms. In 1948, during his second term, Muskie was chosen to be the Democratic minority leader. In 1951, in the middle of his third term, Muskie resigned from the Maine House of Representatives to become district director of the Maine Office of Price Stabilization, a position he occupied for just one year before becoming a Democratic national committeeman in 1952.

In 1954, Muskie decided to run for governor. According to Muskie's biographer David Nevin, he ran not so much out of a strong personal desire as out of a sense of responsibility. He was simply the only possibility for the Maine Democratic Party, and when he won, he was probably even more surprised than the Republicans were. He served two terms as governor, working during both terms with a state legislature that was predominantly (four to one) Republican. However, due to his tact and skills in politics and consensus building, he was able to steer his economic and educational programs through the legislature. He also pushed for legislation that created a state program for the building of water treatment plants and that established a classifica-

tion system for improving the quality of Maine's streams and rivers.

Muskie decided not to seek a third term as governor in 1958. Instead, he ran for the U.S. Senate, defeating the Republican incumbent, Frederick G. Payne, by a wide margin. He was the first Democrat ever to be elected to the Senate by the state of Maine and kept his seat from 1959 until 1980. Muskie was instrumental in bringing about some of the most important environmental legislation in the history of the United States. In a short piece written for *Commonweal*, Abigail McCarthy quotes political commentator Mark Shields, "Before he [Muskie] began his work, there were no national laws and international agreements governing the quality of the country's air and water." Muskie responded to this dearth of environmental legislation, earning himself the nickname "Mr. Clean" for his efforts. He was the chief sponsor and floor manager of the Clean Air Act of 1963 and the Water Quality Act of 1965, and in 1967 he sponsored a $428,300,000 authorization for pollution control efforts. He supported Pres. Lyndon Johnson's Model Cities Act of 1966, which provided $1.2 billion to improve housing, recreation areas, education, and health in economically depressed urban areas throughout the United States. Muskie served as chairman of the Environment and Public Works Subcommittee on Environmental Pollution, and he was the Senate Budget Committee's first chairman from 1974 to 1980, in which post he developed a complex system for tracking federal spending.

In 1968, Muskie gained national prominence as the running mate in Hubert Humphrey's unsuccessful bid for the presidency. Throughout the campaign,

Muskie was described as being Humphrey's greatest asset. In 1972, Muskie entered the presidential race himself. He did not, however, receive the Democratic nomination. In 1980, he resigned his position as senator to serve as Pres. JIMMY CARTER's secretary of state in the final year of the Carter administration.

After Carter failed to be reelected in the 1980 presidential election, Muskie practiced law in Washington. In 1986, he served on an investigative board, headed by Sen. John G. Tower, which examined the role of President Reagan's National Security Council staff in the Iran-Contra "arms for hostages" affair. He also served on a delegation that traveled to Vietnam to explore the lifting of the U.S. trade embargo with that country and chaired the Center for National Policy, a Democratic think tank. He died of a heart attack on March 26, 1996, after being treated at Georgetown University Hospital for a blocked artery in his leg. He was survived by his wife of 48 years, Jane, and five grown children.

BIBLIOGRAPHY

McCarthy, Abigail, "Edmund S. Muskie: Let Us Now Praise Honorable Men," *Commonweal*, 1996; Muskie, Edmund S., *Journeys*, 1972; Nevin, David, *Muskie of Maine*, 1972.

Nabhan, Gary

(March 17, 1952–)
Ethnobiologist, Agricultural and Desert Ecologist, Nature Writer

Gary Nabhan is an award-wining writer and conservationist whose wide-ranging, prolific work has explored such connections as those between cultural diversity and biological diversity, between people and desert wildlife, between wild and cultivated plants, and between poetry and natural science. His second book, *Gathering the Desert* (1985), received the John Burroughs Medal for nature writing. The MacArthur Foundation gave him a "genius" fellowship in 1990, the same year he received a Pew Scholarship on Conservation and Environment. Sicily honored him in 1991 with the Premio Gaia Award for his contributions to "a culture of the environment." Nabhan has focused his projects and writings largely in the Sonoran Desert region of northwestern Mexico and the southwestern United States.

Gary Paul Nabhan was born on March 17, 1952, in Gary, Indiana, one of three children of Theodore and Jerri Nabhan. Lake Michigan's Indiana Dunes, tucked among the wasteland of steel mills and power plants, provided Nabhan's early introduction to the wild outdoors, as recounted in *The Geography of Childhood: Why Children Need Wild Places* (1994), a book he coauthored with longtime friend Stephen Trimble. His Lebanese family and community also set the stage for his interest and success in cross-cultural work. After a short stint at Cornell College, Nabhan attended Prescott College in Arizona, where he graduated in 1975 with a liberal arts degree that combined regional American literature and environmental biology. His postgraduate work at the University of Arizona, Tucson, resulted in a M.S. degree in plant sciences in 1978 and a Ph.D. in arid lands resources in 1983 and in a continuing love for the desert and its peoples.

As a graduate student, Nabhan started collecting seeds from Native American farmers, first as a job for the U.S. Department of Agriculture (USDA) seed bank and then to help a nutrition project introducing vegetable gardening on the Tohono O'odham (Papago) reservation. Organizers learned that while broccoli and spinach were hard to promote, seeds of corn, beans, and squash varieties that their grandparents had grown were often requested—they thrived under desert growing conditions, and the O'odham already had names, recipes, and good memories of them. As Nabhan helped grow native seeds for redistribution, he realized these crops were worth saving not only for the genes they might add to future modern hybrids, but because they still had cultural value in their present form. His first book, *The Desert Smells like Rain: A Naturalist in Papago Indian Country* (1982), is a "sensitive and compassionate portrayal" (according to Rep. MORRIS UDALL) of the Indian people he worked with at this time.

In 1983, Nabhan cofounded Native Seeds/SEARCH, a nonprofit group that has expanded the seed-saving efforts to include Native peoples throughout the southwest and to promote the use of native crops and wild foods. The essays in *Enduring Seeds: Native American Agriculture and Wild Plant Conservation* (1989) and *Cultures of Habitat: On Nature, Culture and Story* (1997) chronicle some of the important projects started by Nabhan and Native Seeds/SEARCH. Nabhan has been especially involved in a program to publicize the benefits of traditional desert plant foods in preventing diabetes, particularly among high-risk indigenous populations. His interest in studying one of the few areas where wild chiles grow in the United States recently led to its designation as a protected botanical area.

In 1986, Nabhan became assistant director of the Desert Botanical Garden in Phoenix. His projects there included establishing a hands-on ethnobotanical trail; creating a database of endangered, useful plants found along the U.S.-Mexico border; and editing a book still widely used, *Desert Wildflowers: A Guide for Identifying, Locating and Enjoying Arizona Wildflowers and Cactus Blossoms* (1988). Nabhan's research projects took him to Organ Pipe Cactus National Monument, where he met his second wife, park naturalist Caroline Wilson. With Wilson he coauthored *Canyons of Color: Utah's Slickrock Wildlands* (1995). His work at this borderland monument, and an association with Conservation International, led to Mexican regulations to stop wholesale destruction of ironwood trees that are used for imitation Seri Indian artwork and to make mesquite charcoal. Nabhan helped form the Ironwood Alliance, a binational coalition that monitors ironwood tree protection efforts and successfully lobbied for the protection of the Ironwood Forest, a 129,000-acre site near Tucson that received national monument status in 2000.

In 1991, Nabhan moved to Tucson to become writer-in-residence at the Arizona-Sonora Desert Museum, where he edited a collection of natural history essays, *Counting Sheep: Twenty Ways of Seeing Desert Bighorn* (1993). He also finished writing a lovely account of his 1990 Franciscan pilgrimage to Assisi, *Songbirds, Truffles, and Wolves: An American Naturalist in Italy* (1993).

As director of science at the Arizona-Sonora Desert Museum, Nabhan cofounded the museum's "Forgotten Pollinator" campaign to raise awareness on the vital but little-appreciated interdependence of plants and the animals that help them reproduce. This project included the book *The Forgotten Pollinators* (1996) that Nabhan cowrote with Stephen L. Buchmann. Another book, *Desert Legends: Re-storying the Sonoran Borderland* (1994), provides some of the basis for another museum program, the Sense of Place Project, which works with small communities on both sides of the U.S.-Mexico border to document, preserve, and celebrate the region's cultural and ecological heritage. In the spring of 2000, Nabhan organized a successful 250-mile cross-country, cross-border, cross-cultural Desert Walk to raise money for Native American internships, to heighten awareness about the epidemic of diabetes among Native American communities, and to promote intergenerational cultural exchanges among the Seri, Tohono O'odham, and Yoeme—an unusual event, yet perhaps typical of Nabhan's thinking.

In 2000, Nabhan became director of Northern Arizona University's Center for

Sustainable Environments, a research center specializing in the sustainable use of natural resources on the Colorado Plateau. Nabhan lives in Flagstaff, Arizona. He is married to Laurie Monti, and has two children, Laura Rose and Dustin Corvus, from his first marriage.

BIBLIOGRAPHY
"Arizona-Sonora Desert Museum," www.desertmuseum.org; Erickson, Jim, "Biologist Nabhan Leaving Desert," *Arizona Daily Star*, 2000; Goldstein, Carol, "Gary Nabhan: Native Seeds," *Omni*, 1994; "Native Seeds/SEARCH," www.nativeseeds.org.

Nader, Ralph

(February 27, 1934–)
Public Interest Lawyer, Founder of Numerous Public Interest Organizations

Thanks to the tireless efforts of consumer advocate Ralph Nader, the United States is a safer place. Automobiles in the United States now have mandatory seat belts, industrial air polluters must conform to the standards of the Clean Air Act, and the Occupational Safety and Health Administration oversees safer and healthier U.S. workplaces. In 1969, Nader founded the Center for Study of Responsive Law, which monitors government regulatory agencies to ensure that they are working honestly and effectively. He has founded numerous other organizations to watch over the government or lobby Congress, including the Public Interest Research Groups (PIRGs) and Public Citizen.

Ralph Nader was born on February 27, 1934, in Winsted, Connecticut, to parents who had emigrated in 1912 from Lebanon. As a child, Nader listened closely to his family's dinner table conversations about social injustice. From an early age he showed an inclination toward law. He spent free time watching trials at Winsted's town hall, and his

recreational reading was the *Congressional Record*. As a student at Princeton University, he became an activist, protesting the spraying of the campus trees with the pesticide dichlordiphenyltrichlor (DDT). He studied law at Harvard University, where a research project on automobile safety initiated his career in consumer advocacy.

Nader was able to continue his work on the topic of auto safety as a staff consultant for Daniel Patrick Moynihan, who was then President Johnson's assistant secretary of labor. Nader's 1964 "Report on the Context, Condition and Recommended Direction of Federal Activity in Highway Safety" was converted into the widely read *Unsafe at Any Speed* (1965). That book deserves much of the credit for the 1966 passage of the National Traffic and Motor Vehicle Safety Act, which mandated seat belts and forced the car industry to install collapsible steering columns and padded dashboards. Nader and his cause gained further public attention when he sued General Motors in the mid-1960s for invasion of privacy. Gen-

Consumer advocate Ralph Nader announces his bid for the Green Party nomination for the U.S. presidency, 21 February 2000 in Washington, D.C. (AFP/Corbis)

eral Motors, threatened by Nader's research, had been making harrassing phone calls to him and sending seductive women his way to lure him into compromising situations. General Motors agreed to pay $425,000 to Nader in order to settle the suit, a sum 30 times more than any previous invasion-of-privacy settlement.

Building on his growing notoriety, Nader used his settlement money to found a network of watchdog and consumer activist organizations. His energetic teams of lawyers and researchers, nicknamed "Nader's Raiders," produced stacks of reports incriminating big business for various social and environmental ills. They documented the lack of effectiveness of company-run workplace hazard policies; these documents became important ammunition in the fight for the Occupational Safety and Health Act (OSH Act; 1970). Once the OSH Act was passed, Nader's Raiders discovered and publicized key problems that the Occupational Safety and Health Administration—the federal agency created by the OSH Act—would have to respond to, such as cancer epidemics in certain chemical plants and the black lung disease in coal mines. The studies by Nader's Raiders of the industrial sources of air pollution aided the passage of the Clean Air Act of 1970.

Nader's Raiders worked through the organizations he founded, such as the Center for Study of Responsive Law, a

Washington, D.C.–based nonprofit watchdog that monitors government regulatory agencies, and both Public Citizen and the U.S. Public Interest Research Group (U.S. PIRG), membership groups for concerned consumers. In 1970, Nader exhorted the students of the University of Oregon to form their own student consumer activist organization. They founded the Oregon Student Public Interest Research Group, which was the first of many student affiliates to U.S. PIRG. The PIRGs, funded by students, with student volunteers and professional staffs in each state where they exist, continue to promote environmental legislation at a statewide level.

Nader continues to direct Public Citizen, leading it on diverse research campaigns that address consumer abuse. Nader is a frequent speaker on college campuses. In his talks, he reminds his listeners that unless monitored by private citizens and the government, industry will exploit its workers and the public and pollute the environment. To ensure that business is not given free rein to abuse and destroy, Nader recommends that people pay close attention to its activities and join organizations that can mobilize quickly in response to problems. Nader has been recognized by *Life Magazine* as one of the 100 most influential people of the twentieth century. He ran for president as the Green Party candidate in 1996 and 2000.

BIBLIOGRAPHY

DeLeon, David, *Leaders from the 1960s*, 1994; Gottlieb, Robert, *Forcing the Spring*, 1993; Nader, Ralph, "Beyond Politics as Usual," (audio recording produced by Alternative Radio), 1996; Nader, Ralph, "Corporate Power: Profits before People," (audio recording produced by Alternative Radio), 1994.

Nagel, Carlos

(February 7, 1931–)
Founder of Friends of Pronatura

As a founder of Friends of Pronatura (FPN), an organization promoting the conservation of natural and cultural diversity in the Mexico and U.S. borderlands, Carlos Nagel has been a leader in the effort to create a sustainable future for the region—environmentally, economically, and socially. Nagel believes that in the same way that a healthy environment depends on diversity, a healthy social dynamic thrives on a diversity of perspectives, and he has thus devoted his career to building bridges between cultures. He works hard to encourage community participation in the process of ecosystem management, bringing together researchers, nongovernmental organizations, representatives of the local communities, and members of the Tohono O'odham Nation and other Native American groups to discuss regional land use issues and increase local awareness of the environment. He is also a founding member and president of the board of the International Sonoran Desert Alliance, an organization that helps foster cross-cultural dialogue regarding ecological, economic, and social issues.

Carlos Nagel was born on February 7, 1931, in Argentina and came to New York with his family as an adolescent. After serving in the U.S. military from 1952 to 1954 in Korea, he attended the University of Washington, earning a B.A. in 1958 in museum administration. Beginning in 1959, Nagel worked for nine years as superintendent of the National Institutes of Health Primate Ecology research facility in Puerto Rico. As administrator, he worked closely with the ecologists who carried out the scientific population dynamics research, and he introduced innovative participatory management techniques. His supervisor, mentor, and friend during his years at the research facility was Carl Koford, who contributed greatly to Nagel's environmental awareness. Nagel later described this period as a time when, without his knowing it, he was being injected with the "antibody" of ecological consciousness—cultivating an appreciation of healthy, diverse ecosystems that came to permeate his outlook on the world.

In 1968 he was named assistant director of the Oakland Museum in California and stayed for one year, leaving in 1969 to become director of the Museums of New Mexico in Santa Fe. In 1974 he was asked to join the Arizona-Sonora Desert Museum in Tucson to establish a comprehensive binational environmental program with Mexico. He left the museum in 1978 to create the Cultural Exchange Service, a consultant service specializing in cross-cultural communications that seeks to find common ground among individuals with different orientations in businesses, environmental groups, and civic organizations.

One of Nagel's best-known achievements began to take shape in 1985, when he and a group of environmentally concerned Tucson residents began looking for a way to help support Pronatura, the premier national conservation organization in Mexico. When the Tucson group approached Pronatura to ask what they could do, the founders of the organization suggested a kind of "reverse imperialism," with the creation of a U.S. branch of the Mexico-based organization. With that, Nagel and the others founded the nonprofit Friends of Pronatura to assist its Mexican counterpart in promoting education, research, and information dissemination on environmental issues affecting southwestern North America. Since its inception, FPN has worked extensively on U.S.-Mexico borderland issues and has supported several of the six chapters of Pronatura in Mexico with grants and promotion in the United States.

In 1988, Friends of Pronatura received a grant from the U.S. Man and the Biosphere Program (U.S. MAB), a project designed to promote ecosystem management programs that incorporate sustainable development, to document global change and biological diversity through monitoring and scientific research, and to organize regional and international cooperation in the effort to resolve complex issues of multipurpose land use. U.S. MAB uses United Nations Educational, Scientific and Cultural Organization (UNESCO)–designated biosphere reserves as sites for implementing management plans, conducting research, and bringing together cooperating institutions. The grant provided to Friends of Pronatura allowed the group to study applications of the biosphere reserve concept in the region. As president of FPN, Nagel used the opportunity to put one of his strongest beliefs to use: that just as the health of an ecosystem stems from its biological diversity, the health of a social environment is promoted by a diversity of opinions, viewpoints, attitudes, feelings, and actions. During the following year, Nagel successfully expanded the discussions on the biosphere reserve to include leaders of the Tohono O'odham and local communities, state and local governments, and nongovernmental organizations. This effort fostered a new willingness to use the biosphere reserve concept as a structure for empowering diverse parties to maintain dialogue and cooperate in addressing shared problems. For example, Nagel planned a symposium to discuss protection of the Pinacate, an ecologically fragile area in northwest Sonora, Mexico. The gathering, held in Hermosillo, Sonora, in 1988, included researchers, representatives from nongovernmental organizations, community members, and members of the Tohono O'odham, a Native American nation whose traditional homelands encompassed much of the area. This was the first time the O'odham were consulted and given an opportunity to express their views on the region, which included sites of critical importance to their culture, history, and identity.

To continue to nurture international relationships and to increase local awareness about the need to protect the western Sonora Desert border region, another forum on land use was held in 1992 in Ajo, Arizona. About 200 people attended from all over the region, including federal, state, county, and tribal government officials; business leaders; and members of the O'odham Nation. The group tackled many complex issues, including the potential environmental and socioeconomic effects of the North American Free Trade Agreement and how to expand and improve communication among agencies and citizens regarding the biosphere reserve concept. As a result of the positive response from the conference, Nagel collaborated with the Sonoran Institute, an organization devoted to community-based conservation, to forge an alliance called the International Sonoran Desert Alliance (ISDA), made up of individuals from the United States, Mexico, and the Tohono O'odham Nation. ISDA has since grown into a large network, and in 1994 it was incorporated as a nonprofit organization in Arizona, and its first board of directors was elected. In 1998 Nagel was asked to join the board of ISDA and in 1999 was elected president. He is also a member of the board of directors of the Sonoran Institute, the Development Center for Appropriate Technology, and the Yonosé Foundation in Tucson.

Nagel's work has involved him in other groups as well. He is a founding member of the following organizations: Environmental Committee of the Arizona-Mexico/Sonora-Arizona Commissions, the Arizona-Mexico Border Health Foundation, and Hands Across the Border (a school exchange program that has involved over 6,000 Sonoran and Arizona students each year since 1982). He has also served as vice president of the board for the Center for Studies of Deserts and Oceans (in Puerto Peñasco, Sonora). He also presents seminars on a broad spectrum of topics ranging from intercultural issues to the relations between the environment and business. In addition, he continues his work for Friends of Pronatura as the current chief executive officer/president. In 1998 the Arizona-Sonora Desert Museum presented Nagel with its Sixth Annual Luminaria Award for his work to increase public knowledge and appreciation for the environment, particularly the Sonoran Desert. Nagel is unmarried, his partner since 1978 having died in 1999; he has five adopted children. He lives in Tucson, Arizona.

BIBLIOGRAPHY

Faulkner, Tina, "Community-Based Groups Set the Agenda for Conservation of the Colorado River Delta," *Borderlines*, 1999; Nagel, Carlos, "Commentary," *Natural Resources Journal*, 1993; "Pronatura," http://www.pronatura.org.mx; Williams, Florence, "On the Borderline," *High Country News*, 1994.

Nash, Roderick

(January 7, 1939–)
Environmental Historian and Ethicist

Roderick Nash's seminal work, *Wilderness and the American Mind* (1967), is considered a classic study of how the concept of wilderness has shaped the American character. Nash, a professor of history and environmental studies (since retired), successfully bridged the gap between academic discourse and popular literature on the environment. He brought a new voice to the debate over wilderness—that of the historian. His exploration of wilderness as an intellectual idea led the way toward a new consideration of the role of philosophy and ethics in the field of environmentalism. *Wilderness and the American Mind* continues to be widely read; now in its third edition, it has remained in print for more than 30 years. Nash has become a well-known and highly respected spokesman for wilderness preservation and environmental education.

Roderick Frazier Nash was born January 7, 1939, in New York City. He is the son of Jay B. Nash, a professor, and Emma (Frazier) Nash. Nash graduated *magna cum laude* with a degree in history and literature from Harvard University in 1960. He pursued graduate studies at the University of Wisconsin, earning his M.A. in 1961 and his Ph.D. in 1964. He taught history at Dartmouth College from 1964 to 1966 before moving to the University of California at Santa Barbara (UCSB) as an assistant professor in the history department. In 1970, he founded UCSB's Department of Environmental Studies; he served as chair of the department for five years.

Nash began his writing career as an academic historian, specializing in intellectual and social history. In 1967, he zeroed in on the subject that would prove to be his forte: the idea of wilderness. His academic background gave him an as yet unexplored perspective from which to examine the subject of wilderness. The mid-1960s were fraught with political battles over what exactly constitutes a wilderness and what uses were consistent with a wilderness designation. Nash removed the discussion from the highly emotional political arena and brought it into the intellectual realm: wilderness as a social construct in the minds of Americans. His hugely successful book, *Wilderness and the American Mind*, had an influence that extended beyond academia. He was invited to appear on television and in Public Broadcasting Service (PBS) and National Geographic documentaries, and he was interviewed in popular magazines.

In the 1970s, Nash wrote many articles for the general public on why wilderness should be valued. "What Is Wilderness and Why Do We Need it?" appeared in the *Daily Idahoan* in Moscow, Idaho, in 1976; it and others were widely reprinted in newspapers, magazines, and textbooks. During this period, Nash also received several grants from the Rockefeller Foundation to research environmental issues. Although he was drawn more and more into the environmental arena, Nash continued to write on other aspects of American history. *The Nervous Generation* (1970) and *From These Beginnings* (1973) focused on the social, intellectual, and biographical elements of American history. In 1973, a revised edition of *Wilderness and the American Mind* was issued. The new edition reflected the evolution of Nash's thought since the original publication. The first edition began with the preconceived ideas of wilderness that immigrants to the United States brought with them from the Old World—attitudes that had been drawn, in turn, from the Judeo-Christian depiction of nature in the Bible. The revised edition looked beyond the Western historical tradition and a purely intellectual idea of nature. In the preface, Nash delved into prehistory to explore the intuitive meaning of wilderness as perceived by early man.

Underlying Nash's evolving philosophy is a fundamental belief that in the United States, understanding of the environment must start with an understanding of the role of wilderness in the nation's imagination. In a 1989 documentary in the PBS series *The American Experience*, he restated the premise he first presented in *Wilderness and the American Mind*. "If there was one thing that shaped our character and culture, one single factor you could point to, it would be wild country." The United States, he claimed, based its national identity on wilderness just as Europe based its on a shared history full of intellectual and artistic traditions. Wilderness made the United States fundamentally different from Europe. Nash traced the transformation of this perception of wilderness from an essentially negative attitude in the early years of the United States to the positive attitude held two hundred years later. Through skillful use of historical examples, Nash conveyed how Puritans saw the wilderness as a hostile wasteland and moral vacuum capable of sucking all the humanity out of man, while later generations came to see it as a benign spiritual force that soothed the often dangerously overstimulated modern mind.

By 1989, Nash had developed an environmental philosophy that centered on the ethics of the human-nature relationship. In *The Rights of Nature: A History of Environmental Ethics*, Nash demonstrates that the concept in the United States of who—or even what—warrants ethical and legal consideration has changed radically from the days of the founding fathers; the circle of inclusion has gradually expanded from White males only to encompass females, Blacks, and other disenfranchised groups. Nash argues that extending legal and ethical rights to all living species—even plants and trees—is a logical extension of our liberal tradition.

Nash's influence has been felt not only through his writing but also through his teaching. Many of his graduate students in UCSB's environmental studies program have become environmental writers. One of these was Calvin Martin, author of *Keepers of the Game*, which won the American Historical Association's Beveridge Award in 1978. Nash's role as a leader in the field of environmentalism was recognized in 1974 when he was selected by the American Academy of Achievement as "one of forty giants of accomplishment from America's great fields of endeavor." The Charles A. Lindbergh Fund named him a Lindbergh Fellow in 1982 for advocating a balance between technological progress and environmental preservation. In 1988, he received the William G. Anderson Award of the American Alliance for Health, Physical Education, and Recreation to honor the contributions he had made to the field of outdoor recreation. Nash has served as a consultant to many government agencies and nongovernmental organizations, including the National Park Service, *National Geographic*, the U.S. Forest Service, the state of Alaska, and the Rockefeller Foundation. He has been associated in a leadership role with various environmental organizations and publications, including the Sierra Club, Friends of the Earth, the Yosemite Institute, *Environmental Ethics*, and *Journal of Environmental Education*.

As an environmental leader at the beginning of the twenty-first century, Nash has often been asked what he thinks the future holds for planet earth. In articles and interviews, he remains unabashedly optimistic about the future, having faith in nature's ability to restore balance. He envisions population "islands" surrounded by undeveloped land and a generation of children brought up with an ethic of sustainability. Nash believes such changes can come about through a new model of environmental education that embraces the study of both the humanities, such as ethics and philosophy, and the sciences.

Although a native New Yorker—he likes to say that he lived for 18 years with a view of a brick wall—Nash became a backcountry enthusiast. After serving as a fishing guide in Ontario and Wyoming in the mid-1950s, he became one of the first professional river guides in the country, running the Snake River in Grand Teton National Park. Over the next 30 years, Nash developed into an experienced riverboatman, running the Colorado River through the Grand Canyon over 50 times and performing first descents of rivers in California, Alaska, and Peru. Nash recently updated his 1978 guide to the best whitewater rafting spots in the West (coauthored with Robert Collins); a revised edition of *The Big Drops: Ten Legendary Rapids of the American West* was published in 2000.

Now retired from UCSB, Nash maintains an active intellectual presence in the field of environmental literature through his erudite book reviews, which appear frequently in such publications as

American Historical Review, New England Quarterly, and *Journal of American History.* Nash and his wife, Lindamel Murray, live in Santa Barbara, California, and Crested Butte, Colorado. He has two grown daughters, Laura and Jennifer.

BIBLIOGRAPHY

Murphy, Pat, "Roderick Nash: Environmental Disciple," Environmental News Network (http://www.enn.com), 2000; Nash, Roderick, *The Rights of Nature: A History of Environmental Ethics,* 1989; Nash, Roderick, *Wilderness and the American Mind,* 1967; Public Broadcasting System, "The Wilderness Idea: John Muir, Gifford Pinchot, and the First Great Battle for Wilderness," (video recording) 1989; "Rod Nash, Author, Professor of Environmental Studies," *Lands of Brighter Destiny: The Public Lands of the American West,* Elizabeth Darby Junkin, ed., 1986.

Nearing, Helen, and Scott Nearing

(February 23, 1904–September 17, 1995; August 6, 1883–August 24, 1983)
Writers, Founders of Back-to-the-land Movement

Helen and Scott Nearing have been described as the grandparents of the back-to-the-land movement. Following Scott Nearing's controversial and subversive academic career, in which he was fired from every university he worked for because of his radical views, the Nearings moved to a farm in Vermont, where they farmed organically and lived simply for 19 years. Then Helen and Scott Nearing moved to Maine, where they did the same thing until their deaths at the ages of 91 and 100, respectively. They were prolific and influential writers, whose books addressed all manner of social issues and whose ideas inspired many others to return to the land, and to live "the good life," the title of their best-known book.

The oldest of six children, Scott Nearing was born on August 6, 1883, in the company town of Morris Run, Pennsylvania, to Louis and Minnie Zabriski Nearing. Scott's grandfather, Winfield Scott Nearing, supervised the Morris Run Coal Company with a union-busting iron hand (he earned the nickname Czar Nearing of Morris Run). Scott's father operated the company fruit and vegetable store. In 1898, when Scott was 15 years old, his family moved to Philadelphia, and he enrolled in Central Manual Training High School. He decided not to enroll in the more academic Central High School, because he sought a practical education, one that combined both theory and practice. He graduated from high school in 1901 and entered the University of Pennsylvania Law School. Scott did not enjoy studying law, as Steve Sherman writes in *A Scott Nearing Reader, The Good Life in Bad Times:* "It was increasingly clear that anything having to do with the gathering of unseemly amounts of money, and defending it as lawyers were trained to do, nettled him." He decided to turn his attention to the social sciences.

In 1903, Nearing entered the Wharton School at the University of Pennsylvania, where he studied economics and oratory at Temple College (now Temple University). He received a B.S. in economics from the Wharton School and a Bachelor of Oratory from Temple College in 1905. That same year, he began work as secretary of the Pennsylvania Child Labor Commission, a position he held until 1907. In 1908, Scott married Nellie Marguerite Seeds. They would later separate in 1925. Nearing earned a Ph.D. in economics at the Wharton School in 1909. He taught economics at the Wharton School as an instructor from 1906 to 1914, before being promoted to assistant professor in 1914. He also taught economics at Swarthmore College from 1908 to 1913.

In 1915, in a nationally publicized academic freedom case, Nearing was dismissed from the Wharton School for his vocal opposition to child labor. The University of Pennsylvania was under pressure to silence Nearing. Powerful textile manufacturers, who employed large numbers of children, exercised a significant degree of control over the amount of state funding the university received, and in the interest of retaining that funding, the University of Pennsylvania fired Nearing after nine years of employment. For the next two years, from 1915 to 1917, Nearing worked at the University of Toledo as a professor of social science and dean of the Colleges of Arts and Sciences. He was fired from the University of Toledo for vocalizing his pacifist views, and he was indicted by a federal grand jury for writing a pamphlet entitled "The Great Madness," which spoke out against U.S. involvement in World War I. He went to trial in 1919 and was acquitted of all charges. In 1917, Nearing became an instructor at the Rand School of Social Science in New York and joined the Socialist Party. He ran for Congress on the Socialist Party ticket in 1918, but lost.

During this period of his life, Nearing was writing copiously. Between 1917 and 1937, he published more than 30 books and pamphlets, including *The Menace of Militarism* (1917), *Dollar Diplomacy: A Study in American Imperialism* (1925), *Whither China? An Economic Interpretation of Recent Events in the Far East* (1927), *Black America* (1929), and *War: Organized Destruction and Mass Murder by Civilized Nations* (1931). He also joined the Communist Party in 1927, only to be expelled three years later after publishing *The Twilight of Empire* (1930), a history of imperialism that challenged the party's official text, *Imperialism*, written by Lenin. In 1929, Scott met Helen Knothe.

Helen Knothe was born February 23, 1904, in New York City to Frank and Maria Obreen Knothe. In a 1994 interview with *Whole Earth Review*, she described her family as "rather out of the ordinary. . . . They were intellectuals, they were musical, they were artistic." Her parents, both vegetarians, were interested in eastern religions and were "quite philanthropic." Helen began studying the violin as a child, and by the time she was 17 she faced a choice: stay in the United States and go to college (Vassar or Wellesley) or travel to Europe and study the violin. She chose Europe. Not only was she a developing concert violinist, she was also a student and practitioner of eastern philosophies. She spent the years 1921 to 1925 with Krishnamurti, an important Hindu spiritual leader. Helen then spent several years traveling around Australia before moving back to the United States permanently. She met Scott

Nearing in 1929; their first date was spent admiring fall foliage in New Jersey.

By 1931, Scott Nearing had been separated from his first wife for six years. He and Helen moved to New York together, where they attempted to piece together a living in the impoverished Great Depression economy. In 1932, they decided that if they were going to be poor, they might as well be poor somewhere where they could at least grow their own food, so they bought a farm near Jamaica, Vermont. They refurbished the farmhouse and rejuvenated the soil through natural, organic means. They lived off of what they produced and sold maple syrup and sugar to pay the taxes. During the winters, they traveled extensively, visiting places such as Germany, Russia, Austria, and Spain. Scott used these trips to gather material for his books, which he continued to produce at a rapid pace.

It was at their Vermont homestead that the couple developed their four-four-four daily schedule. They would spend four hours of each day gardening or otherwise working to provide for their basic needs. They would spend four hours undertaking professional activities (Scott Nearing eventually wrote a total of 50 books, six of them coauthored with Helen). And they would spend four hours working in service to their community, fulfilling their obligations as members of the human race. In 1947, after the death of Scott's first wife, Helen and Scott were married. By the early 1950s, hundreds of visitors were visiting the Nearing farm each year. This popularity, along with the development of a nearby ski resort, impelled the Nearings to move once more in search of solitude. In 1951, they purchased a house in Harborside, Maine.

In Maine, they continued to farm (their cash crop was blueberries rather than maple syrup and sugar), and they wrote what is perhaps their best known work, *Living the Good Life* (1954). This book became a bestseller in the early 1960s with the dawning of the back-to-the-land movement, and the Nearings became mentors for those seeking to escape the materialism and corruption that was becoming prevalent as the United States was beset by racial and social injustices.

In 1972, at the age of 90, Scott Nearing wrote his autobiography, *The Making of a Radical*, and he and Helen began constructing a new stone house, which they finished in 1976. In 1973, Scott Nearing received an honorary degree from the Wharton School of Economics, the institution that had fired him 58 years earlier. He died in 1983, just a few weeks after his one hundredth birthday, purposefully fasting to death. In her *Whole Earth Review* interview, Helen Nearing describes the experience of losing her husband in this way as being completely natural, saying "I wish more people could go as readily and easily and unaffected as he was. . . . He was a model for me in his living and in his going." Helen Nearing continued to live and farm in Maine. She also continued to write, publishing *Simple Food for the Good Life* in 1985 and *Loving and Leaving the Good Life* in 1992. She died in a car accident in 1995 at the age of 91.

BIBLIOGRAPHY

Cole, John N., "Scott Nearing's Ninety-Three-Year Plan," *Horticulture*, 1976; Nearing, Scott, *The Making of a Radical*, 1972; Sherman, Steve, *A Scott Nearing Reader: The Good Life in Bad Times*, 1989; Simon, Tami, "The View from Ninety," *Whole Earth Review*, 1994.

Needleman, Herbert

(December 13, 1927–)
Pediatrician, Lead Researcher

Author of a 1979 study proving a link between low-level lead exposure and impaired mental function, Herbert Needleman has been a long-time advocate for reducing lead levels in the environment. Needleman's work intensified the growing momentum to enforce stricter regulation of lead. Because his findings so clearly showed the need to eliminate the possibility of lead exposure, Needleman became a target for the already-ailing lead industry and had to defend his research from industry-planted complaints. Levels of lead in children's bloodstreams have fallen by 75 percent since 1979, thanks to Needleman's research and the national concern and activism it has spurred.

Herbert Needleman was born on December 13, 1927, in Philadelphia. He graduated from Muhlenberg College with a B.S. in 1948 and earned his M.D. from the University of Pennsylvania in 1952. In order to finance his studies while in medical school, he worked as a laborer at a DuPont chemical plant, where he encountered firsthand the toxic effects of lead. He noticed that one older group of workers kept to themselves and behaved oddly, and when he asked about them, Needleman was told they were from "the house of butterflies." Years before, more than 300 workers had suffered acute lead poisoning at the plant. Four of them died, and others were permanently disabled and unable to return to work. While the group Needleman had noticed was still able to work, they suffered lasting neurological damage, including a tendency to gesture at imaginary insects.

Needleman next encountered lead poisoning in 1957, while completing a pediatric residency at an inner-city hospital in Philadelphia. Needleman diagnosed a three-year-old girl with lead poisoning, acquired from the lead paint on the walls of her family's home. A paint chip as small as a dime can result in acute symptoms, and the girl had probably eaten a small piece. When lead enters the bloodstream, the body mistakes it for calcium. In much of the body this substitution causes no obvious harm, but in the brain, lead causes small blood vessels to leak, resulting in swelling and pressure. The child had classic symptoms of lead poisoning, and Needleman told her mother she could not bring the girl back to their house, because a second exposure would almost certainly cause catastrophic damage. The mother told Needleman that any other house they could afford to rent would also have lead paint. Needleman realized that the problem of lead exposure extended beyond his medical skills. Lead was ever present in the environment, not only in paint but also in gasoline, automobile exhaust, and the pipes that carried drinking water. The solution would have to be political as well as scientific.

In 1974, while teaching at Harvard Medical School, Needleman began a study of the effects of low-level lead exposure on children's behavior and intelligence. Needleman compared groups of children with low and high levels of lead exposure, correlating their intelligence quotients (IQs), educational achievements, and teachers' blind evaluations of their behavioral characteristics. His research demonstrated a clear, direct link between chronic, low levels of lead exposure and impaired mental function. The students with higher lead exposure showed consistently lower IQ scores, weaker academic performance, and higher negative behavior ratings. Needleman published his findings in 1979, at a time when the lead industry was already feeling regulatory pressure. In 1977, lead had been banned from household paint, and the Occupational Safety and Health Act of 1978 contained new, stricter monitoring of lead in the workplace. Needleman's work earned him an influential voice as an adviser to the Environmental Protection Agency (EPA), which phased in a ban on leaded gasoline during the 1980s. The Centers for Disease Control, too, issued stricter guidelines for what could be considered safe lead blood levels.

In 1981, Needleman met Dr. Claire Ernhart while both were serving as consultants to the EPA during revisions to the Clean Air Act. Ernhart, a developmental psychologist who received substantial grant money from the lead industry, charged that Needleman's 1979 study was flawed and that he should therefore be disqualified as an adviser to the government. A panel of outside experts reexamined Needleman's work and confirmed his findings. In 1991, Ernhart again challenged Needleman's study, bringing formal charges of scientific misconduct in a letter to the National Institutes of Health. Needleman was investigated over a period of three years, before being formally cleared of all charges in 1994.

In 1990, Needleman founded the Alliance to End Childhood Lead Poisoning, a public interest group devoted to developing comprehensive strategies to end lead poisoning. The alliance's mission is "to frame the national agenda, formulate innovative approaches, and bring critical resources to bear—scientific and technical knowledge, law and public policy, economic forces, national allies, and community organizations and leaders—to prevent childhood lead poisoning." Needleman has also been active in efforts to reduce exposure to pesticides, many of which lead to brain damage and other neurological impairment. In 1992, he was awarded the National Wildlife Federation's Conservation Achievement Award and in 1995 the Heinz Award in the Environment.

Needleman's work has contributed to a dramatic reduction in the lead hazard in the United States. Levels of lead in children's blood fell by more than 75 percent during the 20 years following the publication of his 1979 study, and his advocacy played no small role in the reduction. He continues to be active in the struggle to remove lead-based paint from homes in poor neighborhoods in urban communities. Herbert Needleman is a pediatrician and child psychiatrist at the University of Pittsburgh Medical Center.

BIBLIOGRAPHY

"Alliance to End Childhood Lead Poisoning," http://www.aeclp.org; "Dr. Herbert Needleman," The Heinz Awards http://www.awards.heinz.org; Lewis, Thomas, "The Difficult Quest of Herbert Needleman," *National Wildlife*, 1995.

Nelson, Gaylord

(June 4, 1916–)
Governor of Wisconsin, U.S. Senator from Wisconsin

Wisconsin native Gaylord Nelson spent 32 years of his life as a Democratic politician, including ten years as a state senator, four years as Wisconsin's governor, and 18 years in the U.S. Senate. During his career he led the fight to pass such landmark environmental legislation as the Clean Air Act, the National Pesticide Control Act, the Water Quality Act, the Natural Lakes Preservation Act, and the Wild Rivers Act, but he is perhaps best known as the father of Earth Day.

Born on June 4, 1916, Gaylord Nelson attributes his deeply ingrained environmentalism to his upbringing in the small town of Clear Lake, Wisconsin. Clear Lake was surrounded by lakes and marshes, which Nelson explored throughout his boyhood. After being inspired by a visiting politician, Nelson undertook his first political project: asking the town council to line the roads that led into town with elm trees. The council declined to take on the project, but Nelson did not lose his interest in politics. Nelson earned a bachelor's degree at San Jose State College in California in 1939 and graduated from the University of Wisconsin law school in 1942. Immediately he was drafted into the military and served for the remainder of World War II. Upon his return to civilian life, Nelson married Carrie Lee and was elected to Wisconsin's state senate, an office he retained for ten years.

When Nelson was elected governor in 1958, he reread one of his favorite books, ALDO LEOPOLD's *Sand County Almanac*, which urged a view of nature as a community that we humans must respect and conserve. Nelson convinced the state legislature to set aside $50 million to buy and conserve wildlands in the state. During his governorship, Wisconsin became the first state to regulate the use of the harmful, nonbiodegradable laundry detergents that were polluting lakes and rivers. The state also passed strict laws against trash dumping and littering.

Nelson was elected U.S. senator in 1962, bringing a previously nonexistent environmental consciousness to Congress. He successfully convinced Pres. John F. Kennedy to undertake a national conservation tour during the summer of 1963. With 80 reporters signed up for the trip, Nelson thought the tour would be a giant success and would awaken the American public to environmental problems. However, newspaper editors were not interested, and the tour went completely unpublicized. Still, Nelson continued his work quietly.

After a visit to Santa Barbara, California, where he witnessed the destruction wrought by a huge oil spill, an idea occurred to him. He was familiar with the "teach-ins" that were being organized around the Vietnam War and issued a call for a country-wide environmental teach-in. Once the idea caught on, he enlisted Harvard Law student DENIS HAYES to coordinate the event. Earth Day, April 22, 1970, was celebrated at 2,000 colleges and 10,000 elementary and secondary schools by 20 million Americans, who spent the day picking up trash, dredging rivers, closing streets to cars, and a variety of other activities. Public enthusiasm for Earth Day carried a lot of weight with Nelson's colleagues in Washington. President Nixon had dedicated his 1970 State of the Union speech to the cause of environmentalism, and in the decade following Earth Day, much environmental legislation was passed, including the Water Quality Act, the National Pesticide Control Act, the Clean Air Act, the Natural Lakes Preservation Act, and the Wild Rivers Act. Nelson took a leadership role in the passage of much of this legislation.

Nelson left public office in 1980 and went to work for the Wilderness Society, which his intellectual mentor Aldo Leopold had helped found. Nelson was the honorary chair of the 1990 Earth Day and has won the Conservationist of the Year award from the National Wildlife Federation and the United Nations Environmental Program's Environmental Leadership Award. He is married to Carrie Lee Nelson, and together they have three children.

Gaylord Nelson (Courtesy of Anne Ronan Picture Library)

BIBLIOGRAPHY

"Joint Resolution to Designate April 22, 1990 as Earth Day and to Set Aside the Day for Public Activities Promoting Preservation of the Global Environment," Public Law 101-186, 1989; Motavalli, Jim, "Founding Father: Gaylord Nelson on Earth Day's Past, Present and Future," *E Magazine*, 1995; Mowrey, Mark, and Tim Redmond, *Not in Our Back Yard*, 1993; Shulman, Jeffrey, and Teresa Rodgers, *Gaylord Nelson: A Day for the Earth*, 1992.

Norton, Bryan

(July 19, 1944–)
Philosopher

Author of numerous works on ethics and the environment, Bryan Norton is an influential voice in the fields of environmental policy and philosophy. Much of his work has been directed at showing that environmentalism, as a movement, is more philosophically coherent and po-

Brian Norton viewing Peregrine Falcons on Half Dome in Yosemite National Park, California (Galen Rowell/Corbis)

litically unified than is usually credited, an argument made most clearly in his 1991 book *Toward Unity among Environmentalists*. In particular, Norton has argued that philosophical differences among environmentalists should not be allowed to interfere in efforts to effect change. Norton's work aims to unite theory and action, in part through his role as a consultant to government agencies and other environmental groups, including the Environmental Protection Agency, the U.S. Forest Service, and Zoo Atlanta.

Bryan Norton was born on July 19, 1944, in Marshall, Michigan, where his family ran a farm. Norton later credited his love of nature to his early years on the farm. He studied government and philosophy at the University of Michigan in the 1960s, where he absorbed the value of social activism. He graduated with a B.A. in 1966 and stayed at the university to earn a Ph.D. in philosophy in 1970. He published his first book, *Linguistic Frameworks and Ontology*, in 1977 while teaching at the University of South Florida in Sarasota.

Norton's professional interest in environmentalism began in 1981, when he was invited by the University of Maryland's Center for Philosophy and Public Policy to study the Endangered Species Act of 1973. The study was funded in part by the National Science Foundation's Ethics and Values in Science and Technology Program. Norton began looking at the act during the heyday of the snail darter controversy. The case was one of the first to test the power of the Endangered Species Act, when a population of the obscure, endangered fish was found in the Tennessee River where a large dam project was in development. The battle between environmentalists and the powerful Tennessee Valley Authority received national press coverage and focused attention on the question of whether all species deserve unlimited protection to exist. (The case eventually drew to a close without a clear victory on either side, when a population of snail darters was found on another branch of the Tennessee River.) While reviewing this and other cases, Norton became interested in a question that was to hold his attention for much of the next decade: On what philosophical grounds can one argue for the preservation of biological diversity? Norton's first consideration of the question was published in *Why Preserve Natural Variety?* (1987), a book that was the direct result of his work at the University of Maryland. The book examines the various rationales behind the Endangered Species Act and reviews a number of pressing practical concerns, such as how to prioritize species for protection, when time and resources are limited. Should we, for example, focus on those species that are most at risk for immediate extinction, even if these efforts might be futile, or should we instead focus our efforts on less critical cases, where species might be kept from reaching near-extinction?

Toward Unity among Environmentalists attempts to step back from these debates and find a solution to what Norton calls "the environmentalists' dilemma" of reconciling human-centered arguments for environmental protection with demands that species and ecosystems be protected for themselves. During 1985 and 1986, Norton was the Gilbert White Fellow at Resources for the Future, where he studied a variety of environmental organizations. His research convinced him that environmentalists need a more thoroughgoing discussion of values. He argues that environmental debates are often structured on a false dichotomy between utilitarian approaches, which call for environmental protection on the basis of economic value, and moral or ethical approaches, which argue that biodiversity is intrinsically valuable apart from human needs. Norton bridges this gap with a policy-centered approach, focusing on unity of purpose and action rather than differences in ideas. He foregrounds the context of environmental debates to show that while humans are ever-present, we are just one part of the larger picture, dependent upon the very ecosystems we seek to preserve. In the epilogue to the book, he writes, "We must value nature from our point of view *in a total context* which includes our cultural history and our natural history. Nature must be valued, from the ecological-evolutionary viewpoint of environmentalists, in its full contemporary complexity and in its largest temporal dynamic."

Norton has gone on to develop his ideas in numerous essays and further books, including the edited collections *Ecosystem Health* (1992), with Benjamin Haskell and Robert Costanza, and *Ethics on the Ark* (1995), with Michael Hutchins, Elizabeth Stevens, and Terry Maple. The latter collection came out of a workshop held in Atlanta, Georgia, in March 1992, that brought wildlife conservationists, biologists, philosophers, animal rights activists, and zoo professionals together to explore differences among individuals and groups and to come to consensus wherever possible to make policy recommendations for the role of zoos in wildlife preservation. In this and in his role as adviser to the Environmental Protection Agency and other governmental agencies, Norton works to put his ideas into action, facilitating practical efforts at environmental conservation today. Norton is on the editorial boards of *Environmental Ethics, Environmental Values, Ethics and the Environment*, and *Ecosystem Health*. He is professor of philosophy at Georgia Institute of Technology in Atlanta, Georgia.

BIBLIOGRAPHY

"Bryan Norton," Georgia Institute of Technology, http://www.spp.gatech.edu/faculty/bnorton.htm; Kim, Ke Chung, and Robert Weaver, *Biodiversity and Landscapes*, 1994; Preston, Christopher, "Epistemology and Intrinsic Values: Norton and Callicott's Critiques of Rolston," *Environmental Ethics*, 1998; Steverson, Brian, "Contextualism and Norton's Convergence Hypothesis," *Environmental Ethics*, 1995.

Noss, Reed

(June 23, 1952–)
Ecologist, Conservation Biologist

Reed Noss, president of the Society for Conservation Biology, has been a prominent leader in the fields of conservation biology and conservation planning and in the effort to raise awareness of the value and present decline of global biodiversity. His own career has paralleled the advent of the mission-oriented approach toward science, and he has advanced and promoted large-scale ecosystem management and protection. He has authored and edited numerous books, journal articles, and research publications, and has served as a consultant, adviser, and professor for universities, nonprofit organizations, government agencies, and private firms.

Reed Frederick Noss was born in Dayton, Ohio, on June 23, 1952, the son of James F. and Margaret J. Noss. At a young age, he experienced and learned to treasure natural areas around his home only to see them destroyed by development. This, in turn, sparked a personal and professional interest in conservation. Noss graduated from the School of Education at the University of Dayton, Ohio, in 1975 with a B.S. He then began graduate study in outdoor recreation at Antioch College in Yellow Springs, Ohio, before moving on to the University of Tennessee in Knoxville and receiving an M.S. in ecology in 1979. He then worked several years for the Ohio Department of Natural Resources, followed by a year with the Florida Natural Areas Inventory (a project of The Nature Conservancy). He earned his Ph.D. in wildlife ecology from the School of Forest Resources and Conservation at the University of Florida in 1988.

While working toward his doctorate in Florida, Noss conducted community-level studies and surveys of endangered species, including the red-cockaded woodpecker and gopher tortoise. Noss also developed a proposal for establishing a statewide network of core reserves, connecting corridors, and buffer zones in Florida. While this effort in conservation planning was considered crude and radical at the time, it was later revised and now serves as the base for the land acquisition program in Florida and a model for large-scale ecosystem preservation in other places as well.

After receiving his doctorate, Reed Noss took on a variety of positions as researcher, adviser, and consultant. His career has centered on efforts to protect ecosystems on large scales, enhance existing preserves of wildlands, and educate people of the importance of biodiversity and the value of conservation planning. Noss has had an atypical career in the sense that he has never served as a full-time academic, although he has worked in some capacity for Oregon State University as well as the University of Idaho, Stanford University, and other schools throughout the 1990s. In 2000, he was named an adjunct professor of biology at the University of Oregon, where he teaches conservation biology. He has also worked on behalf of such nonprofit organizations as the World Wildlife Fund and The Nature Conservancy and government agencies such as the U.S. Environmental Protection Agency, but primarily he has been a self-employed consultant.

Noss's work examines the benefits of approaching conservation from a larger scale rather than species-by-species, advances the vital relationship between species and habitat, and promotes the protection and restoration of native biodiversity and ecological integrity. His case studies have identified and assessed core reserves, migration corridors, and at-risk areas in Ohio, Florida, the Blue Mountains of Oregon and Washington, as well as in such ecosystems as the Greater North Cascades, the Greater Yellowstone, the Utah/Wyoming Rockies, and the Klamath/Siskiyou. Noss has also performed as a consultant and adviser outside of the country, specifically in Chile.

In 1991, Noss cofounded with DAVE FOREMAN (a founder of Earth First!) the Wildlands Project, an organization that specifically advocates a connected network of wildlands reserves through conservation science. Promoting the mission-oriented approach of conservation biology and planning, Noss served as science director for the group through 1996.

In 1993, he was named the editor of *Conservation Biology*, the journal of the Society for Conservation Biology, a position he held until 1997. Over his career, Noss has published more than 175 scientific and semitechnical papers and been a regular contributor and peer reviewer for journals such as *Ecology*, *Ecological Applications*, and *Journal of Wildlife*, as well as *Conservation Biology*. His articles and reports stress the value of keystone species, including large carnivores, in monitoring ecosystem and bioregional health, as well as the crucial role of wildlife corridors and connectivity in maintaining biodiversity and healthy wildlife populations. Also in 1993, Noss was awarded a three-year Pew Fellowship in Conservation and the Environment.

His first book, *Saving Nature's Legacy*, coauthored with Allen Cooperrider, was published in 1994. Derived from an instructional series the two men conducted for the U.S. Bureau of Land Management, the book provides guidelines on setting conservation priorities, inventorying biodiversity, and establishing and monitoring reserve networks. The book specifically defines conservation biology as "science in the service of conservation" with the "fundamental belief . . . that biodiversity is *good* and should be conserved." It notes the growth of the field of conservation biology as a direct response to the decline of biodiversity and differentiates it from traditional fields of science in that it is "cross-disciplinary and depends on the interaction of many different fields (e.g., geography, economics, political science in addition to wildlife biology, forestry, ecology)." *Saving Nature's Legacy* received the annual Conservation Community Award for Outstanding Achievement in the Field of Publications from the Natural Resources Council of America in 1995. Also in 1995, Noss received the Edward T. LaRoe Memorial Award, the highest award of the Society for Conservation Biology.

Noss's second book, *The Science of Conservation Planning*, coauthored by Michael A. O'Connell and Dennis Murphy, was published in 1997. It focuses on the successful management of Habitat Conservation Plans as laid out in the Endangered Species Act. His third book, *The Redwoods Forest: History, Ecology, and Conservation of the Coastal Redwoods*, which he edited and to which he contributed several chapters, was published in 1999.

In 1999, Noss and his wife began a private consulting firm, Conservation Science, Inc., of which he is president and chief scientist. In this same year, he was named president of the Society for Conservation Biology. At this post, he continues to confront the challenge of making the general public and media more aware of the countervailing trends of the continuing decline of biodiversity and the growing field of conservation science.

Noss presently directs research in the Rocky Mountains, as well as other areas, assessing habitat suitability and promoting the crucial role of large carnivores in protecting the biodiversity and ecological integrity of wildlands. He continues to concentrate on the application of natural science to conservation planning at regional to continental scales. He coedited a fourth book, *Ecological Integrity: Integrating Environment, Conservation, and Health*, which was published in 2000. Another book, *Restoration of Large Mammals*, is due out in 2001.

Reed Noss currently lives with his wife, Myra Wilson Noss, and their three children outside of Corvallis, Oregon.

BIBLIOGRAPHY

Noss, Reed F., *A Citizen's Guide to Ecosystem Management* (booklet published by the Biodiversity Legal Foundation), 1999; Noss, Reed F., "Indicators for Monitoring Biodiversity: A Hierarchical Approach," *Conservation Biology*, 1990; Noss, Reed F., "A Regional Landscape Approach to Maintain Diversity," *Bioscience*, 1983; Noss, Reed F., and Allen Y. Cooperrider, *Saving Nature's Legacy*, 1994; Noss, Reed F., and Dennis D. Murphy, "Species and Habitat Are Inseparable," *Conservation Biology*, 1995; Soulé, Michael, and Reed Noss, "Rewilding and Biodiversity," *Wild Earth*, 1998.

Odum, Eugene

(September 17, 1913–)
Ecologist

Eugene Odum is known worldwide among ecologists for his groundbreaking *Fundamentals of Ecology*, the only textbook on ecosystem ecology for ten years after it was first published in 1953. Odum is unquestionably the father of modern ecology; as a result of his dogged quest to develop a discipline that studies the biological and physical components of the natural world as a system, *ecology* has become a household word.

Eugene Pleasants Odum was born in Lake Sunapee, New Hampshire, on September 17, 1913, the first of two sons of Howard and Anna Louise (Kranz) Odum. As a young boy he moved to Chapel Hill, North Carolina, and spent hours in the expansive wilderness that surrounded the small southern town. The young Odum would wake early and disappear into the woods to watch birds. At first he was interested in different species of birds, then he began to wonder about birds and their environment, and eventually, he became more and more interested in the whole environment, or ecosystem, itself. Odum pursued his interest in college, receiving a bachelor's degree in biology in 1934 and a master's in 1936, both from the University of North Carolina at Chapel Hill.

Odum left his beloved southern home in 1936 to pursue a doctoral degree in zoology with a minor in ecology at the University of Illinois. During graduate school, Odum began developing his idea of ecology as an integrative science that includes not only the study of the biologi-

cal aspects of a community but also its physical, chemical, and geographic components. Ecology, he believed, was much more than a subdiscipline of biology. He graduated in 1939, and after a yearlong stint as a resident biologist at the Edmund Niles Huyck Preserve near Albany, New York, he returned to the South to pursue an academic career that would lead to the development of a new discipline of science.

Odum became an assistant professor of zoology at the University of Georgia, Athens, in 1940. He made $1,800 a year, which at that time was enough to support himself and his young wife, Martha, who was an artist. His career was interrupted during World War II when he was called upon to teach anatomy and physiology to nurses and premedical students. In 1945, he and his colleagues in the biology and zoology departments held a meeting to discuss what courses undergraduate students should be required to take for their major. Odum suggested ecology. "They looked at me like, what's that?" Odum recalls. He took the question as a challenge, and the young assistant professor set out to establish the principles upon which a scientific discipline of ecology could be established.

Five years later, Odum articulated the principles he had developed in a textbook, *Fundamentals of Ecology*, so that the principles of ecology could be taught to students in the same fashion that biology or physics or chemistry is taught. The only problem was that universities did not yet offer courses on ecology.

Without much of a market for his book, Odum had a difficult time getting it published. When he was finally successful in 1953, it became the only published textbook on ecology, and for the first several years the book was in print, very few copies sold. Then in the early 1970s, when environmental consciousness began to develop in the United States, sales of his book increased dramatically. Today, the book has been published in three editions and translated into numerous languages. It continues to be used worldwide, and Odum is currently working on the fourth edition.

Back in Georgia, Odum was successful in integrating ecology courses into the university curriculum; in 1951, he also began an ecology research program at the Savannah River Atomic Energy Site in Aiken, South Carolina. The site has become the Savannah River Ecology Laboratory, and the 300-square-mile area is now a national environmental research park. In 1954, Odum led the way in establishing the University of Georgia's Marine Institute on Sapelo Island to study Georgia's coastal marsh ecosystem, and in 1961, he became the founding director of the university's Institute of Ecology. He served in that capacity until 1984 and continues as director emeritus.

Throughout his distinguished career, Odum has written extensively on ecosystem ecology, with 12 books and countless journal papers to his credit. His *Ecology* book series, first published in 1963, has been printed in five editions; the 1997 edition was *Ecology: A Bridge between Science and Society*. His most recent book is *Ecological Vignettes: Ecological Approaches to Dealing with Human Predicaments*, published in 1998. As reflected in its title, Odum's current ecological focus is on the impact of human beings on ecosystems, which he terms *human ecology*.

In *Ecology: Our Endangered Life Support System* (1993), Odum argues for the extension of market economics to include ecosystem services. Market economics is not sustainable, he says, because it is based on the increasing consumption of goods with no monetary cost calculated in for the air, water, and soil that are used or destroyed to produce the goods. Odum believes that in order to solve our environmental problems, the emerging discipline of human ecology must include not only the natural sciences, but the social sciences and humanities as well. "Science alone will not solve any of our problems," Odum says. "Science can help, but it's not going to save the world from environmental deterioration because the problems and solutions deal with people and the nonscience disciplines."

Using an analogy from ecology, Odum argues that as a society, we have developed beyond the "pioneer" stage of development, when survival depends on producing numerous offspring, to a "mature" stage of development, when maintenance of the quality of life is more important than growth in size. As in any ecosystem, growth beyond a certain level requires very high maintenance and begins to produce diminishing returns. "We now suffer from too many good things," says Odum. "We have to change somewhat the way we live."

Odum has received numerous awards throughout his career, including three prestigious international honors: the Institut de la Vie Prize from the French government in 1975; the Tyler Ecology Award, presented at the White House by Pres.

JIMMY CARTER in 1977; and the Crafoord Prize from the Royal Swedish Academy of Science in 1987. The Crafoord is considered the Nobel Prize of ecology. Odum shared two of these awards with his brother, Howard T. Odum, a professor of environmental engineering at the University of Florida and an ecosystem ecologist. Odum has received honorary degrees from the Hofstra University, Ferum College, the University of North Carolina–Asheville, Universidad del Valle in Guatemala, the Ohio State University, and Universidad San Francisco de Quito in Ecuador.

Although Odum retired as a professor in 1984, he continues to be involved in the development of ecology at the University of Georgia, and under his leadership, the Institute of Ecology will soon become part of the university's new College of the Environment. The new college will seek to promote interdisciplinary study of environmental issues to seek solutions to real world problems.

In addition to continuing his writing on human ecology, Odum is working on a book of his wife's paintings with his added commentary, entitled *Essence of Place, Watercolor Sketches by Martha Odum and Ecological Commentaries by Eugene Odum*. Odum and his wife were sometimes invited to give seminars together, featuring her landscape paintings and his commentary on ecology. Martha Odum died in 1995 of cancer. Odum also survives his son, William, who was a marine biologist and director of the School of Environmental Science at the University of Virginia, and another son, Daniel, who died in childhood. Odum lives in Athens, Georgia.

BIBLIOGRAPHY

Chaffin, Tom, "Father Ecology Marches On," *Georgia Magazine*, 1998; Edwards, Lorraine, "Father of Ecology," *Georgia Journal*, 1983; Ezzard, Martha, "Ecologist Odum Still Planting Ideas," *Atlanta Journal*, 1998; Shearer, Lee, "Father of Ecosystem Ecology," *Athens Magazine*, 1996; Williams, Phil, "Eugene Odum: An Ecologist's Life" (video produced by University of Georgia), 1997.

Olmsted, Frederick Law, Sr.

(April 26, 1822–August 28, 1903)
Landscape Architect

Frederick Law Olmsted Sr. designed Central Park in New York City and many other city parks, in an attempt to humanize and make more healthy the expanding industrialized urban centers of the mid-nineteenth century. He was also a respected journalist and social thinker; he served as general secretary of the U.S. Sanitary Commission, the agency in charge of health services for soldiers during the Civil War; and he helped inspire the National Park Service and the U.S. Forest Service.

Frederick Law Olmsted Sr. was born on April 26, 1822, in Hartford, Connecticut, a city that his ancestors had helped found seven generations before. He had relatives throughout the region and from a very

Frederick Law Olmsted Sr. (Bettmann/Corbis)

early age was allowed to explore the countryside surrounding Hartford. When he was nine years old, he took his six-year-old brother on a journey by foot to visit an aunt and an uncle 16 miles away. His parents also took their eight children on many long trips by horse-drawn cart through rural New England. Olmsted attended country schools until he was a teenager, at which time he boarded with clergymen who tutored him privately.

After a few false starts in careers that did not suit him (business, the merchant marine, farming), Olmsted began to travel and write. In 1850, he toured England and parts of continental Europe on foot to explore agriculture and landscape architecture, and in 1852 he published the book, *Walks and Talks of an American Farmer in England*. Next, he took three long investigative trips through the American South, publishing his observations of slavery and its effects on the South in serial form in the *New York Daily Times*. (In 1861, these were collected in a three-volume set called *The Cotton Kingdom*.)

Olmsted became widely recognized as an influential progressive thinker and was offered the position of superintendent of Central Park in New York City in 1857. Olmsted saw the post as an opportunity to improve public health in an increasingly dirty and crowded city, especially for the poor, who were not able to escape the city as the wealthier classes could. Central Park, at that time a squatter's settlement in the rural area of Manhattan, was in the process of being developed as a recreational area, and the park's board of commissioners held a contest for a park design. Prominent architect Calvert Vaux invited Olmsted to collaborate on a design, and they won the competition. Vaux and Olmsted devoted the next four years to the park's construction. During this period, in 1859, Olmsted married his late brother's widow, Mary, and adopted her three children. In 1861 Olmsted had to take a leave of absence from landscape architecture in order to direct the U.S. Sanitary Commission, which attended the wounded soldiers of the U.S. Civil War. He stayed with that post until 1863 and was lauded for his skill in organizing his employees in the brand new field of sanitation.

After his stint with the Sanitary Commission, Olmsted moved his family to California, where he managed a mining estate. Having been previously depleted of all its gold, the mine failed. While in California, however, Olmsted visited and was deeply impressed by the Yosemite Valley. He was appointed to Yosemite's

Board of Commissioners, which convinced California's congress to declare Yosemite a state park in 1864. Twenty-six years later, in 1890, Yosemite became a national park. Olmsted used two justifications to argue for the protection of Yosemite: the tourism it would attract would be good business, and the government should purchase the land before the wealthy elite did, in order to assure access to it for all citizens.

Upon his return to New York in 1865, he and Vaux became partners and were contracted to design several more parks, including Brooklyn's Prospect Park, Washington and Jackson Parks in Chicago, and Seaside Park in Bridgeport, Connecticut. In all, he and his partners (Vaux, until 1872; then later his stepson, John C. Olmsted; his son, FREDERICK LAW OLMSTED JR.; Charles Eliot; and Henry Codman) designed some 80 public parks and 13 college grounds. In each case, the Olmsted approach was to humanize the area and make it a healthier place to live. The design for the Riverside subdivision of Chicago, for example, included streets that followed natural curves in the land and left the most attractive areas of the site as commons, where neighbors could recreate together. Olmsted and Vaux were called to Buffalo to help that growing city avoid the hygiene problems faced by other cities. Their solution included over 600 acres of new tree-filled parks that served both as recreation grounds and buffers between neighborhoods. When in 1881 Boston city officials asked Olmsted's firm to help with the pestilent Back Bay area, a flood-prone marsh filled with the sewage from Boston's suburbs, which drained into the Charles River, Olmsted designed a system of parks to surround Back Bay and protect it from

pollution and at the same time to protect nearby neighborhoods from flooding. The Capitol area of Washington, D.C., was another Olmsted project; he landscaped the grounds with an eye for shade and year-round greenery and proposed the marble staircase and terrace addition to the capitol that distinguishes it today.

Another of Olmsted's accomplishments of the 1880s was the protection of Niagara Falls from industrial development and a tourist-oriented sprawl right up to the river's banks. Olmsted's organizational skills helped convince the New York and Ontario governments to found a binational park around the falls in 1883. Olmsted and Vaux were hired to landscape the new park in 1886.

Olmsted became interested in forestry in the early 1890s while designing Biltmore, the Vanderbilt family's large country estate in Asheville, North Carolina. When young GIFFORD PINCHOT approached Olmsted in 1892 to ask to manage the estate's forests according to the scientific forestry principles he had just studied in Europe, Olmsted gave his consent. Olmsted later supported Pinchot in his quest to establish a national Division of Forestry to manage the nation's national forests. For this support, Olmsted is credited for providing impetus for the establishment of the U.S. Forest Service.

Although Olmsted paid close attention to recreational and health considerations in his designs, his private clients often cut corners in the implementation, in order to cut cost. Olmsted also spent years fighting with commissioners and regulatory committees, because his designs did not conform to all of the rules of the day. This adversity took a toll on his physical and emotional health, but his inherent idealism gave him the strength he needed to continue until his retirement in 1895. Olmsted died on August 28, 1903, in Waverly, Massachusetts.

BIBLIOGRAPHY

Beveridge, Charles E., *Frederick Law Olmsted: Designing the American Landscape*, 1995; Fein, Albert, *Frederick Law Olmsted and the American Environmental Tradition*, 1972; Roper, Laura Wood. *FLO: A Biography of Frederick Law Olmsted*, 1973; Stevenson, Elizabeth, *Park Maker: A Life of Frederick Law Olmsted*, 1977; Strong, Douglas H., *Dreamers and Defenders: American Conservationists*, 1988.

Olmsted, Frederick Law, Jr.

(July 24, 1870–December 25, 1957)
Landscape Architect

Primed by FREDERICK LAW OLMSTED SR., who was not only Frederick Law Olmsted Jr.'s father but also the father of landscape architecture, Frederick Law Olmsted Jr. was an important landscape architect himself. He is also remembered as an effective public servant and a dedicated conservationist.

Henry Perkins Olmsted was born on July 24, 1870, in New York City and was immediately nicknamed "Boy." His father decided early on that this son would eventually take over his prestigious landscape architecture firm, and so when "Boy" was four years old, his father had the child's name changed to Frederick Law Olmsted Jr. Sharing a name would aid continuity in the family business, Olmsted Sr. believed. The young Olmsted grew up steeped in landscape architecture. His father worked at home, and prominent people in the field often visited. There were maps and surveys all over the house. Olmsted Jr. was once found scanning a map of New York City while sleepwalking.

Olmsted Jr.'s apprenticeship with his father began early. At the age of 16, he accompanied his father to California, to help with the design of Stanford University. He attended Harvard University and supplemented his education there with piles of landscape architecture readings heaped on by his father. Upon graduation in 1894, Olmsted Jr. accompanied his father to Biltmore, the country estate of the Vanderbilt family. This was to be the last major project of Olmsted Sr. and the first for Olmsted Jr.

When senility claimed Olmsted Sr., Olmsted Jr. joined his stepbrother, John Olmsted, as a partner in the family firm. His major early assignment was to continue the work in Washington, D.C., begun by his father in the 1880s. As a member of the Senate Park Commission, Olmsted designed the grounds of the White House, the Jefferson Memorial, the Washington Monument, the Shrine of the Immaculate Conception, the National Arboretum, and Lafayette and Rock Creek Parks. Indeed the parklike appearance of much of Washington, D.C., today is due in great part to two generations of the Olmsted aesthetic.

Olmsted's firm also drew up the plans for numerous suburban developments and cities and towns, including Rochester, New York; Boulder, Colorado; New Haven, Connecticut; and Pittsburgh, Pennsylvania. He helped found the American Society of Landscape Architects and the American Institute of Planners.

In addition to his work on urban planning and city parks, Olmsted also was an advocate of national parks. He campaigned actively against the Hetch Hetchy dam that submerged an area compared by JOHN MUIR to the world's most inspiring cathedrals. He also worked with Horace McFarland of the American Civic Association to draft an unsuccessful bill to create a national park bureau in 1910.

Six years later, Olmsted collaborated with STEPHEN MATHER, director of the national parks, to write the National Park Service Act. Olmsted's contribution consisted of the policy statement for the new National Park Service: "To conserve the scenery and the natural and historic objects and the wildlife . . . and to provide for the enjoyment of same in such manner and by such means as will leave them unimpaired for the enjoyment of future generations." Thanks to powerful lobbying by Mather and his team, this bill made it through Congress and was signed by President Wilson in 1916.

Olmsted stood up for the parks in the early 1920s when various bills were introduced to Congress to allow livestock grazing within parks or construction in national parks of dams for hydroelectricity or irrigation. He, along with Mather and National Park Association director ROBERT STERLING YARD, successfully lobbied the Senate against these bills.

Olmsted helped further park development in 1928 when he drew up a master state park plan for California. The plan recommended parks in redwood groves, coastal and desert areas, in the mountains, and at historic sites. California voters then approved a bond issue to raise six million dollars to start purchasing land for state parks. Olmsted was invited to serve on the Yosemite Planning and Policy Committee, which he did from 1928 to 1956. Olmsted's entry in the *Dictionary of American Biography* comments that this must have seemed to Olmsted Jr. a satisfying post, given that Olmsted Sr. had been the first commissioner of the park in 1864, and its protection as a park has generally been attributed to him.

As a gift for his 83rd birthday, a group of Olmsted's friends bought and dedicated the Frederick Law Olmsted Redwood Grove in the Prairie Creek Redwoods State Park in Humboldt County, California. Olmsted died in Malibu, California, on December 25, 1957. Olmsted and his wife, Sarah, had one daughter, Charlotte.

BIBLIOGRAPHY

Fein, Albert, *Frederick Law Olmsted and the American Environmental Tradition*, 1972; "Frederick Law Olmsted," *Dictionary of American Biography, Supplement VI*, 1980; "Frederick Law Olmsted, 1870–1957, An Appreciation of the Man and His Works, Landscape Architecture," *Landscape Architecture*, 1958; Roper, Laura Wood, *FLO: A Biography of Frederick Law Olmsted*, 1973.

Olson, Molly Harriss

(July 11, 1960–)
Sustainability Consultant

Sustainability expert Molly Harriss Olson has worked with the Great Barrier Reef Marine Park Authority, the World Conservation Union (IUCN), Wildlife Australia, Greenpeace, and the Natural Step (TNS) to promote conservation and sustainability. She served as inaugural executive director of President Clinton's Council on Sustainable Development, and currently she directs the Australian company Eco Futures Pty Ltd. to develop innovative strategies to accelerate the adoption of sustainable practices and facilitate collaboration among leaders in business, government, and civic organizations on achieving a sustainable society.

Molly Harriss Olson was born on July 11, 1960, in Alliance Nebraska, and moved with her family at an early age to Palo Alto, California. Her mother, who was very interested in Native Americans, told stories to young Olson about the strong connections Native Americans had with the earth. Her father, who loved his work as a music teacher, provided her with an example of how one could make a living by pursuing a passion. Molly Olson's own passion for nature started early with a keen interest in kelp beds, coral reefs, and scuba diving. As a teenager, she learned to dive in Pacific Grove, California, not far from her home, and that passion eventually took her to the Great Barrier Reef and Australia.

Olson pursued Environmental Studies at the University of California at Santa Cruz (UCSC). During a summer internship in Washington, D.C., Olson studied the Alaska Native Claims Settlement Act and realized that economics would have more to bear on the future of this magnificent area than its ecological significance. Upon her return to UCSC the next fall, Olson added economics to her study program. In 1982 she graduated from UCSC with a joint B.A. degree in economics and environmental studies. Her honors thesis, which she researched in Australia, was entitled "The Great Barrier Reef: A Legal and Economic Analysis." The paper was all about sustainability long before the term came into vogue.

Continuing her education, Olson became a Bates resident scholar at Yale University's School of Forestry and Environmental Studies, earning a master's degree in environmental policy there in 1985. At the same time that she was becoming more interested in sustainability and how it benefited conservation efforts, she remained interested in coral reef habitat, especially the Great Barrier Reef. During the late 1980s, Olson authored a number of reports on conservation in Australia for the Australian government, local communities, and the World Conservation Union, and she edited *Wildlife Australia* magazine. Olson headed Greenpeace's Ocean Ecology department from Australia, leading campaigns on the famous Greenpeace vessel, *The Rainbow Warrior*. She conducted scientific research on the Great Barrier Reef and served as a delegate to numerous international meetings and negotiations. As a member of the IUCN's Commission on Parks and Protected Areas, Olson helped organize the IUCN's General Assembly in Perth in 1990. By 1991, Olson had become an adviser to the environmental minister of Australia and conducted a major review of the Australian government's funding for the environment.

During the first Clinton campaign for U.S. president, Olson returned to the United States to serve as deputy national coordinator of Environmentalists for Clinton-Gore, 1992, setting up committees of prominent pro-Clinton environmentalists in every state. Upon the victory of the Clinton-Gore team, Olson was appointed executive director of the President's Council on Sustainable Development (PCSD), a 25-member council whose distinguished members represented industry, government, and environmental, labor, and civil rights organizations. The council worked for three years to develop a national strategy for sustainable development, articulated in the 1996 publication *Sustainable America: A New Consensus*. The goals addressed the interdependent issues of health, environment, economic prosperity, equity, conservation, stewardship, sustainable communities, civic engagement, population, international responsibility, and education and were accompanied by policy recommendations and indicators of progress. Before completing her work at the PCSD, Olson ensured that an implementation phase for the report would follow. This implementation culminated with a national town meeting with Vice President AL GORE in April 1999. The council disbanded in September 1999.

Continuing to use her expertise in sustainability, in 1995 Olson became the first executive director of the U.S. branch of the Natural Step, an international nonprofit educational organization that promotes sustainability for corporations, organizations, and individuals. The Natural Step, which was founded in 1992 by Swedish oncologist Karl-Henrick Robèrt and brought to the United States by environmental industrialist PAUL HAWKEN, cites scientific principles and laws in defining how societies must function if they are to be sustainable. Olson's work during her two years at the Natural Step attracted the attention of the World Economic Forum, which in 1995 selected her for its Global Leaders for Tomorrow Program, comprising leaders born after 1950 who have distinguished themselves as recognized leaders in the world community.

Currently, Olson and her husband, Australian environmental leader Philip Toyne, direct a new Australian company, Eco Futures Pty Ltd., which advises business, government, and civic leaders on strategies for achieving sustainability. She and Toyne live on a farm in the village of Gundaroo, New South Wales, with their son, Atticus, and Toyne's son, Jamie.

BIBLIOGRAPHY

"The Natural Step," http://www.naturalstep.org/; Olson, Molly Harriss, "Accepting the Sustainable Development Challenge," *Willamette Law Review*, 1995; Olson, Molly Harriss, "Shaping a Path to Sustainability," *UNEP Our Planet*, 1996; The President's Council on Sustainable Development, *Sustainable America: A New Consensus for the Future*, 1996.

Olson, Sigurd

(April 4, 1899–January 13, 1982)
Nature Writer, President of the National Parks Association, Conservationist

Sigurd Olson was a nature writer and activist whose contemplative essays celebrated the beauty and spiritual nature of wilderness, with a particular focus on the lake country of northern Minnesota. He held several influential positions in the national conservation movement throughout the 1950s and 1960s, serving as president of the National Parks Association and the Wilderness Society Council. In 1962, he was appointed consultant to Secretary of the Interior STEWART UDALL; in this capacity, he was instrumental in crafting the Wilderness Act of 1964. He was also a key figure in the establishment of the Boundary Waters Canoe Area Wilderness along the Minnesota-Ontario border.

Sigurd Ferdinand Olson was born to the Reverend Lawrence J. and Ida May (Cedarholm) Olson in Humboldt Park, Illinois, on April 4, 1899. His parents were both Scandinavian immigrants: His father was from Sweden, his mother from Denmark. They moved to Wisconsin when Sigurd was seven and his brother Kenneth was 11. His family settled in Ashland, Wisconsin, in 1912. Olson was fascinated by nature from an early age and spent much time in such solitary pursuits as fishing, trapping, and hunting. Olson attended Northland College in Ashland for two years. In the fall of 1918, he transferred to the University of Wisconsin at Madison not long before Congress lowered the age of draft eligibility to 18. Olson immediately joined the Students Army Training Corps, but he served for only eight weeks before the armistice was signed. He returned to the University of Wisconsin and graduated in 1920 with a degree in animal husbandry. Olson found the agriculture courses dry, a far cry from the romance and beauty he had experienced on the farm; nevertheless, he accepted a job teaching high school agricultural classes in northern Minnesota. Although Olson was to continue in the field of education for 27 years, he found teaching unfulfilling. He spent every weekend camping in the woods or canoeing on the region's many lakes, forming a deep attachment to the area known as "canoe country." His first extended canoe trip also yielded his first published writing, an article for the *Milwaukee Journal.*

Olson spent summers working on the Uhrenholdt farm near Seely, Wisconsin. He married Elizabeth Uhrenholdt, the farmer's daughter, in 1921. Olson sought to join his academic and outdoor interests through a career as a field geologist. He enrolled in a graduate program in geology at the University of Wisconsin but became disillusioned as he realized that most geologists were employed by mining companies, whose devastation he had seen firsthand in the iron ore region of northern Minnesota. After one semester, he left the university and took a job as a high school biology teacher in Ely, Minnesota, in the midst of his beloved "canoe country." Olson's sons, Sigurd T. and Robert K., were born in Ely in 1923 and 1925, respectively. Again, he changed his focus—this time embarking on the study of ecology. The Olsons soon moved to Champaign, Illinois, so that he could attend the University of Illinois. At a time when predators were seen as pests and wolves were the target of predator control programs, Olson set about investigating the role wolves play in an ecosystem; his thesis on timber wolves is considered a pioneering work in the field. Olson received his master's degree in ecology in 1932.

Upon his return to Ely, Olson began teaching full-time at Ely Community College. He later became dean of the college, a position he held for 11 years. Throughout his academic career, Olson continued writing nature essays, despite active discouragement from agents and editors, who assured him they would never sell. Persevering, Olson sold a few articles and wrote a syndicated newspaper column for a brief period before World War II. He eventually came to the attention of Alfred Knopf, who expressed interest in publishing a set of Olson's essays. *The Singing Wilderness* was published in 1956 and appeared on the *New York Times* bestseller list. Olson went on to publish seven more books over the next 24 years.

Olson deliberately wrote in a simple, nonintellectual style so that he could reach a wide range of readers. He appealed to rural sportsmen because he enjoyed hunting and fishing and to conservationists because he advocated wilderness preservation. His personal philosophy—espoused throughout his work but most apparent in such later works as *Reflections from the North Country*—centered on the idea that modern man retains a racial memory of the wilderness, to which the human spirit responds. Olson believed that without a life that allows an intuitive, receptive response to nature, individuals—and thus communities—suffer. The solution, according to Olson, lies in a human ecology that addresses the need to preserve wilderness as well as the need to provide economic sustainability.

Sigurd Olson (Wilderness Society)

The success of his first book allowed Olson to resign as dean of Ely Community College. Realizing that his fledgling writing career would not support him, Olson accepted a position as spokesman for the movement to protect the Quetico-Superior area from road and dam building and increased air traffic. He gained national recognition for his success in a high-profile campaign to ban aircraft in the wilderness. Olson helped negotiate an international treaty governing management of the area along the Minnesota-Ontario border that was later to become the Boundary Waters Canoe Area Wilderness.

Olson went on to serve as the president of the National Parks Association from 1953 to 1958 and as president of the

Wilderness Society Council from 1963 to 1968. As a member of the secretary of the interior's Advisory Committee on Conservation from 1960 to 1966, he advised Secretary Stewart Udall on wilderness and national parks issues. His work had a direct influence on the creation and passage of the Wilderness Bill in 1964. He was elected to the Izaak Walton League Hall of Fame in 1963.

Olson's efforts to bridge the divide between conservationists and rural people dependent upon resource extraction sometimes led to conflict. His belief in hunting as a legitimate means of getting close to nature was harshly criticized by his fellow conservationists. They also pressured him to accept nothing less than full wilderness protection for the Boundary Waters Canoe Area. On the other hand, his rural supporters in northern Minnesota accused him of hypocrisy for failing to advocate logging in the wilderness area. At one point, he was hanged in effigy in his hometown of Ely.

Olson became a trustee of Northland College in 1970 and continued an active involvement with the college until his death 12 years later. The Sigurd Olson Environmental Institute was founded in 1972 to serve as the environmental outreach program of the college. The institute works with government agencies, businesses, and citizens on both sides of the U.S.-Canadian border to address environmental problems affecting the Quetico-Superior region. In 1974, Olson was awarded the John Burroughs Medal, the highest honor in nature writing. He received the Robert Marshall Award of the Wilderness Society in 1981.

When he was 80 years old, Olson underwent surgery for cancer of the colon; the surgery, although successful, left him in a weakened condition. He still had enough internal strength, however, to refuse the U.S. Department of the Interior's Conservation Service Award, as a protest against the department's receptivity to oil and gas exploration under the new secretary, James Watt. On January 13, 1982, Olson suffered a fatal heart attack while snowshoeing near his home in Ely. His final written words, later discovered on a sheet of paper left in his typewriter, were: "A new adventure is coming up and I'm sure it will be a good one."

Ten years after Olson's death, the Sigurd Olson Environmental Institute established the Sigurd F. Olson Nature Writing Award, given biennially to recognize writing that both conveys the spirit of the north and raises awareness about the need to preserve wilderness for future generations. In 1998, Olson's sons donated the family cabin near Ely, Minnesota, to the newly created Listening Point Foundation. The cabin is the center of a wilderness education program that carries on the author's ecological philosophy. The preservation of the 26-acre property on Burntside Lake fulfills Olson's wish, as expressed in his book *Listening Point*, that it be "a place reserved for vistas and dreams and long thoughts." The 100th anniversary of Olson's birth was recognized with a U.S. Senate tribute by Sen. Russ Feingold of Wisconsin on August 3, 1999.

BIBLIOGRAPHY

Backes, David, *A Wilderness Within: The Life of Sigurd F. Olson*, 1997; Olson, Sigurd, *Reflections from the North Country*, 1976; Olson, Sigurd, *The Singing Wilderness*, 1956; "The Sigurd F. Olson Web Site," http://www.uwm.edu/Dept/JMC/Olson/; "Two to Remember [Conservationists Sigurd Olson and Aldo Leopold]," *Field & Stream*, 1998.

Orr, David

(January 10, 1944-)
Environmental Studies Professor, Writer

David Orr, chair and professor of environmental studies at Oberlin College in Ohio, is an influential figure in educational reform. He has written prolifically on topics including the responsibility of educational institutions to promote "ecological literacy" (an understanding of how humans fit into the earth's ecological web) and how architectural design on college campuses can affect the success of environmental curricula.

David Wesley Orr was born on January 10, 1944, in Des Moines, Iowa, and was raised in New Wilmington, Pennsylvania. He studied history at Westminster College in New Wilmington, receiving his B.A. in 1965. He earned an M.A. from Michigan State University in 1966 and a Ph.D. in international relations from the University of Pennsylvania in 1973. Orr taught at the University of North Carolina at Chapel Hill and has been on the faculty of Oberlin College since 1990.

Orr's first book, which he coedited with Marvin Soroos in 1979, was entitled *The Global Predicament: Ecological Perspectives on World Order.* It is a collection of essays on topics such as growth and development from an ecological perspective, food alternatives for the future, resource scarcity, and environmental management of such international resources as oceans.

Orr's current work focuses on educational reform and how ecological design can contribute to it. He believes that the current ecological crisis is in part a result of the shortcomings of the educational system. Rather than teaching students to respect the interconnectedness of life, which might have prevented much of the environmental destruction we now have to deal with, students in traditional institutions have been taught to compartmentalize life. They have been encouraged to separate themselves from the rest of the ecological web, separate their emotions from their intellects, and separate practice from theory. Although colleges and universities are pumping out some of the best-informed graduates in their history, the students nonetheless lack many areas of knowledge and, most of all, ecological wisdom.

Instead of preparing students to compete in a global economy, which is what

David Orr (Courtesy of Anne Ronan Picture Library)

most academic institutions currently see as their mission, Orr advocates teaching students to work to make life on earth more environmentally and socially sustainable. He believes that colleges should become models of sustainability and earth stewardship. In his 1994 book *Earth in Mind: On Education, Environment and the Human Prospect,* Orr advocates a new ratings system for colleges that would include such criteria as consumption/discards per student, management policy on waste, recycling, purchasing, landscaping, energy use, and how graduates contribute to helping the world become more sustainable.

Orr spent the late 1990s overseeing the Adam Joseph Lewis Center for Environmental Studies building project at Oberlin College, designed by WILLIAM MCDONOUGH of William McDonough and Partners. From this experience, Orr shared recommendations for those embarking on similar projects in his 1999 article entitled "The Architecture of Science" for *Conservation Biology.* Students and faculty should participate in the design process along with the architects and engineers. The building should reflect an environmental ethic in terms of what materials are used, how energy-efficient it is, and how alternative sources of energy power it. Building data such as energy performance, energy production, water quality, indoor air quality, and emissions should be on display on site.

Orr is the education editor for *Conservation Biology* and serves on the editorial advisory board of *Orion Nature Quarterly.* His articles frequently expand on situations he encounters at Oberlin, such as proposals for new campus car garages or the college's electronic communication network. He examines them from the perspective of well-known ecological thinkers such as GARRETT HARDIN or WENDELL BERRY and draws parallels with similar phenomena worldwide. Orr also has critiqued recent well-publicized challenges to the environmental movement with reviews that point out the inconsistencies and fallacies of such antienvironmentalist books as *A Moment on Earth* (1995) by Gregg Easterbrook and Martin Lewis's *Green Delusions* (1992).

Orr has received numerous awards for his work. He was awarded a National Conservation Achievement Award by the National Wildlife Federation in 1993, a Lyndhurst Prize in 1992 "to recognize the educational, cultural, and charitable activities of particular individuals of exceptional talent, character, and moral vision," and the Benton Box Award from Clemson University for his work in environmental education (1995).

Orr lives in Oberlin, Ohio, with his wife and their two sons.

BIBLIOGRAPHY

Orr, David W., *Earth In Mind: On Education, Environment, and the Human Prospect,* 1994; Orr, David W., *Ecological Literacy: Education and the Transition to a Postmodern World,* 1992; Orr, David W., and David J. Eagen, *The Campus and Environmental Responsibility,* 1992.

Osborn, Fairfield

(January 15, 1887–September 16, 1969)
Writer, Founder of the Conservation Foundation, President of the New York Zoological Society

Fairfield Osborn alerted readers of the mid-twentieth century to the dangers of unbridled population growth and consumption of natural resources through his popular books *Our Plundered Planet* (1948) and *The Limits of the Earth* (1953). Through the New York Zoological Society (NYZS), which he headed for almost 30 years, Osborn helped found the Conservation Foundation, an organization that served as a research and policy institute on natural resource issues.

Henry Fairfield Osborn Jr. was born in Princeton, New Jersey, on January 15, 1887, to Henry Fairfield and Lucretia (Perry) Osborn. His father was a well-known biology professor at Princeton and Columbia Universities, president of the American Museum of Natural History, and a founder and president of the Bronx Zoo. Young "Fair" Osborn grew up in a world of wealth and social prestige and early in his life demonstrated an abiding interest in the natural world. He was allowed to keep his own private menagerie at his family's Madison Avenue brownstone and took one of his pets, a nocturnal flying squirrel, to school with him every day, curled up in his pocket. Osborn atttended Groton School in Groton, Connecticut, and Princeton University, where he earned an A.B. in 1909. He spent one year at Trinity College, Cambridge University, for graduate study in 1909–1910. After his formal education, he spent a few years working in the San Francisco freight yards and laying railroad track in Nevada. In 1914, he married Marjorie Mary Lamond, with whom he would have three daughters and enjoy a long marriage.

After service in the American Expeditionary Force during World War I, Osborn spent the first two decades of his professional life working in business: for Union Oil Company, as treasurer for a label-making company, and then as a partner in the Redmond and Company investment banking firm. He retired from business in 1935.

Despite his immersion in the business world, his heart had remained with animals. In 1922, he had been named a trustee of the New York Zoological Society, which ran the Bronx Zoo and Coney Island's New York aquarium. He served as the treasurer of the NYZS executive board from 1923 to 1935, and as secretary from 1935 to 1940. When he became president of the NYZS in 1940, a position he would hold until just before his death in 1969, "the zoo began to change," recounted LAURANCE ROCKEFELLER, a friend of Osborn and co–board member of the NYZS, in an elegy to Osborn published in *Reader's Digest.* Osborn, who still maintained a deep love for animals and reportedly was able to detect signs of poor emotional and physical health before even the zookeepers could, did away with cages whenever possible and placed animals instead in open areas that recreated as best as possible their natural habitats. His intuitive approach increased survival rates for gorillas, which until then usually died in captivity. He advised that zookeepers give gorilla babies more physical affection, that they hold and embrace them. According to Rockefeller, survival rates turned around after that. Osborn also raised millions of dollars for the NYZS from his many affluent friends. His *New York Times* obituary reported that "he could go to the very affluent and, without batting an eye, ask for a million dollars for zoo and aquarium improvements; he said that if he asked for a million dollars he was confident he would 'get at least $25,000.'"

At the same time that Osborn devoted himself to improving the welfare of captive animals and the public's ability to learn from and enjoy them, he tried to raise awareness of worsening environmental conditions worldwide. He wrote *Our Plundered Planet* in 1948 about a war that Osborn says in the introduction to the book could potentially result in far more destruction than the recently ended World War II: man's war against nature. The book decries the rampant destruction of the natural environment by an ever-increasing, hungry population and presents "only one solution: Man must recognize the necessity of cooperating with nature. He must temper his demands and use and conserve the natural living resources of this earth in a manner that alone can provide for the continuation of civilization." If the population rate increase did not slow, and people did not amend their relationship with the earth, he warned, "Nature holds the trump card." His next book, *The Limits of the Earth*, published in 1953, provided further information about the devastating impact on water, agriculture, and other life-sustaining resources. Both books, written for nonspecialists, were translated into several languages, and as the first such warnings sounded, they in-

spired many to join efforts to work toward population control and conservation of natural resources. Osborn served on the Conservation Advisory Committee of the U.S. Department of the Interior from 1950 to 1957 and on the Planning Committee of the Economic and Social Council of the United Nations.

Following the publication of *Our Plundered Planet*, Osborn and other colleagues at the NYZS founded the Conservation Foundation in 1948, a research institute that produced books, reports, and films about natural resources, flood control, and endangered species. Osborn served as the organization's first president until 1962 and then became chairman of its board. The Conservation Foundation was originally part of the NYZS, but affiliated with the World Wildlife Fund (WWF) in 1985 and in 1990 was formally consolidated with the WWF.

Osborn is also known for his work with conservation philanthropist Laurance Rockefeller to establish the Jackson Hole Wildlife Park near Moran, Wyoming, on land owned by the Rockefeller family. Together they founded the Jackson Hole Biological Station to facilitate studies by NYZS biologists there.

Osborn received many awards and honors during his life, including the Medal of Honor from the Theodore Roosevelt Memorial Association in 1949, another Medal of Honor from the city of New York in 1960, and the Gold Medal of the New York Zoological Society in 1966. He continued working actively with the New York Zoological Society and the Conservation Foundation until shortly before his death. Osborn died in New York City on September 16, 1969, after a series of strokes.

BIBLIOGRAPHY

Bridges, William, *A Gathering of Animals: An Unconventional History of the New York Zoological Society*, 1974; "Fairfield Osborn, the Zoo's No. 1 Showman, Dies," *New York Times*, 1969; Goddard, Donald, ed., *Saving Wildlife: A Century of Conservation*, 1995; Rockefeller, Laurance, "My Most Unforgettable Character," *The Reader's Digest*, 1972; Stroud, Richard, *National Leaders of American Conservation*, 1985.

Owings, Margaret

(April 29, 1913–January 21, 1999)
Artist, Founder of Friends of the Sea Otter

An artist with a profound appreciation of nature, Margaret Owings fought to protect the wildlife and wild habitat in her home state of California and became known as the leader of a broad range of environmental causes. While serving as a commissioner of the California State Parks Commission, she led a successful opposition to the construction of a freeway through Pacific Creek Redwoods State Park. She crusaded for some of the threatened species that lived near her, waging battles to eradicate both bounty and sport hunting of mountain lions in the state and founding Friends of the Sea Otter, an organization committed to protecting the California sea otter and its habitat. Owings also established the Rachel Carson Memorial Fund shortly after Carson's death, to ensure that her work to educate the public about hazardous chemicals would continue.

Margaret Wentworth was born on April 29, 1913, in Berkeley, California, to Frank W. and Jean (Pond) Wentworth. She was aware of the natural world from a very early age and enjoyed spending time in the family's rock and live oak garden as she grew up. Her father was a trustee at Mills College where she later went to school, graduating with a bachelor's degree in art in 1934. She completed postgraduate work in art at the Fogg Museum at Harvard University in 1935. From her formal training in art she gained greater sensitivity to life and the natural world, and found that for her, art and nature seemed to go hand-in-hand. She married Malcolm Millard in 1937 and later had a daughter with him. Her marriage took her to Chicago for ten years, then to Carmel, California, before it ended in divorce. At about this time, in 1947, she became interested in environmental activism and joined the Point Lobos League, a small group of people who opposed the mining of sand from a beautiful beach and the construction of a housing development on the coast south of Carmel. The group garnered support and raised funds and was then able to buy a mile and a half stretch of coastline and preserve it as a state park. Many years later Owings still looked back on the group's accomplishment with satisfaction.

On December 30, 1953, she married Nathaniel Owings, a founding partner in Skidmore, Owings & Merrill, one of the country's preeminent architectural firms. Nathaniel Owings undertook plans to design and build a house for them in Big Sur on the coast of California, on a site projecting out from a cliff and affording striking views of the ocean and mountains around it. He built the house in accordance with their desire to live in a structure that had an active presence on the landscape while still respecting the character of the land and its delicate ecology.

By this time, Margaret Owings was becoming well known as an artist, and her paintings had been featured in one-woman shows at the Santa Barbara Museum of Art (1940) and the Stanford Art Gallery (1951). She and Nathaniel continued to work on their spectacular cliff-top house at Big Sur, but they were both increasingly drawn to the landscape and people of New Mexico, particularly near the Rio Grande valley. Together they started building an adobe farmhouse there in 1956, with high-walled courtyards, ancient woodwork, and a studio for Owings to paint in. The couple moved seasonally between the two houses for the next 30 or so years until Nathaniel Owings's death in 1984.

In 1962, Owings was dismayed when a mountain lion was killed for bounty near their Big Sur home. Since the late 1800s, hunter had killed thousands of mountain lions for rewards, following the view that fewer mountain lions meant more deer, cattle, and sheep. Owings opposed the practice and began a crusade to stop it, enlisting such famous conservationists as RACHEL CARSON and ANSEL ADAMS to speak with her in front of legislative committees. In 1963 the state legislature abolished the mountain lion bounty, though they could still be legally hunted for sport. But after several years, when concerns were raised over continuing declines in the mountain lion population, Owings again came to their defense, promoting a moratorium on all hunting of the animal. In 1971 she succeeded, and sport hunting of mountain lions was abolished in California.

By now, Owings was dedicating more and more of her time to conservation causes and had begun serving as a commissioner on the California State Parks Commission in 1963, a position she held until 1969. During that time she led a campaign to prevent the construction of a freeway through prime alluvial flats in Pacific Creek Redwoods State Park, site of the world's tallest trees. Also in 1963, Owings attended the ceremony for Rachel Carson when she received the Audubon Medal for her book *Silent Spring*. At the time, Carson was dying of cancer and remarked that she worried that no one would carry on her work after she was gone. Owings took those words to heart and resolved to keep Carson's legacy alive. Shortly after Carson's death in April 1964, Owings raised $14,000 and founded the Rachel Carson Memorial Fund, administered by the National Audubon Society (now administered by Environmental Defense), to continue Carson's work of combating toxins in the environment.

Out of the broad spectrum of environmental concerns in which Owings was active, she is probably best known for her work with the California sea otter, a threatened marine mammal that she was fond of watching from her cliff-top home at Big Sur. She championed their cause in 1968 by cofounding the Friends of the Sea Otter organization with Dr. Jim Mattison, at a time when the southern sea otter population numbered about 650. In the early days of the organization, it operated on a volunteer basis, and often the meetings were held in the Owingses' home. Through sheer determination, Owings, who served as president from its beginning until the early 1990s, kept the group functioning—rallying scientists, other conservationists, and educators; speaking to legislators both in Sacramento and Washington, D.C.; and helping to establish environmental policy to protect the sea otter. Friends of the Sea Otter has now grown to 4,600 members in 45 states and 13 countries, and the southern sea otter population has grown in number and range and now includes about 2,400 otters along the central California coast.

Owings filled many other roles as well. She was a trustee of Defenders of Wildlife (1969–1974), a board member with the National Parks Foundation (1968–1969), the first woman to be a board member of the African Wildlife Leadership Foundation (1968–1976), and a trustee of the Environmental Defense Fund (1972–1982, then as an honorary trustee). She was recognized with many awards and citations, including the prestigious National Audubon Society Medal in 1983 and the United Nations Environmental Program's Gold Medal Award in 1988. In 1993 she received an honorary doctorate from Mills College, and in 1998 the National Audubon Society named her one of the top 100 most influential conservationists of the century.

Before she died, Owings spent months compiling a collection of her artwork and writings, which she put together into a book titled *Voice from the Sea: Reflections of Wildlife and Wilderness* (1998). The book covers five decades of her writing and emphasizes her lifetime commitment to wildlife conservation. She died of heart failure at her home in Big Sur, California, on January 21, 1999.

BIBLIOGRAPHY

Goldberger, Paul, "Revisiting Big Sur's Wild Bird," *Architectural Digest*, 1996; LaBastille, Anne, "The Crusader of Big Sur," *Living Wilderness*, 1980; Valtin, Tom, "Margaret Owings," *Environmental Defense Fund Newsletter*, 1986.

P

Packard, Steve

(March 28, 1943–)
Restoration Ecologist, Organizer

Through a combination of sheer determination and rare vision, Steve Packard has succeeded in restoring tallgrass prairie, savanna, and oak woodland communities in the midwestern United States—some of the rarest and least-understood ecosystems in the country. While he has had little formal training in biology, he has a talent for finding new and effective approaches to restoration ecology using common sense and intuition. His unorthodox methods initially ruffled feathers in the academic world, yet his years of work along the North Branch of the Chicago River have produced beautiful results, turning weedy remnants into rich native prairies and savannas. Part of his secret lies in his ability to inspire groups of people to help with the physically and intellectually demanding work of restoring degraded land—he created a whole network of volunteers that came to serve as a model for a new kind of organized environmental activism.

Stephen Packard was born on March 28, 1943, in Worcester, Massachusetts. His father was a self-made businessman who had grown up in the slums of Worcester. His mother overcame physical handicaps to develop great spiritual strength and generosity. Steve was the oldest of seven children and even as a young child had a love of nature in his blood. In 1952 his family moved from industrial Worcester to rural Shrewsbury, Massachusetts, and he was able to further develop his consuming interest in the natural world. His fourth-grade teacher had an enlightening influence on him, and he spent many hours with her and her husband, learning about nature's wild things. As a sixth grader he turned his backyard into a nature center, putting up bird feeders and transplanting trees.

He went to Harvard in 1961—the first from his family to attend college—but rebellious tendencies made him restless, and he left after two years. He moved to New York City, where he spent a year in a community of artist and filmmakers, then returned to Harvard to finish his degree in social psychology and anthropology, graduating in 1966. For the next few years he was absorbed in antiwar, civil rights, and related movements. He joined a newsreel collective, which made propaganda films for activist groups, and this eventually took him to Chicago. When the antiwar movement subsided, he was 30 years old and at a loss. At this point, he turned back to the passion of his youth—nature. He came across a ROGER TORY PETERSON guide to wildflowers and decided to learn all the flowers in the Chicago area. This led him to seek out wild places, and what he found were many degraded remnant prairie sites along the North Branch of the Chicago River. His interest in nature fully reawakened, he took a job in 1975 (for $95 a week) with the Illinois Environmental Council, a fledgling lobbying group, and committed himself to saving what was left of the prairies. The experience he gained from his activist days served him well; he had a thorough understanding of politics and a flair for organizing people and keeping them involved. This was the beginning of the North Branch Prairie Project—later renamed the North Branch Restoration Project—whose goal was to restore sites as closely as possible to the condition they were in when European settlers found them.

Many obstacles were in the way of the North Branch project. The land the volunteers wanted to save belonged to the Forest Preserve District, whose administrators were reluctant at first to let a band of volunteers help manage their sites. Also, the prairie sites themselves were in very poor shape—filled with trash and overrun with brush that had long ago choked out the native prairie species in many areas. Packard was left to decide what to plant and how to convince the Forest Preserve District to give his group increasing responsibility. With a little political finesse, lots of determination, and some common sense, the group was soon able to begin their experiment. They cleared brush by hand and attempted many plantings, not knowing at first what would work and what would not. And although it took years to see results, they did not give up.

By 1979 Packard was working as a field representative for the Natural Land Institute, a private, nonprofit environmental group. He also continued working in all his spare time on the North Branch Restoration Project, which was starting to show progress. He and his volunteers had made some headway clearing brush but were now ready for a more natural method of brush control—fire. Their proposal for controlled burning was met with even more bureaucratic balking, but they did succeed in gaining the necessary permits, and they quickly found fire to be an effective and irreplaceable tool. The techniques used have paid off. Twenty years after their project began, native flowers again bloom by the hundreds of thousands on the North Branch prairies and savannas, and birds and other species that had abandoned the area have returned.

Part of the legacy of Packard's early work is that he challenged rigid academic views. Without formal scientific training he was able to take a more open-minded approach to find innovative solutions. His efforts have encouraged recognition of the important contribution restoration practitioners have made through in-the-field experimentation. Also, the success of the North Branch Restoration Project awakened others in the conservation world to the ability of volunteers to take on major responsibilities. At the Illinois Nature Conservancy, where Packard worked from 1983 to 1998, he established a Volunteer Stewardship Network. It has since grown to 8,000 volunteers on 202 Illinois restoration sites. The network has come to serve as a model for similar work in many other states and countries.

Starting in the mid-1980s he broadened his focus to include building a culture of conservation for the Chicago region. A 13-county metropolitan area with over eight million people, the Chicago region paradoxically harbors the greatest concentration of biodiversity in the midwestern United States. He designed and organized a massive collaboration, the Chicago Region Biodiversity Council ("Chicago Wilderness"). It now includes 98 public and private agencies with many scores of scientists, hundreds of land management and education professionals, and thousands of volunteers working to protect and restore 200,000 acres of core natural areas such as corporate campuses, rights-of-way, golf courses, and wooded neighborhoods. The goal is to build a positive relationship between people and nature in an advanced metropolitan culture. Chicago Wilderness has been described as a model for the planet, an important one considering the shift of the world's populations to cities. Currently Packard directs the Chicago Wilderness Program for the National Audubon Society. He resides in Chicago.

BIBLIOGRAPHY

Packard, Stephen, "Rediscovering the Tallgrass Savannas of Illinois," *Proceedings of the Tenth North American Prairie Conference*, Denton, Texas, 1986; Packard, Stephen, "Restoring Oak Ecosystems," *Restoration & Management Notes*, 1993; Packard, Stephen, and John Balaban, "Restoring the Herb Layer in a Degraded Bur Oak 'Closed" Savanna," in *Proceedings of the North American Conference of Savannas and Barrens*, James S. Fralish, Roger C. Anderson, John E. Ebinger, and Robert Szafoni, eds., 1994; Packard, Stephen, and Cornelia F. Mutel, eds., *The Tallgrass Restoration Handbook: For Prairies, Savannas, and Woodlands*, 1997; Stevens, William K., *Miracle under the Oaks: The Revival of Nature in America*, 1995.

Palmer, Paula

(June 8, 1948–)
Sociologist, Director of Global Response

Sociologist and writer Paula Palmer lends her pen to the environmental causes of indigenous peoples throughout the world. Unlike conservationists who see humans as innately destructive of "wilderness," Palmer recognizes that traditional indigenous cultures have refined ways of living on the land without destroying valuable ecosystems. Palmer's career—working with indigenous and Afro-Caribbean residents of the Talamanca Coast of Costa Rica and, since 1996, directing Global Response, an urgent action network that responds to environmental crises affecting indigenous people—has been devoted to helping indigenous people carry out their own goals for a continued sustainable lifestyle.

Born in Valparaiso, Indiana, on June 8, 1948, Palmer studied at Valparaiso University; at the University of Colorado–Boulder, where she received a B.A. in English literature in 1970; and at Michigan State University, where she earned an M.A. in sociology in 1989. During her undergraduate years, Palmer edited a civil rights newspaper and worked in the student movement to end the war in Vietnam. A 1973 backpacking trip to Costa Rica became a turning point in her life. Via train and boat, she made her way to the roadless Talamanca region, intending only to spend a short time there. It was rainy season—there was little to do besides play with the village children and listen to elders tell stories. She became fascinated with the history of the English-speaking Black villagers, who had come to Costa Rica from the Caribbean islands during the 1800s. When they invited her to stay and establish an English-language school to carry on their cultural traditions, she eagerly accepted.

Since there were no textbooks for her to use, Palmer decided to make her own. She tape-recorded and transcribed interviews with the oldest people in the coastal villages, asking them to talk about every aspect of their lives. These transcripts became the children's reading primers. As villagers became more involved in remembering and thinking about their history as a unique people, they encouraged Palmer to put their recollections together in a book. *What Happen* (1977; rev. ed., 1993) has become Costa Rica's classic oral history of Talamanca's Afro-Caribbean community.

Palmer went on to two other major writing projects in Talamanca. As director of the Talamanca Community Research Project, she taught high school students how to do oral history research; her indigenous and Afro-Caribbean students became authors of articles that Palmer published in a series of magazines called *Nuestra Talamanca Ayer y Hoy*. Costa Rica's Ministry of Education later reprinted these magazines as a textbook about the unique cultures of Talamanca. In a region that lacked electricity, telephones, and paved roads, the magazines generated community pride and increased appreciation and cooperation among Talamanca's different ethnic groups. Young people learned to value the traditional knowledge and skills that sustained their communities as well as the natural resources of the forests and the sea.

With two indigenous coauthors, Juanita Sánchez and Gloria Mayorga, Palmer also published *Taking Care of Sibö's Gifts*, an oral history of the KéköLdi Indigenous Reserve in Talamanca. This book is a product of a method of sociology called participatory action research, in which a researcher helps members of a community analyze their own culture, history, economy, and ecology and then use what they learned when making decisions about their future. As a result of their research with Palmer, the Kekoldi indigenous community began a project to breed endangered green iguanas and opened a community center where elders teach their language and their traditional knowledge and practices to young people. Once they saw that outsiders were interested in their culture, they began to lead educational tours through parts of their forest home. As a result of this work, the Kekoldi people were awarded the Richard Evans Schultes prize for ethnobotany and biodiversity conservation.

Palmer returned to the United States in 1993 and is currently director of Global Response, an international network for environmental action and education that helps communities around the world organize effective campaigns to stop or prevent environmental destruction. At the request of local communities, Global Response publishes action alerts that describe specific environmental threats. Global Response members—adults, teens, and children—are asked to write letters to specific officials, urging them to make environmentally sound decisions. The organization has a history of successful campaigns. For Palmer, it is a way to create a world community of ordinary citizens, young and old, who work together to preserve the health and abundance of life on earth.

In addition to her work at Global Response, Palmer is the environment and education editor of the American Indian quarterly *Winds of Change*. She is also on the adjunct faculty in the Environmental Studies Department of the Naropa Institute.

BIBLIOGRAPHY

Palmer, Paula, "Empowering Indigenous Peoples to Preserve Their Forests and Cultures," *The Forum for Advancing Basic Education and Literacy*, 1994; Palmer, Paula, "We Can Survive and Prosper," *Cultural Survival Quarterly*, 1992; Palmer, Paula, and Corrine Glesne, *Coastal Talamanca: A Cultural and Ecological Guide*, 1993; Palmer, Paula, Juanita Sánchez, and Gloria Mayorga, *Taking Care of Sibo's Gifts: An Environmental Treatise from Costa Rica's KéköLdi Indigenous Reserve*, 1993.

Parkman, Francis

(September 16, 1823–November 8, 1893)
Historian

Author of *The Oregon Trail* and the seven-volume *France and England in North America*, Francis Parkman chronicled the early history of the colonization of the northern United States and Canada and created a vivid description of the wilderness encountered by the early settlers. Parkman was an avid outdoorsman and naturalist, as well as an expert on the cultivation of roses. He was an early advocate for the preservation of nature, particularly the great tracts of unspoiled forest in the West, and created through his writing an appreciation for the importance of the wilderness in the United States and the lives of its citizens.

Francis Parkman was born on September 16, 1823, on Boston's Beacon Hill, to a life of privilege. His parents were both from prominent New England families, and from his mother's family Parkman inherited enough money that he never had to earn his own living. He was a frail child and at age eight was sent to live with his maternal grandfather, Nathaniel Hall. Hall had retired to the country near Medford, Massachusetts, and while in his care Parkman learned the joys of a life in the woods. During the four years he lived in Medford, Parkman explored the wilderness of the Middlesex Fells, which bordered his grandfather's land, and became a close and enthusiastic observer of nature. When his family considered him sufficiently strong, Parkman was returned to Boston, where he went to preparatory school before entering Harvard in 1840. There he studied with the historian Jared Sparks, who helped inspire the project that was to be Parkman's life work: the history of the French and Indian wars in New England.

Parkman graduated from Harvard in 1844 and went on to take a law degree in 1846. Parkman had no intention of practicing law, however, and while at Harvard began publishing prose sketches in *Knickerbocker*, covering historical topics and his adventures in the White Mountains. In 1846 Parkman set out for a journey to the West with his cousin Quincy

Shaw, the journey chronicled in *The Oregon Trail*. Parkman undertook the trip in part to observe Native Americans, who were central in his history of New England, but with whom he had had little direct experience. With Shaw and Henry Chatillon, a guide they met in Missouri, Parkman journeyed across the eastern half of the Oregon Trail and part of the Santa Fe Trail. In Wyoming they met a band of Oglala Sioux. Parkman stayed several weeks in the company of the Indians and was deeply affected by the experience. He saw in the Oglala Sioux both a model for the way the wilderness tests and molds men and a reflection of the racist stereotypes of the time. *The Oregon Trail* is marred by Parkman's racist comments on the inferiority of Native Americans, though Parkman also later argued for more equitable treatment of native peoples as the United States expanded its borders westward. The principal drama of the book is that of the masculine adventurer, testing himself in the wilds, mastering the obstacles of the land through strength and bravery, and conquering wildlife with his trusty rifle. Like THEODORE ROOSEVELT, who read and admired *The Oregon Trail*, Parkman wished to preserve nature so that men could continue to prove their manhood through hunting.

Upon returning from his journey, Parkman's always-precarious health completely collapsed. He was stricken with arthritis, and his eyesight nearly failed. For much of the rest of his life he wrote through dictation. In spite of these obstacles, Parkman began publishing *The Oregon Trail* in serial form in *Knickerbocker* in 1847 and in book form in 1849. He then set to work on the pieces of *France and England in North America*, which he called "a history of the forest," because the main subject of the work is the encounter of French, English, and Native Americans with the forests of New England. Parkman's history is filled with close description of the forest and detailed accounts of the adventures of soldiers and trackers during the disputes over control of Canada and the northern United States. Parkman knew the land forms, plants, and animals of his setting, and his work preserves a record of the experiences of the European settlers as they met the frightening, foreign wilderness of North America. One of Parkman's contributions to the creation of an environmentalist sentiment in the United States is his recognition that this wilderness was not an enemy to be conquered, but a treasure worth preserving. For Parkman, the wilderness played a crucial role in developing the strength of the nation and its people, through testing and molding the national character, and *France and England in North America* records both the events of the early chapters of North American history and Parkman's love of nature.

The completion of this history took most of the rest of Parkman's life, concluding with the last volume, *A Half-Century of Conflict*, in 1892. Parkman also wrote numerous other works, including essays in support of CARL SCHURZ's policies of forest preservation. Parkman witnessed the disappearance of ancient forests in New England and hoped that similar destruction could be avoided in the West. Parkman was a good naturalist: He understood that destroying forests affected all parts of the surrounding ecosystem and wrote detailed descriptions of the delicate balance among land, water, plants, and animals. Parkman also became heavily involved in horticulture, particularly of roses and lilies. His 1866 *Book of Roses* was for many years considered the definitive work on the subject of rose cultivation. Parkman was appointed professor of horticulture at Harvard and served as president of the Massachusetts Horticultural Society. He died on November 8, 1893, of peritonitis, at his home in Brookline, Massachusetts.

BIBLIOGRAPHY

Gale, Robert, *Francis Parkman*, 1973; Jacobs, Wilbur, "Francis Parkman—Naturalist—Environmental Savant," *Pacific Historical Review*, 1992; Townsend, Kim, "Francis Parkman and the Male Tradition," *American Quarterly*, 1986.

Perkins, Jane

(April 18, 1948–)
Labor Organizer, Director of Friends of the Earth

During her career, Jane Perkins has worked both as a labor organizer and as an environmental activist. Currently, she combines the two, working within the American Federation of Labor–Congress of Industrial Organizations (AFL-CIO) as its liaison to the environmental movement. Her mission is to bridge the perceived gap between labor and environmentalists, defining goals that both camps can support in a quest for a healthy environment and economically secure workers.

Jane Perkins was born in Pottsville, Pennsylvania, on April 18, 1948. From an early age she learned of the hazards of unsafe work environments and the potential of unions to make a difference in the lives of workers. Both of her grandfathers contracted black lung disease and died in their forties as a consequence of working in coal mines without protection. Her mother sewed piece goods in factories and was a member of the International Ladies Garment Workers Union. She participated in several strikes both as a striking worker and a scabbing worker, crossing the picket lines to substitute for strikers. Watching her mother in these difficult situations made a deep impression on Perkins.

Perkins was the first member of her family to attend college. She earned a B.A. in humanities from Pennsylvania State University in 1971. After being unfairly fired from her first job after college and feeling firsthand the frustration of injustice without union representation, Perkins sought work in the labor movement. She took a job as business agent in the Harrisburg office of the Pennsylvania Social Services Union, a local of the Service Employee's International Union (SEIU). She spent six years helping workers negotiate contracts, handling grievances and arbitrations, and organizing new workers, before moving on to assume the elected office of secretary-treasurer of the union local.

The Three Mile Island nuclear power plant accident in 1979 was the catalyst for Perkins's transformation into an environmentalist. She and other family members were evacuated from their Harrisburg, Pennsylvania, homes when the power plant melted down. Because of her activity in the labor movement, Perkins immediately sensed the dilemma for those concerned with both environmental safety and job security. A well-known labor activist by this time, Perkins was invited by the local environmental group Three Mile Island Alert to represent the voice of labor within their group. Perkins agreed to join and soon organized the ad hoc Greater Harrisburg Labor Committee for Safe Energy and Full Employment, which in turn organized a labor march on the issue in Harrisburg. Perkins's work on the march caused problems for her with some unions. She was accused of causing dissension within the labor movement and faced internal union charges. Other unions, notably the Machinists' Union, the United Mine Workers, and the Furniture Workers, had come out against nuclear power and supported Perkins during the controversy.

The nuclear accident and her burgeoning environmentalism propelled Perkins into politics. In 1981 she was elected to Harrisburg's city council, where she created and chaired an ad hoc Three Mile Island committee, as well as presiding over committees dealing with other city affairs. She ran for state senate as a Democratic candidate in 1984, on a platform of fair taxes, personal integrity, and government spending priorities. Although she lost the election, she was endorsed by farmers', citizens', labor, women's, consumers', and environmental groups. The year 1985 took her to Washington, D.C., where she served as the first director of the New Populist Forum (now called Democrats 2000), which was founded by Iowa senator Tom Harkin, Texas agricultural commissioner Jim Hightower, and Illinois congressman Lane Evans. This organization serves as a bridge for debate between political progressives and conservative populists and as a network for progressive populists elected to office.

By this point in her career, Perkins was recognized as an expert in organizing people, motivating them to participate in projects and activities, solving organization-wide problem issues, and doing this all with limited budgetary resources. The newly reorganized environmental organization Friends of the Earth (FOE) recruited her in 1991 to repair its ailing management systems, eliminate its debt, and iron out problems with its recent merger with the Environmental Policy Institute and the Oceanic Society. After helping FOE reorganize and refocus, Perkins spent one year as a private-sector consultant and then two more as a consultant to the AFL-CIO. In 1997 she joined the staff of the AFL-CIO, where she works as its environmental liaison.

Perkins's mission at the AFL-CIO is to help the labor movement integrate the goal of a healthy environment with its advocacy for workers. She has assembled a group of labor activists and environmentalists interested in the goal, and they have met since 1997 to develop a set of principles for collaboration on emissions issues. So far, Perkins reports, the group agrees that reducing emissions—be they carbon, toxics, or other industrial byproducts—is important and that no worker can be sacrificed in the pursuit of a healthier environment, a concept the unions call Just Transition. The struggle to find common ground continues; workers and environmentalists have painful memories of past conflicts, and establishing trust is not easy. The next step in this process is to apply the principles to appropriate situations in communities outside of Washington, D.C.

Perkins is married to Andrew Stern, the current president of the SEIU. They have two children, Matt and Cassie.

BIBLIOGRAPHY

"AFL-CIO," http://www.aflcio.org/home.htm; Dougherty, Laurie, "Between the Devil and the Deep Blue Sea: Workers in the Global Environment," *Dollars and Sense Magazine*, 1999; Harwood, Jon, "Perkins Seeks Water-bond Review," *Evening News*, 1983; "Jane Perkins Is Selected Woman of Year," *The (Harrisburg) Patriot*, 1983; Moberg, David, "Brothers and Sisters Greens and Labor: It's a Coalition That Gives Corporate Polluters Fits," *Sierra*, 1999; Perkins, Jane, "Recognizing and Attacking Environmental Racism," *Clearinghouse Review*, 1992.

Peterson, Roger Tory

(August 28, 1908–July 28, 1996)
Ornithologist, Painter, Writer

Bird-watchers and other nature enthusiasts all over the world know Roger Tory Peterson as the inventor of field guides. By simplifying the techniques used in the field, he made bird identification accessible to everyone, not just scientists and museum curators. Birding became a hobby and a passion for millions, awakening them to the natural world. Peterson was active in many ornithological and conservation organizations and is well known for his paintings of birds and other wildlife.

Born on August 28, 1908, in Jamestown, New York, Roger Tory Peterson was the son of European immigrants. His parents both came to the United States as young children, his mother from what was then eastern Germany and his father from Sweden. He grew up roaming the fields and hills around his home, often neglecting household chores. A dreamy thinker, he did not fit in well at school and was nicknamed "Bugs Peterson" because of his affinity for everything from snakes to skunks. But at the age of 11 he found inspiration. His seventh-grade teacher, who was interested in birds, started a Junior Audubon Club in which she passed out pamphlets on birds and had the children color on bird sketches. Peterson was hooked. He began bird-watching in the field and painting birds instead of doing homework.

Peterson's first job after high school was at Union Furniture Company, painting Chinese scenes on expensive lacquered cabinets, a job that allowed him some creativity and a chance to keep painting. He could not afford college, but after a few years he did move to New York City to study art at the National Academy of Design for three years. Later, from 1931 to 1934, he worked as a teacher in the science department at the Rivers School in Brookline, Massachu-

Roger Tory Peterson (Courtesy of Anne Ronan Picture Library)

setts. It was there that he developed the idea that would change the course of nature study forever.

He had decided to reinvent bird identification in the field and began by simply combining his two loves: bird-watching and painting. The tradition for anyone studying birds up until this point was to head to the fields with a shotgun. Identifying birds in a less destructive way was not easy. "Bird guides" at the time were either hefty scientific texts or skimpy leaflets with little helpful information. Peterson's goal was a marriage of the two: a small, lightweight guide, organized to show birds as they appear in the field; sometimes distant, sometimes aloft, sometimes in poor light. This could be done by observing "field marks": easy to spot characteristics that distinguish one

species from another. Peterson did all the illustrations and descriptions himself, a painstaking labor of love met with huge skepticism from publishers. Five turned him down before Houghton Mifflin, in 1934, prompted by *Bird Lore* editor WILLIAM VOGT, cautiously printed 2,000 copies, insisting that Peterson receive only half a share of royalties. They need not have worried. The print run sold out in one week and launched what is today a multi-billion-dollar industry. The original book, *Peterson's Field Guide to Eastern Birds*, has never been out of print. The original book expanded into a whole series of over 45 guides that help nature enthusiasts identify anything from butterflies to seashells, from reptiles to mammals, from minerals to the weather, all of which he edited until his death.

It is hard today to appreciate what a brilliant and influential idea Peterson had with his field guide. The impact of the original book and all the field guides that have followed has been tremendous. At the time of publication it brought a new awareness of birds and their world. People began to value birds as soon as they could name them, identify them, and enjoy them. And consequently people began to care about habitat and the destruction of it. Many credit Peterson for paving the way for the environmental movement because he got people to pay attention to what they were seeing.

Following the success of his field guide, Peterson served as education director for the National Audubon Society and art editor for *Audubon Magazine* from 1934 to 1943. In 1936 he married his first wife, Mildred Washington; though they shared a passion for the natural world, their marriage was turbulent and ended after six years. Very shortly after that he married his second wife, Barbara Coulter, in 1943. The same year Peterson was drafted into the Army Corps of Engineers, stationed near Washington, D.C., where he served until 1945. During this time, he and Barbara had two sons, Tory and Lee.

After his army term Peterson continued writing and editing field guides, working on paintings, and photographing and filming birds all over the world. He remained involved in many organizations and agencies until the time of his death. He was art director for the National Wildlife Federation's conservation stamp program from 1946 to 1975 and served on the board of directors of various agencies such as the National Audubon Society, the World Wildlife Fund, and Hawk Mountain Sanctuary Association. He gave the keynote address and was a panelist at the Earth Care Conference at the United Nations in New York in 1975 and gave numerous lectures throughout the country emphasizing all phases of conservation from wildlife preservation to pollution and population control. In 1964 he addressed a U.S. Senate investigative committee hearing on dichlordiphenyltrichlor (DDT), recommending that it be banned and lobbying for strict control on all new chemicals.

After 33 years together, his marriage to Barbara crumbled apart. They were divorced in 1976, and once again Peterson quickly remarried, this time to Virginia Westervelt.

By the end of his career he had accumulated many honors and awards, including the Audubon Conservation Medal from the National Audubon Society, the Gold Medal from the World Wildlife Fund, and in 1980 the Presidential Medal of Freedom, and he was twice nominated for the Nobel Peace Prize. Roger Tory Peterson died at 87 years of age in his home in Old Lyme, Connecticut, on July 28, 1996.

BIBLIOGRAPHY

Devlin, John C., and Grace Naismith, *The World of Roger Tory Peterson*, 1977; Gordon, John Steele, "Inventing the Bird Business," *American Heritage*, 1996; Graham, Frank, Jr., "Field Guide for America: Roger Tory Peterson Taught This Country to See Its Birds—and Much More," *Audubon*, 1996; Leo, John, "He Was a Natural," *U.S. News and World Report*, 1996; Zinsser, William, "A Field Guide to Roger Tory Peterson," *Audubon*, 1992.

Peterson, Russell

(October 3, 1916–)
Governor of Delaware, President of National Audubon Society

Russell Peterson defies conventional stereotypes—he has been a nearly lifelong Republican and has worked for the chemical industry, yet he is widely recognized as a powerful and influential proponent of environmental causes. As governor of Delaware he enacted strict legislation protecting the state's coastline from development by the oil industry, despite the billions of dollars of rapid industrial growth it would provide. Following his term as governor, he held a succession of high-level appointed positions in state and federal administrations and nonprofit organizations, including chairman of the President's Council on Environmental Quality and president of the National Audubon Society. In each position, he worked his way to the forefront of the environmental movement, always battling industry's resistance to change and promoting a belief that a healthy environment leads to a healthy economy.

On October 3, 1916, Russell Wilbur Peterson was born in Portage, Wisconsin, to John Anton and Emma (Anthony) Peterson, the seventh of eight sons. Neither of his parents had any formal education past fifth grade. During his elementary school years Peterson held a number of jobs, delivering newspapers and working in restaurants. In high school he proved his competitive spirit and perseverance on the football and basketball teams. He graduated from high school as president of his class in 1934, in the midst of the Depression. His family was in financial straits and tried to talk him into staying home to help, but he was determined to go to college. He began his studies at the University of Wisconsin in 1934, and by working two or more jobs he paid his way through school. At the end of his junior year in 1937, he married Lillian Turner—a marriage that lasted until her death 57 years later. He graduated the next year with a B.S. in chemistry and immediately entered the university's Ph.D. program in chemistry, receiving his doctorate in 1942.

The DuPont Company offered him a job at their Experimental Station in Wilmington, Delaware, and he accepted. For the next 26 years he worked for DuPont, where he was frequently promoted and eventually groomed for top management. But although his career was flourishing, he grew disillusioned with some of their policies. In 1968 he left DuPont and turned his attention toward civic activism and politics; he quickly became a prominent figure in the local Republican Party. That same year he ran a successful campaign for governor of Delaware, and in January 1969 he was sworn in. Almost immediately he began reforming Delaware's archaic justice system—abolishing the public whipping post and the debtor's prison and introducing concepts such as out-of-prison punishment as an alternative to building more prisons.

Peterson continued to make waves the following year when he unveiled his plan to protect the Delaware coastal zone from industrial development. The coastline had come under serious threat from the planned construction of a Shell Oil refinery, which would likely become the stimulus for a complex of other industrial plants. Governor Peterson issued a moratorium on all new industrial development in Delaware coastal areas, pending a report from his new task force on coastal affairs. Throughout the year while the task force gathered data, there was little public debate on the issue and almost no opposition from heavy industry—probably because they were reassured by Peterson's background as a Republican and former official at DuPont. But when the report came out and Peterson started pushing the Coastal Zone Act, which prohibited all new development on the coast, it was too late for the oil companies to fight it. With the support of the people of Delaware behind it, the state legislature voted for the protection of their coastal areas and the Coastal Zone Act has remained intact, despite frequent assaults from heavy industry.

In 1973, Governor Peterson lost his bid for a second term, which he attributed to a financial problem. Less than ten months later, he was sworn in as chairman of the President's Council on Environmental Quality, intent on continuing his conservation work on a national scale. In his three years on the council, he tackled issues such as nuclear waste and ozone depletion and countered the argument that environmental regulations are bad for the economy. Following his work on the Council of Environmental Quality, he held a number of appointed positions with state and federal administrations, nonprofit organizations, and the United Nations—continuing his dedicated commitment to environmental causes.

In 1979, the National Audubon Society, looking for someone with visibility and political influence, named Peterson as its new president. At the time, the Audubon Society had 350,000 members but had yet to become an effective lobbying force. The organization was operating with a one-million-dollar deficit and lacked a clear organization of its environmental themes. Under Peterson, Audubon showed a 19 percent per year increase in revenue, its membership increased to 550,000, it gained potent political influence, and it launched a major effort toward reaching each of the five goals that define the Audubon Society: conserving plants and animals, fostering careful use of land and water, promoting rational strategies for energy use, protecting life from pollution, and seeking

solutions for global problems involving population, resources, and the environment. Peterson strongly believed that overpopulation is the single biggest threat to the global environment and broadened Audubon's focus to global issues. Under his leadership, Audubon established an international program, which played a role in United Nations conferences on the environment, in the World Conservation Union, and in the International Council for Bird Preservation. He also reestablished a youth education program called the Audubon Adventures Club, which introduced millions of schoolchildren to birding and nature study.

In the early 1980s, during President Reagan's first term, with Secretary of State James Watt leading an antienvironment crusade, Peterson was an ideal foil. As a Republican formerly involved with the chemical industry, he was proof against Reagan's and Watt's charge that the environmental movement was a liberal cabal. He participated in the Group of Ten, a gathering of ten leaders of national environmental groups who strategized responses to Reagan's antienvironmental policies.

Peterson left Audubon in 1985 to spend more time writing and speaking about environmental issues. He taught as a visiting professor at Dartmouth College, Carleton College, and the University of Wisconsin–Madison over the next several years and continued to serve on boards of directors for various environmental organizations. He has won many awards, including the Gold Medal from the World Wildlife Fund in 1971, the Audubon Award in 1977, and the Robert Marshall Award from the Wilderness Society in 1984, and was named Conservationist of the Year by the Wildlife Federation in 1972. In the mid-1990s, as Republicans became more and more conservative, Peterson realized there was little he could do to turn the party around, and in October 1996 he became a Democrat. He lives with his second wife, June, in Wilmington, Delaware.

BIBLIOGRAPHY

Gottlieb, Robert, *Forcing the Spring: The Transformation of the American Environmental Movement*, 1993; Graham, Frank, Jr., *The Audubon Ark: A History of the National Audubon Society*, 1990; Holden, Constance, "Peterson Leaving a Changed Audubon Society," *Science*, 1984; Peterson, Russell W., *Rebel with a Conscience*, 1999.

Pinchot, Gifford

(August 11, 1865–October 4, 1946)
Forester, Founder of U.S. Forest Service

The first professional forester in America, Gifford Pinchot introduced the methods of scientific forest management to the United States. He believed that forests were like crops that, if properly cultivated, would yield a profit and would continue bearing indefinitely. Pinchot became the first director of the U.S. Forest Service, which he established during Pres. THEODORE ROOSEVELT's term in office.

Gifford Pinchot was born on August 11, 1865, in Simsbury, Connecticut. His wealthy parents raised him at their estate, "Grey Towers," in Milford, Pennsylvania, and in France, where he lived between the ages of six and nine. His father, businessman James Pinchot, was one of the few American men at that time concerned about the depletion of the nation's forests. When his son was still a boy, James Pinchot decided that Gifford should do something to solve the problem. Young Pinchot agreed and was happy to follow his father's guidance. Knowing that forestry was a field that Eu-

ropeans had been practicing for many years, Pinchot's father urged his son to study forestry in Europe. After graduating from Yale with a B.A. in 1889, Pinchot traveled to Germany, where he met the great Prussian forester Detrich Brandis, who became his mentor. Brandis's first recommendation was that Pinchot attend the French National School of Forestry in Nancy, France. After a semester there, Pinchot visited the managed forests of central Europe with Brandis and a group of his students. They saw forestry being practiced in such a way that no part of the tree was wasted, not even the sawdust. There was no clear-cutting; to the contrary, only selected trees were cut, and the direction in which they fell was determined carefully to prevent them from damaging other trees on their way down. This was in clear contrast to the clear-cutting that was then standard forestry practice in the United States. Pinchot returned after 13 months of study, eager to visit the vast forests of the West (he was embarrassed to admit that

Gifford Pinchot (Corbis)

he had never traveled west of the Allegheny mountains) and to put what he had learned in Europe into practice.

In 1892, Pinchot was offered a job at Biltmore, the Vanderbilt family's 5,000-acre forested estate in Asheville, North Carolina. The estate had been designed by the eminent landscape architect FREDERICK LAW OLMSTED SR., who along with Vanderbilt was interested in helping Pinchot develop the field of managed forestry in the United States. At Biltmore, Pinchot successfully proved that scientific forest management would fulfill three objectives: the forest would yield a profit to its owner; there would be a constant annual yield of timber that would allow a permanent workforce to remain employed year-round; managed forestry would actually improve the forest rather than degrade it. Pinchot prepared an exhibit for the 1893 Chicago World's Fair that demonstrated these objectives; this was the first of the many publicity campaigns that he undertook to promote managed forestry during his life.

As the country's only trained forester, and as the scion of an elite family who boasted personal friendships with all of the U.S. presidents who served during his lifetime, Pinchot's skills and reputation quickly became known in government circles. When Secretary of the Interior Hoke Smith asked the National Academy of Science in 1896 to make a survey of the country's forests, Pinchot was asked to serve as secretary of the National Forest Commission. The commission, headed by dendrologist CHARLES SPRAGUE SARGENT, spent the summer of 1896 surveying the major forests between Montana and Arizona. Everyone on the commission was concerned about the health of forests in the United States, but there was a major rift about how best to protect them. Sargent and his allies were in favor of a purely preservationist strategy: locking up the forests and employing the military to protect them from destruction. Pinchot believed that forests were a national resource to be managed for a sustained yield. Despite differing opinions, the commission filed a single report recommending that Pres. Grover Cleveland set aside over 21 million acres of forested land. When Cleveland signed them into the national reserve in 1897, westerners reacted angrily, believing that this would set back the economic development of the region. Pinchot learned an important lesson from this experience, that communication with the public would be crucial in raising public support for any further forestry policy.

In 1898, Pinchot became head of a tiny new office in the Department of Agriculture, the Division of Forestry. He invited a small group of his friends and colleagues, mostly classmates from his Yale days, to work with him there. Although the Division of Forestry did not have direct control of the forest reserves, which were held by the Department of the Interior, it offered management advice for state-owned and private forests. As requests for help grew, so did the need for more trained foresters. Pinchot convinced his family in 1900 to endow his alma mater with $150,000 to found the Yale Forestry School and to host summer classes at Grey Towers.

Once Theodore Roosevelt became president in 1901, Pinchot became one of the most influential men in Washington. He was a member of Roosevelt's informal "tennis cabinet," Roosevelt's closest friends and advisers, who were as likely to be found stripped to their shorts and boxing with the president as discussing policy during long walks in the countryside with him. Roosevelt was a committed conservationist, interested both in preserving spectacular sites from any human intervention and in sustainable management of natural resources. Roosevelt trusted Pinchot to the extent that Pinchot wrote most of his conservation speeches, and Roosevelt insisted that Pinchot be allowed to review any policy having to do with natural resources.

Pinchot organized two major conservation conferences during Roosevelt's tenure, both of which served to publicize his approach to conservation. The American Forest Congress in 1905 was organized shortly before a Pinchot-inspired bill arrived in Congress. Influential politicians, lumbermen, miners, stockmen, and railroad executives were invited to discuss the nation's forest policy. The Forestry Division bombarded the national press with press releases proclaiming that if timber-cutting continued at the current rate for another 60 years, all forests in the United States would disappear. A month after the conference, Congress passed a bill consolidating the management of all 86 million acres of U.S. forest reserves and handing them to Pinchot's newly renamed

U.S. Forest Service (USFS). The bill allowed the USFS to issue timber-cutting permits and collect usage fees. Under Pinchot's decentralized management plan, rangers in the field were given the power to make arrests, which made illegal cutting easier to control. Three years later, in 1908, Pinchot and Roosevelt organized the Governor's Conservation Conference, intended to make conservation a priority for state governments. Pinchot's ideas permeated the conference. He wrote many of the speeches given by Roosevelt and some of the governors, and he purposely excluded preservationists like Charles Sprague Sargent and JOHN MUIR, whose viewpoints he was not interested in publicizing.

Shortly before Roosevelt left office, Congress began to rebel at what it perceived as too much emphasis on conservation. Under Pinchot's influence, Roosevelt had established 130 new national forests, covering 173 million acres, and had increased funding for the USFS one hundred-fold. In 1907, an amendment was tacked onto an important agriculture appropriations bill that took away the president's power to declare new national forest reserves. Pinchot and Roosevelt responded by declaring 16 million acres of forest reserves in the two weeks before Roosevelt signed the bill.

This sneaky tactic turned politicians against Pinchot. That, plus problems that Pinchot had with a new secretary of the interior, Richard Ballinger, caused President Taft to dismiss Pinchot, then 45 years old, in 1910. Upon his departure from Washington, Pinchot founded the National Conservation Association, which fought for his approach to conservation.

Pinchot married suffragette Cornelia Bryce in 1915, at the age of 49. She accompanied him through the political career that occupied the rest of his life. During the 1920s and 1930s, Pinchot served as governor of Pennsylvania for two terms and as commissioner and secretary for the Pennyslvania Department of Forests and Waters. Although he was no longer involved formally in Washington, D.C., affairs, Pinchot served as an adviser to many later national parks and national forest officials.

Pinchot died in Milford, Pennsylvania, on October 4, 1946, after a long bout with leukemia.

BIBLIOGRAPHY

McGeary, M. Nelson, *Gifford Pinchot: Forester Politician*, 1960; Pinchot, Gifford, *Breaking New Ground*, 1947; Strong, Douglas, *Dreamers & Defenders: American Conservationists*, 1988.

Plotkin, Mark

(May 21, 1955–)
Ethnobotanist, Founder of Shaman's Apprentice Program and Ethnobiology and Conservation Team

Ethnobotanist Mark Plotkin is known for his adventures with the indigenous peoples of the Amazon rain forest and his attempts to help them conserve their ecosystem and their way of life. Soon after Plotkin began visiting the Amazon in 1977, he found that young indigenous people were losing interest in their peoples' ancient traditions, attracted instead to Western ways. To combat what he worried would be a tragic loss of knowledge, Plotkin initiated the Shaman's Apprentice Program in the mid-1980s, which encourages young tribespeople to study with elderly shamans. Plotkin also collaborates with the innovative Shaman Pharmaceuticals, which sends ethnobotanists to study with shamans and reciprocates by funding health and development projects for the communities that host their expeditions.

Mark J. Plotkin was born on May 21, 1955, in New Orleans, the son of Helene, a schoolteacher, and George, who owned a shoe store in New Orleans's French Quarter. As a boy, his parents took him frequently to the Audubon Zoo, and he was allowed to visit the swamps around New Orleans to collect reptiles. He kept his animals in the basement; his mother moved the washing machine into the kitchen to avoid disturbing her son's basement zoo.

Plotkin entered the University of Pennsylvania in 1973 to study biology but was disappointed when he found that the faculty there focused on molecular and cellular biology. So he dropped out and moved to Cambridge, Massachusetts, where he found a job as curatorial assistant at Har-

vard's Museum of Comparative Zoology. He took night classes with Harvard's extension program and enrolled in one given by eminent ethnobotanist RICHARD EVANS SCHULTES. Watching Schultes's slides and listening to his accounts of decades of Amazon research, Plotkin was seduced. He realized that his dreams of being a jungle explorer could become reality, and from then on he worked to make it happen.

During his undergraduate work at Harvard, Plotkin studied chameleons in Haiti and other lizards in Venezuela, French Guiana, and the island of Guadeloupe. He earned an A.B. degree *cum laude* in 1979 and immediately went on to Yale University's Forestry School for a master's degree in wildlife ecology, which he received in 1981. After that he attended Tufts University and earned a Ph.D. in biological conservation in 1986.

Plotkin took the first of his many trips to South American rain forests in 1977, when he accompanied primatologist and zoologist RUSSELL MITTERMEIER on a search through French Guiana's jungles for the elusive black caiman. Later in 1979, he traveled to Surinam, where he studied how the Saramacca Maroon people— rain forest–dwelling descendants of escaped slaves—used native medicinal plants. Plotkin continues to work primarily in Surinam because of its largely intact forest and because the national forest service is dedicated to conservation.

Convincing native shamans to reveal their knowledge to him about medicinal plants is a skill that Plotkin developed over time. He recalled in a 1989 *Smith-*

sonian article that at first, the native Tiriós called him *panankiri*, which meant "alien." While crossing a river with a group of Tiriós tribespeople during that first 1982 visit, Plotkin clumsily slipped on a rock and fell into a river, and one shaman continues to call him "whiteman-who-fell-on-his-ass." But Plotkin's cheerful, friendly manner, his repeated visits over the years, his respect, his willingness to try even their hallucinogenic preparations, and his sincere interest in plants eventually charmed his shaman hosts, who tell him what they know about each plant and allow him to take botanical samples.

From the beginning, Plotkin knew that countless medicinal plants were powerful healers. He had observed many cures, where the shaman combined spiritual work with potent botanical preparations. Plotkin says he has also experienced several cures first hand. Once shamans cured him of a persistent fungal infection of the skin; on another occasion he was attacked by a swarm of wasps and some crushed bark applied by a shaman made the pain disappear almost immediately; and an old muscular injury was healed with a salve and clouds of smoke from an shaman's pipe. But Plotkin's interest in learning about rain forest medicine and his concern that the knowledge and the forest be conserved have never been motivated solely by a desire to bring powerful medicines back to the United States.

A hallmark of Plotkin's work has always been to insist that rain forest conservation depends upon the survival of the rain forest's indigenous people and their traditions. Throughout his career he has worked to strengthen the native cultures he has visited. During the 1980s, Plotkin noticed that with each visit to native villages, native culture seemed to be a little less present; Western dress, Walkmans, and outboard motors for their canoes were becoming more and more prevalent. Plotkin became concerned that young people were losing interest in their culture's shamanic traditions and that aging shamans had no one who wanted to inherit their wealth of knowledge. Plotkin quickly realized that individually he was not capable of preserving the knowledge, thirsty for it as he was, since he was just an occasional visitor to many different tribal villages. So, with resources from the World Wildlife Fund, where he worked as plant conservation coordinator from 1985 to 1990, Plotkin initiated the Shaman's Apprentice Program. The program works with the Tiriós tribe and three others and encourages the young to study directly with the elderly shamans. As of 1998, Plotkin reported during an Internet chat session, there were 60 or 70 apprentices in the Amazon, and he expected the interest among other tribes to grow quickly.

Plotkin has collaborated with LISA CONTE's Shaman Pharmaceutical Company, which is unique in two ways. It sends ethnobotanists to work with shamans in gathering promising substances, and it tries to reciprocate by helping the communities that host the ethnobotanists. Fifteen percent of the expenses of each scouting expedition are donated to the host community for such projects as demarcation of traditional lands, building health centers, and installing potable water systems. Plotkin is insistent that indigenous inhabitants of the rain forest are the best conservationists and that the best strategy to conserve the rain forest habitat is to help indigenous rain forest dwellers strengthen their communities.

In addition to his work with World Wildlife Fund, Plotkin has held a series of conservation-oriented positions at academic and environmental organizations. He worked as researcher associate in ethnobotanical conservation at the Harvard Botanical Museum from 1981 to 1990, taught Harvard's first classes on tropical rain forest conservation from 1983 to 1985, served as the vice president for plant conservation at Conservation International, and has acted as consultant to Shaman Pharmaceuticals since 1989. In 1995, Plotkin founded the Ethnobiology and Conservation Team, an association of ethnobotanists who work with their indigenous colleagues to promote greater ethnobotanical knowledge and more effective conservation. Plotkin is married to Costa Rican conservationist Liliana Madrigal, and has two daughters, Gabrielle and Anne. He and his family live near Washington, D.C.

BIBLIOGRAPHY

"Ethnobiology and Conservation Team," http://www.ethnobotany.org; Hallowell, Christopher, "In Search of the Shaman's Vanishing Wisdom," *Time*, 1999; "Heroes for the Planet: Mark Plotkin, Modern Medicine Man," http://cgi.pathfinder.com/time/reports/environment/heroes/live/; Jackson, Donald Dale, "Searching for Medicinal Wealth in Amazonia," *Smithsonian*, 1989; Plotkin, Mark, *Tales of a Shaman's Apprentice*, 1993; Reed, Susan, "Sorcerer's Apprentice," *People*, 1993.

Poe, Melissa

(September 5, 1979–)
Founder of Kids F.A.C.E.®

In 1989, when Melissa Poe was nine years old, she founded Kids F.A.C.E.®, a children's organization for environmental action. Kids F.A.C.E.® has grown from one chapter in Nashville, Tennessee, with six members to a membership of 300,000 in all 50 states and more than 20 foreign countries and has sponsored such events as the "One in a Million" campaign to plant one million trees throughout the United States and the creation of a huge Earth Flag with patches from more than 20,000 children worldwide.

Melissa Poe was born on September 5, 1979, in Nashville, Tennessee, to Patrick and Patricia Poe. At the age of nine, she and her mother were watching *Highway to Heaven* on television, a program about what would happen to the world if natural resources were not protected and pollution continued unabated. She was horrified, but when actor Michael Landon concluded the program by saying that it was not too late and that people who cared would do something to prevent disaster, Poe immediately decided that she would be one of those who cared and would "do something." Her parents supported her in this conviction, telling her that it was right to take action if she felt strongly about a problem. Poe thought that if she wrote to President Bush about conser-

Melissa Poe (Courtesy of Anne Ronan Picture Library)

vation and pollution, he would surely recognize the urgency of the problem and respond with all of his power. But, she recalled in a 1997 article for *Out of the Blue*, "It took the President twelve weeks to write me back. Actually, he never did write me back—all I received from his office was a form letter telling me to stay in school and not to do drugs." She also wrote to other political leaders, and she wrote commentaries for the local newspaper and television station. And she called the major environmental groups of the country but was told that children her age could not do very much. At this point she decided to found a children's club to protect the environment. She recruited her brother

Mason as her first member and five others from her elementary school.

The club's original activities were to promote recycling, plant trees, pick up trash, and recruit other members. Poe excelled in publicity and outreach. During the club's first year, in 1990, she recruited a board of directors composed of both children and adults, requested nonprofit status from the Internal Revenue Service, solicited support from the Wal-Mart foundation to sponsor members and publish a newsletter (Kids F.A.C.E.® Illustrated), and arranged for the placement of 250 billboards across the United States with an enlarged copy of her letter to President Bush. After she appeared on the *Today* show that year, a group of children in Hattiesburg, Mississippi, formed the first chapter of Kids F.A.C.E.® The following year, in 1991, she developed a program to train youth in public speaking and filmed a public service announcement for television, underwritten by Wal-Mart, about the club as a way for children to become involved in the environmental movement. During 1991 she collaborated with Children's Alliance for the Protection of the Environment (now defunct) and the U.S. Forest Service to create the Children's Forest concept for special areas within national forests that are designed by and for children. That year she also served as the Tennessee chair of Global Releaf, a reforestation campaign.

Poe remained busy throughout the 1990s. Through her preadolescence and adolescence, she maintained an active public life, speaking at Rio de Janeiro, Brazil, for the 1992 Earth Summit, on many television shows, and at schools and community meetings nationwide. With Kids F.A.C.E.®, she followed through with the Children's Forest program, proposing one for Mt. Hood and Gifford Pinchot National Forest and helping to establish the first one in San Bernardino National Forest in California; she coordinated the sewing of a huge Kids Earth Flag, 100 feet high by 200 feet long, with panels from 20,000 children from around the world, and took it first to Washington, D.C., for its official unveiling on Earth Day 1995 and later across the country on a year-long tour. The flag is still available for showings through F.A.C.E.®

In 1993 at a youth environmental summit in Ohio, Poe met Tara Church of El Segundo, California, who in 1987 had established with 12 other Girl Scouts a tree-planting organization called Tree Musketeers. Church and Poe began collaborating on tree-planting projects, and in 1997 launched Kids F.A.C.E.®'s and Tree Musketeers's One in a Million campaign to plant a million trees before the end of the century. Currently, more than 700,000 trees have been planted by that many children nationwide.

Poe resigned as leader of Kids F.A.C.E.® in 1997, when she was 18 years old, because she felt it was important that the organization continue to be run by children. Co–child executive officers Ashley Craw and Rachel Jones were elected to replace her. Poe has received more than 20 awards and recognitions, ranging from the National Junior League's 1999 Elizabeth Award to Disney's International Eco Hero in 1998, from *E-Magazine*'s Young Eco-Hero in 1995 to GI-Joe Real American Hero in 1991. Poe, now a student at Wake Forest University in North Carolina, continues to mentor to activist children and organizations.

BIBLIOGRAPHY

Black, James, "Melissa Poe Planted the Seed," *Southern Living*, 1999; Hansen, Mark Victor, Barbara Nichols, and Patty Hansen, *Out of the Blue: Delight Comes into Our Lives*, 1997; "Kids F.A.C.E.®," http://www.kidsface.org/; McGrath, Kim, "Green FACE," *Wake Forest Magazine*, 1999.

Popper, Deborah, and Frank Popper

(August 18, 1947– ; March 26, 1944–)
Cocreators of the Buffalo Commons Concept

Frank and Deborah Popper have brought their knowledge of land use planning and geography to the study of the American Great Plains, where heavy agriculture and extractive industries have exploited the land and led to large population losses. The Poppers first wrote about the Great Plains in 1987 in the magazine *Planning*, where they articulated the concept of the Buffalo Commons, a proposal for restoring large parts of the plains to their presettlement state, with bison instead of cattle and native shortgrass instead of cultivated wheat, corn, and cotton. The idea provoked heated responses and sparked a vigorous debate, and the Poppers accomplished in one article what few academics achieve in a lifetime: They caught the attention of the nation and engaged an entire region in a dialogue about its future. From the various responses came new ideas for the Great Plains based on preservation and sustainable industries such as tourism, and many of the Poppers's original proposals, such as more buffalo production, are springing to life on the ground.

Deborah Epstein was born on August 18, 1947, in New York City to Irving (a lawyer) and Fay (Falkowsky) Epstein. She grew up in New York and attended Bryn Mawr College, graduating with a bachelor's degree in history in 1969. In August 1968 she and Frank Popper were married, and they moved to Boston after her graduation. Deborah then took a job as a neighborhood worker in the Boston Model Cities program, a government-funded service that tried to integrate redevelopment of inner city neighborhoods with a wide range of social services and job opportunities. In 1971 the Poppers moved to Chicago, where their children were born: Joanna in 1972 and Nicholas in 1977. Deborah divided her time between her children and library school at Rosary College (now Dominican University) in River Forest, Illinois. She received her master's degree in library science in 1977. She held library positions first in Montgomery County, Maryland, in 1983, and then in Mountainside, New Jersey, from 1984 to 1985. She then began work on a second master's degree from Rutgers University, and in 1986 she began working there too, as a research assistant in the Department of Urban Studies and teaching assistant in the Department of Geography. At Rutgers she completed her master's degree in geography in 1987 and her doctorate in geography in 1992. After teaching at New York University, she is now an associate professor of geography at the College of Staten Island/City University of New York, where she has worked since 1994.

Frank J. Popper was born on March 26, 1944, in Chicago, Illinois, to Hans (a physician) and Lina (Billig) Popper. He received his bachelor's degree in psychology from Haverford College, graduating in 1965 with high honors. During the following year he did graduate work at the Massachusetts Institute of Technology and then went to Harvard University to complete his master's degree in public administration, which he received in 1968. He began working that year as a research associate at the Twentieth Century Fund (now the Century Fund) in New York City, staying for one year. He continued his studies at Harvard and earned a Ph.D. there in political science in 1972. In the meantime he had moved to Chicago in 1971 and become a staff associate with the Public Administration Service, where he worked until 1973. At that point he switched to the American Society of Planning Officials (now the American Planning Association), where he worked as a senior research associate until 1974. Between 1975 and 1981 he directed the Twentieth Century Fund's Project on the Politics of State Land-Use Regulation. In 1979 and 1980 he also worked as a senior associate at the Environmental Law Institute in Washington, D.C. He then held a Gilbert White fellowship at Resources for the Future in Washington before becoming an associate professor in the urban studies department at Rutgers University in New Brunswick, New Jersey, in 1983 and a full professor in 1991.

In the December 1987 issue of *Planning*, a magazine for urban planners, the Poppers wrote "The Great Plains: From Dust to Dust." The fragile semiarid grassland ecosystems of the U.S. Great Plains, which extend from Montana and North Dakota in the north to New Mexico and Texas in the south, had suffered decades of abuse from heavy-handed farming, too many roads, overgrazing, and too many people. The Poppers argued that ecological impoverishment was showing up in declines of farm and ranch economies, increasing dependence on federal subsidies, dropping water tables, rising wind erosion, and long-term depopulation. They suggested that because many rural residents were already giving up trying to make a living from the current agricultural system, the Great Plains could feasibly be restored to early-nineteenth-century conditions, before settlement and overly aggressive agriculture took over. The Poppers called their restoration approach the Buffalo Commons and proposed that a federal agency be established to administer the removal of fences, the replanting of native shortgrass prairies, and the restocking of native animals such as bison. The native prairie ecosystem would reestablish itself. The Buffalo Commons would become the world's largest historic and wildlife preservation project.

After the Poppers's article came out, the Associated Press wire service picked up the story, and it was printed in more than 150 newspapers, magazines, and journals, creating a buzz of responses that ranged from supportive interest to outrage. The Poppers were inundated with mail and speaking invitations at a range of forums: planners, environmentalists, farmers, ranchers, local communities, agricultural economists, and businesspeople, among others. Many farm and ranch families in plains communities scorned the Buffalo Commons, seeing it as a threat to their way of life and an assault on their ancestry. Some of the Poppers's Great Plains appearances required armed guards, and a talk in Montana in 1992 had to be canceled because of death threats. But contrary to many local perceptions, the Poppers were not threatening the forced removal of people living on the Great Plains but instead were offering the Buffalo Commons as a metaphor for the alternative future of a region in economic and ecological decline. Once this point became clear, a dialogue often became possible. The Poppers's idea touched off a national debate on the region's problems. The Buffalo Commons caused policy makers and residents alike to think about economic possibilities for the Great Plains that place more emphasis on preservation and ecotourism and less on agriculture and extraction.

The Poppers still study the region. Deborah Popper's doctoral dissertation explored why some Great Plains counties have maintained stable populations while others have decreased. The Poppers have written numerous journal and newspaper articles exploring the future of the region's rural areas. The Buffalo Commons remains controversial and visionary, but it has shown surprising impact and accuracy. Many of the Poppers's proposals and predictions are materializing across the plains as traditional agriculture continues to decline. In 1992 a coalition of tribes formed the Intertribal Bison Cooperative to reintroduce buffalo as an alternative to cattle and as a culturally significant part of Native American identity. Many cattle ranchers, realizing that buffalo are better adapted to the Great Plains and are more cost efficient to raise, have switched to bison and prospered. Federal agencies are allowing more buffalo to graze on public lands. The Nature Conservancy, the Sierra Club, and other private groups are buying up plains land, restoring buffalo on it and otherwise promoting buffalo.

In 1997 the Poppers received the American Geographical Society's Paul Vouras Medal for their work. Frank Popper continues his work as a land use consultant to numerous agencies, corporations, and nonprofit groups and as a teacher in the Rutgers urban studies department. He received the university's Presidential Award for distinguished public service in 1997. He has served on the governing boards of various land use and planning councils and on the editorial boards of several journals. Deborah Popper continues to teach at the College of Staten Island in the Department of Political Science, Economics, and Philosophy and serves on the advisory boards of Ecocity Builders and the Rural Americans Project and on the editorial board of the *Journal of Rural Communities.* The Poppers live in Highland Park, New Jersey.

BIBLIOGRAPHY

Manning, Richard, "The Buffalo Is Coming Back," *Defenders of Wildlife,* 1995–1996; Matthews, Anne, *Where the Buffalo Roam,* 1992; Popper, Deborah, and Frank Popper, "The Bison Are Coming," *High Country News,* 1998; Popper, Deborah, and Frank Popper, "The Buffalo Commons, Then and Now," *American Geographical Society's Focus,* 1993; Popper, Deborah, and Frank Popper, "The Reinvention of the American Frontier," *The Amicus Journal,* 1991.

Postel, Sandra

(1956–)
Founder and Director of the Global Water Policy Project, Writer

A leading authority on water usage and conservation, Sandra Postel is founder and director of the Global Water Policy Project in Amherst, Massachusetts. She is also a senior fellow at the Worldwatch Institute, where she served as vice president of research and coauthored more than a dozen State of the World reports. A Pew scholar and international lecturer, she has authored several books and numerous articles on global water issues. She received media attention for her 1992 book *Last Oasis*, which was adapted for PBS television, and for *Pillar of Sand*, her 1999 book about the challenges and risks of global dependence on irrigated agriculture.

Sandra L. Postel was born in Hollis, New York, in 1956 to Harold and Dorothy Postel. She grew up on the south shore of New York's Long Island. After earning her bachelor's degree in geology and political science at Ohio's Wittenberg University in 1978, she went on to study at Duke University, where she was awarded a Master of Environmental Management, Resource Economics, and Policy degree in 1980. At Duke she also gained what she has called the "principal passion" of her life—a compelling interest in water issues.

Postel found employment as a natural resources consultant with a private firm in Menlo Park, California, but after three years she moved on to the nonprofit Worldwatch Institute. She began as a researcher and worked her way up to the level of vice president. From 1988 to 1994, she codirected the institute's State of the World annual reports, which have been used as textbooks in nearly 600 U.S. colleges and universities. Her own alma mater recognized her efforts in 1991, when she was awarded the Duke University School of Forestry and Environmental Studies' Distinguished Alumni Award.

Compiling research on the world's water supply, Postel wrote her first book under the Worldwatch imprint. *Last Oasis* (1992) documented a pending shortage in clean, available water and argued for conservation. Noting that several of the world's major rivers, including the Colorado, Nile, and Yellow Rivers, are used up completely before they reach the ocean, Postel outlined the need for better management and for water usage rates that encourage farms and industries to conserve water. The book was still popular five years after publication. It was reprinted and made into a television documentary that aired on PBS in the four-part Cadillac Desert series in 1997–1998. To date, *Last Oasis* has been published in nine languages.

Postel has lectured at many universities, including Duke, Stanford, the Massachusetts Institute of Technology, and Yale. From 1994 to 1996 she was adjunct professor of international environmental policy at Tufts University in the Fletcher School of Law and Diplomacy. Although she remained a senior fellow at Worldwatch, in 1995 she left to found the Global Water Policy Project in Amherst, Massachusetts, which focuses on international water issues and strategies. In 1995, she was named a Pew scholar in conservation and the environment, promising to work to change the policies that deplete and waste the world's fresh water.

In her latest book, *Pillar of Sand* (1999), Postel calls for a Blue Revolution. She takes on the practice of irrigated farming, noting that 40 percent of the world's food comes from irrigated lands that are increasingly salinized and hemmed by urban growth. To combat water scarcity, she advocates protecting vital ecosystems and using new technology, such as drip irrigation systems and sprinklers. If we cannot irrigate more efficiently and moderately, she predicts, our civilization will deteriorate as did the great empires of the past.

Postel has published a wide array of articles and chapters in popular and scholarly publications, including *Science, Natural History, Environmental Science and Technology, Ecological Applications, The Sciences, Technology Review,* and *Water International.* She has written op-ed features for more than 30 newspapers in the United States and abroad, including the *New York Times* and the *Washington Post.* She has also appeared periodically on radio and television, serving as commentator on CNN's *Futurewatch.*

Her voice is especially strong in international water policy. She has served as a consultant to the World Bank and the United Nations Development Program and has addressed the European Parliament on environmental issues. She has served on the board of directors of the International Water Resources Association and the World Future Society and has acted as adviser to the Environmental Media Association and the Global 2000 program founded by Pres. JIMMY CARTER. She has been a council member in

Sandra Postel (Courtesy of Anne Ronan Picture Library)

the UK-based Forum for the Future and has served as senior adviser to the World Commission on Water.

Currently a visiting senior lecturer in earth and environment at Mount Holyoke College as well as director of the Global Water Policy Project, Postel lives in Amherst, Massachusetts.

BIBLIOGRAPHY

Motavelli, Jim, and Elaine Robbins, "Sandra Postel: The Coming Age of Water Scarcity," *E,* 1998; Postel, Sandra, *Last Oasis: Facing Water Scarcity,* 1992; Postel, Sandra, *Pillar of Sand: Can the Irrigation Miracle Last?* 1999; "Worldwatch Institute," http://www.worldwatch.org.

Pough, Richard

(April 19, 1904–)
Cofounder of the Nature Conservancy

Richard Pough cofounded The Nature Conservancy in 1951 to facilitate the purchase and preservation of ecologically valuable land and was president when the organization made its first acquisition in 1955. His vision—to preserve sanctuaries that represent all existing ecosystems—contrasted with the approach of conservationists at that time, which was to focus on preserving individual endangered species or places of spectacular beauty. Pough went on in 1957 to found the Natural Areas Council, which fomented the creation of local preservationist groups throughout the United States and successfully brokered the purchase or donation of land to be set aside as preserves.

Richard Hooper Pough was born on April 19, 1904, in Brooklyn, New York. His parents, Frances Harvey and Alice (Beckler) Pough, were both graduates of the Massachusetts Institute of Technology (MIT). The family spent summers on Block Island, off of Rhode Island, where Pough developed a passion for birds. At the age of ten, his first summer there, "a sandhill crane landed on the island, and when some hunters shot it I was so upset I ran at them, screaming," he told Frank Graham of *Audubon*. He studied at MIT, where he spent free time birding, and graduated with a B.S. in chemical engineering in 1926. After a year at Harvard University studying Oriental art, Pough obtained a job in Port Arthur, Texas, at a sulfuric acid plant. He worked at night so that he could spend his days birding along the Gulf of Mexico. He later traveled to Europe to study birds and art, then worked at a foundry in St. Louis, and in 1932, with his brother Harold, bought and rebuilt a bankrupt photographic equipment company in Philadelphia.

Pough continued to devote his leisure time to birds. One weekend, he invited the young woman he was courting, Moira Flannery, on a birdwatching expedition. They spent a day on the Jersey shore, during which he was proud to point out the droppings of many different species of birds of prey. As they drove home, Moira pointed at a pile of manure on the streets of Newark and pronounced it "elephant." Indeed, a couple of blocks later they came up behind a troop of elephants walking from a circus train. The couple's affection was sealed, and soon they married.

In the early 1930s, Pough read a study on goshawks that cited data from a large number of the dead birds collected from Hawk Mountain in eastern Pennsylvania. Curious about why there would be so many dead goshawks from that area, he went to investigate and found that hunters came from around the region to shoot hawks as they migrated south along the Kittatinny Ridge. He wrote an article and was invited to talk at a New York City meeting of the Audubon Society and other bird protection groups. Heiress and conservationist ROSALIE EDGE was in the audience, and soon she decided that the only way to stop the slaughter was to buy the mountain and preserve it as a wildlife sanctuary. Pough helped facilitate the deal in 1934.

Audubon Society leaders, with whom he worked on the Hawk Mountain sale, were impressed with Pough and offered him a position on the organization's research staff in 1936. He remained with the Audubon Society for 12 years, focusing on "persecuted species," especially hawks and owls. He was insistent that predator control programs were misguided; "Do away with predators and you destroy the balance of nature," he told the *New Yorker.* Pough was instrumental in prosecuting illegal traffickers of wild bird feathers and strengthening regulations on the use of these feathers by the millinery industry—an issue that the Audubon Society activists had first tackled in the early 1900s and that resurged as plumed hats became popular again in the late 1930s. He was also one of the early voices to decry dichlordiphenyltrichlor (DDT), which was first used during World War II. He told the *New Yorker* in 1945 that "if DDT should ever be used widely and without care, we would have a country without freshwater fish, serpents, frogs, and most of the birds we have now." In 1946, Pough published the first of what would be his three Audubon bird guides (*Audubon Bird Guide*, 1946; *Audubon Water Bird Guide*, 1951; and *Audubon Western Bird Guide*, 1957), concise field guides that show how each bird fits into its ecosystem.

In 1948, Pough became chair of the Department of Conservation and General Ecology of the American Museum of Natural History. He was charged with creating an exhibit called the Hall of North American Forests, the preparation of which allowed him to visit forests throughout the continent. He convinced the museum to buy Great Gull Island off of Connecticut, for sale by the federal government, and to establish a research center for terns there. While at the mu-

seum, he also came up with a way to help the Cahow (also known as the Bermuda petrel), birds that were so rare they were thought to be extinct. Cahows were found on the islets off of Bermuda, but they were threatened by larger tropicbirds, which killed the cahows' young and took over their nests. Pough suggested a device that would exclude birds larger than the cahows from entering the nests; the concept was put into practice and proved successful. Pough remained with the museum until 1956, when he was fired for his attempts to involve the museum in conservation battles, including the collaborative conservationist effort to stop the Echo Park Dam in Dinosaur National Monument and another fight against the construction of a highway through Van Cortlandt Park.

During the early 1950s Pough raised funds to buy the Corkscrew Swamp in Florida as one of the Audubon Society's system of private sanctuaries. This and his previous experiences with Great Gull Island and Hawk Mountain had convinced Pough of the efficaciousness of preserving ecologically valuable parcels of land by purchasing them. Neither the Audubon Society nor the museum, however, was interested in adding land acquisition to their missions. Pough began promoting the idea among garden clubs and the Ecologists' Union, and it caught on quickly. He was appointed chair of the Ecologists' Union's Committee on the Preservation of Natural Conditions. He convinced the Ecologists' Union in 1951 to change its name to The Nature Conservancy, the name of a similar organization in Great Britain, and he successfully solicited a donation of $100,000 from *Reader's Digest* founder Lila A. Wallace to seed a revolving fund that would provide down payment loans to local preservationists who needed to act quickly to buy ecologically desirable land.

Pough became president of The Nature Conservancy in 1953. He served in this capacity until 1956, guiding it through its first purchase of land in 1955, the Sunken Forest on Fire Island, for which ecologist OAKLEIGH THORNE II raised much of the necessary funds. This was the first in what has become the largest privately owned system of nature preserves in the world. Pough urged The Nature Conservancy to remain nonpolitical and steer clear of controversial issues, so that it could focus solely on the purchase of private lands for the purpose of conserving habitats and wildlife that represent the diversity of life on earth.

Pough remained involved with The Nature Conservancy as he founded the Natural Area Council in 1957, an umbrella organization for preservation groups. Kathrine Ordway, heiress of the Minnesota Mining and Manufacturing Company, provided the initial funds for another organization that he founded, the Open Space Institute, which worked to convince wealthy landowners to donate ecologically rich plots of 20 acres or more for preservation. Pough's organizations approached their goal creatively, discouraging large-scale development even when they did not have the funds to purchase entire areas, by "checkerboarding"—buying interspersed plots. Over the years Pough raised millions of dollars and participated in the preservation of thousands of acres of land.

During the late 1960s, Pough participated in the Thorne Ecological Institute's Seminars in Environmental Arts and Sciences. He is credited with convincing Army Corps of Engineers chief William Cassidy to give greater consideration to the environmental impact of its projects. The Corps subsequently established an Environmental Advisory Board, to which Pough was appointed.

Pough has served many other conservation organizations as well. He has been a member of so many organizations that in 1975 he paid about $1,500 per year in dues alone. He has held high-level offices for many of these organizations, including the U.S. section of the International Council for Bird Preservation, World Wildlife Fund–U.S., the Linnean Society, National Parks and Conservation Association, Defenders of Wildlife, the American Ornithologists' Union, Thorne Ecological Institute, and the Association for the Protection of the Adirondacks. He has received numerous recognitions and awards, including the silver medal of the Federation of Garden Clubs of New York State, the Conservation Award of American Motors Corporation, the Horace Marden Albright Scenic Preservation Medal of the American Scenic and Historic Preservation Society, the Frances K. Hutchinson Medal of the Garden Club of America, and the National Audubon Society Medal in 1981.

Pough resides in Chilmark, Massachusetts.

BIBLIOGRAPHY

Boyle, Robert, "An Earth-Saving Bulldozer that Runs on Money," *Sports Illustrated*, 1975; "Corner-Cornerer," *New Yorker*, 1954; Graham, Frank, "Dick Pough: Conservation's Ultimate Entrepreneur," *Audubon*, 1984; Stroud, Richard, *National Leaders of American Conservation*, 1985.

Powell, John Wesley

(March 24, 1834–September 23, 1902)
Explorer, Director of U.S. Geological Survey

Explorer John Wesley Powell made the country's first official surveys of the Rocky Mountain region and is especially remembered as leader of the first river trip through the Grand Canyon. The carefully drawn maps and surveys that his teams produced were the first of the topographic maps that the U.S. Geological Survey is known for today. Powell tried to promote orderly and appropriate settlement of arid western lands by designating certain mineral-rich areas for mining and others, with better access to water, for irrigated agriculture. Today, Powell is cited as one of the original promoters of environmentally appropriate land use.

John Wesley Powell was born on March 24, 1834, in Mount Morris, New York. His father, Joseph Powell, was a Methodist preacher whose strong abolitionist sentiments made the family a target for violent advocates of slavery. Powell was kept home from school to protect him. This seclusion, plus the family's frequent moves—always westward, finally ending up in Illinois—forced Powell to educate himself as he could. While the family lived in Ohio, an elderly neighbor introduced Powell to the marvels of natural history. Inspired, Powell began to rove the land, observing plants and animals and collecting specimens. Once he turned 18 years old, Powell became a country schoolteacher and enrolled at Wheaton College in Illinois. Always more knowledgeable about scientific matters than his professors, Powell also took courses at Oberlin and Jacksonville colleges but never stayed at any school long

enough to earn a degree. Powell did exert an influence on academia, however; historians credit him with introducing scientific curricula to college-level education. His teaching jobs suited him well, for they allowed him long vacations for explorations of the midwestern prairies and the Mississippi and Ohio Rivers. Powell joined the Illinois State Natural History Society when he was 20 years old and became its secretary at age 27.

In 1861, Powell joined the Illinois Volunteer Infantry, mobilized for the Civil War. He took a one-week leave of absence in 1862 to marry Emma Dean, who accompanied her new husband to the

John Wesley Powell (Fotos International/Munawar Hosain/Archive Photos)

warfront and stayed with him until the war ended. Major Powell was an enthusiastic, effective military leader but also found time to explore and collect fossils. Powell returned from the war one-armed, having lost his right arm during the Battle of Shiloh.

This handicap did not dissuade him from renewing his natural history work. Powell became a lecturer in geology at Illinois Normal University and was named curator of the Illinois Natural History Museum in 1865. In the summer of 1867, he organized the first of his many expeditions: he took a group of 16 students and amateur naturalists to the Rocky Mountains. The United States Army provided the group with rations, and the Smithsonian Institution lent them equipment in return for data collection. Later expeditions continued to receive support from the Smithsonian and from other academic institutions.

Powell's most famous trip took place in 1869. He and a group of 11 men were the first Whites to descend a 900-mile stretch of the Colorado and Green Rivers. They spent almost three months in the Grand Canyon, victims of wild torrents, nearly starving in the arid heat. Three of their party climbed out of the three-quarter-mile-deep canyon but were killed, just as they reached the canyon's edge, by Indians who mistook them for a band of at-large rapists. When Powell's small ragged group emerged from the canyon at the end of the summer, they became famous, and Powell had no trouble obtaining funds for later expeditions.

Powell's travels through the West convinced him that the arid West was a delicate place that could be ruined if settlement was not planned carefully. The 1862 Homestead Act encouraged westward expansion by granting settlers titles to plots of 160 acres, which they could use as they saw fit. Powell saw the beginnings of a disaster: land inappropriate for agriculture was being ruined by erosion, and the lives and dreams of western immigrants were being broken when farms and ranches failed owing to lack of rain and erosion of topsoil. Powell, named director of the U.S. Geographical and Geological Survey of the Rocky Mountain Region in 1875, recommended officially that no land in the West (except for a few areas in California, Oregon, and Washington) be farmed except when irrigation was possible. Powell's 1878 *Report on the Lands of the Arid Region of the United States* explained his findings and recommendations. Powell asked Congress to pass bills calling for limits to the size of farms and ranches and for the formation of cooperative irrigation associations. Although his intention was to protect westerners from monopolization by large land-owners and assure community control of irrigation, these ideas scared western landowners, who feared government intrusion.

Yet Powell continued to find favor in Washington. The National Academy of Science asked Congress in 1878 to fund extensive land use surveys of the West, and Congress assented. Powell's friend and colleague, Clarence King, was appointed director of the new U.S. Geological Survey (USGS), and Powell became its director when King died in 1884. Under Powell, the USGS began publishing the topographical maps it still puts out today. It also insisted on applying its scientific findings in land use policy.

After a terrible year for western farmers and ranchers in 1886, in which droughts parched crops and then tremendous blizzards killed livestock, Congress in 1888 passed a bill written by Powell calling for a survey of irrigable lands. In order to prevent wealthy speculators from grabbing all the potentially irrigable lands, Powell also succeeded in blocking grants of titles to all claims filed after the 1888 Irrigation Survey Act. But Powell's success was short-lived. He had alienated many different parties in the West: large landowners who did not like his proposal to limit ranch size, small landowners because Powell had said that only large ranches would be economically viable in arid regions, and real estate speculators, among others. By the early 1890s, Powell's proposals for careful, well-planned land-development had been defeated by his many enemies in the West.

In 1894, Powell retired from the USGS and returned to direct the Smithsonian's Bureau of Ethnology, where he had served as director from 1878 till 1884. As well known as Powell was for his revolutionary ideas on land use, he was also acclaimed for his many studies of North American Indian ethnology. Powell died in Haven, Maine, from a cerebral hemorrhage on September 23, 1902.

BIBLIOGRAPHY

Darrah, William, *Powell of the Colorado*, 1951; Stegner, Wallace, *Beyond the Hundredth Meridian: John Wesley Powell and the Second Opening of the West*, 1954; Strong, Douglas, *Dreamers & Defenders: American Conservationists*, 1988.

Pritchard, Paul C.

(August 27, 1944–)
President of National Parks and Conservation Association, Founder and President of National Park Trust

Paul C. Pritchard has been instrumental in the establishment of more than half of the national parks in the United States and has served as an international adviser on park formation. Founder and first president of the National Park Trust, he also served for 17 years as president of the National Parks and Conservation Association. His achievements in environmental conservation include, among others, creation of the Climate Institute and Friends of China's National Parks and the addition of the Alaskan national parks, as well as extensive writing on the environment. He draws his commitment and energy from the environmental challenges currently facing humanity worldwide. He believes that people "must act both locally and globally, just as we treat our entire bodies with one sense of respect."

Paul C. Pritchard was born on August 27, 1944, in Huntington, West Virginia, son of Eason G. Pritchard and Stella N. Pritchard. His love of nature developed during a boyhood in the hills of Huntington and later on the plains of Kansas City, where nature, he has said, "always seemed essential to our daily life." His family regularly made time for camping, hiking, a flower garden, and Boy Scout activities, and his grandparents' farm provided a classroom in life. He attended the University of Missouri at Columbia, where he earned a B.A. in humanities in 1966. From February 1967 to December 1968, he served as an Intelligence Officer in the United States Army. He returned to college and earned a master's degree in the science of planning from the University of Tennessee, Knoxville, in June 1971.

In 1971, after completing his graduate studies, Pritchard worked for the Office of the Governor of the State of Georgia, first as state transportation coordinator, later as chief of natural resources planning. From that time on, his career increasingly focused on the conservation of natural resources. In 1974, he joined the U.S. Department of Commerce as Pacific region coordinator for the National Oceanic and Atmospheric Administration (NOAA), where he was responsible for preservation grants. There he also supervised research for the five Pacific states and two territories for coastal zone management programs. That responsibility included the funding and creation of the nation's first coastal zone program, in Washington State tidal waters, and of the first estuarine sanctuary at Coos Bay, Oregon.

Pritchard left NOAA and returned to his native Appalachia in 1975 to become the first full-time executive director of the Appalachian Trail Conference, a confederation of over 50 hiking clubs that built the country's first recreation footpath, originally laid out in 1925. Though it is now part of the National Park system, the Appalachian Trail, extending from Maine to Georgia, is still maintained by the conference. During his term as executive director responsible for coordination of trail management, he raised over $1 million for acquisitions and established a citizen membership of some 80,000 members.

From 1977 to 1980, Pritchard was deputy director of the U.S. Department of Interior, overseeing the Bureau of Outdoor Recreation/Heritage Conservation and Recreation Service. Pritchard influenced the National Park system's acquisition of the Alaskan parks, including Denali and Glacier Bay. His dedication to the preservation of the land led him to the National Parks and Conservation Association (NPCA), where he served as president from 1980 through 1996. NPCA is a citizen-supported organization dedicated to protecting and enhancing the National Park system of the United States for present and future generations. Its objectives are realized mainly through public education and outreach. While guiding NPCA, he increased the association's membership from some 23,000 to over 500,000, at the same time expanding its budget to nearly $19 million annually. In 1990, he was instrumental in organizing the first Earth Day March for Parks, which remains an annual NPCA event. At its height, NPCA encouraged the development of more than 1,000 park events around the world, including a Russian march.

In 1983, Pritchard founded the National Park Trust (NPT) to acquire holdings from willing sellers who own land inside national and state parks or refuges. An organization of citizen members, which neither seeks nor accepts federal funds, it is dedicated exclusively to preserving and protecting the country's park lands, wildlife, and historic monuments. State and federal agencies contact the trust with projects that have no federal funding, and NPT selects those it is able to help. It is the only charitable group authorized by Congress to own an entire unit of the National Park system, the Tallgrass Prairie National Preserve in Kansas.

Concurrent with his work at both NPCA and NPT, Pritchard established the Climate Institute, serving as its founding chairman from 1986 until 1988. He remains a board member and chair emeritus. The institute's mission is to protect the balance between climate and life on earth. It is an international leader, interfacing with scientists and policy leaders concerned with global climate change and protection of the ozone layer. In his work for the institute, he has played a significant role in building cooperation among U.S. environmental organizations, through such groups as the National Resources Council of America, which he chaired from 1987 to 1989.

Author of more than 100 articles on the environment, Pritchard contributed to the NPCA book, *National Parks in Crisis*, published in 1982. He edited the 1985 NPCA book *Views of the Green*, a collection of articles by European and North American park experts on the challenges they were addressing. He was asked to write the definition of national parks for the Houghton-Mifflin *Encyclopedia of the Environment*. An adviser to National Geographic Society Books, he has two books slated for publication in 2000 by the society: *Parks of the World* and *The Philosophy of America's Parks*, which he edited.

Pritchard has actively supported the development of state heritage programs. During the course of his career, he also has been influential in the acquisition of some 200 National Park Service units, which include national monuments, battlefields, seashores, and historic sites as well as parks. He is a member of the Council of Editorial Advisors of the National Center for Nonprofit Boards. Among his international projects, Pritchard has consulted on national park formation and management for Canada, Ireland, and China. He is a founder of the Friends of China's National Parks, a private group of U.S. citizens. The recipient of numerous awards, in 1986 he was honored as the first American to receive the Albert Schweitzer Prize for Humanitarianism, awarded by Johns Hopkins University and the Alexander Von Humbolt Foundation of Hamburg, Germany.

In 1999, he was appointed to the Jefferson County, West Virginia, Public Service District Board of Directors. Pritchard lives on a farm near Shepherdstown, West Virginia, with his wife, Susan, and their two youngest sons, Stephen and Christopher. He also has two grown children, Robin and Marcus.

BIBLIOGRAPHY

"Park Service Report Finds Agency Ailing," *National Parks*, 1992; "Paul C. Pritchard, President, National Park Trust," http://www.park-trust.org/paul01.html; Pritchard, Paul C., "U.S. National Parks," *Encyclopedia of the Environment*, Ruth Eblen and William Eblen, eds., 1994; Stroud, Richard, *National Leaders of American Conservation*, 1985.

Pulido, Laura

(January 26, 1962–)
Geographer

Laura Pulido's career has focused on issues of race, social movements, and social justice, especially environmental racism and justice. As a geography professor at the University of Southern California, she studies structures and origins of oppression and how people organize themselves to resist exploitive situations. She describes her research as "very regional," covering the Southwest and southern California in particular. She is active in the Los Angeles community, having served as the Los Angeles city commissioner for the Department of Environmental Affairs. She is also a longtime member of the Labor/Community Strategy Center. Over the last few years she has been involved with labor struggles and has been active in a statewide California network to create a Latino Left.

Laura Pulido was born on January 26, 1962, in Los Angeles, California, to Louis and Berta Pulido. Growing up in Los Angeles, she expected to become a housewife because that was all she saw the women around her doing. She became dissatisfied with her high school and left without graduating but later earned her equivalency diploma. Ambitious and intellectually curious, she became a first-generation college student at Golden West College. She majored in social work, a career path that she had seen other women take, and received her associate's degree in 1983. She then enrolled at California State University at Fresno, still a social work major. However, while at Cal State she took a geography class, and suddenly she was hooked.

She was fascinated that the field of study provided her with answers to questions she had always had, questions of society and environment—Why were all the African Americans in Watts and all the Mexicans in East LA? Why were there no forests in LA? What forces were shaping Los Angeles and the lives of its residents?

Pulido changed her major from social work to geography, and she excelled in her studies. After receiving a B.S in 1985, she went to the University of Wisconsin at Madison. Far from home, deep in winter darkness and snow, she was homesick and determined to get back to Los Angeles and "never leave again." However, among the benefits that came from attending the University of Wisconsin was the influence of Dr. Diana Liverman, who urged her to continue her studies in urban planning at the University of California at Los Angeles (UCLA). After receiving her master's degree from Wisconsin in 1987, she enrolled in the doctoral program at UCLA. She received her Ph.D. in urban planning in 1991.

At UCLA Pulido grew to be passionate about politics, especially the issue of racism and how that shapes the city. Rather than studying specific incidents or events as a historian would, she says she looks for a broader understanding of what racism is, what economic exploitation is, and "how the experience is played out on the landscape." For example, the urban structure of Los Angeles has "clean suburban areas and Black and Brown spaces" in the industrial zones. Suburbanization was essentially a massive subsidy for Whites to move out of the city, while nonwhites were refused mortgages and equal housing opportunities even up to 1970. Even now, Pulido says, nonwhites do not choose to live in industrial areas of Los Angeles, but economic forces and racism force them to stay in that area.

This has led her to explore how groups react to oppression and White privilege. The development of a "people of color" identity is crucial in the fight against environmental racism. It empowers people to understand what is going on and why, when their communities are neglected or degraded by government policies. However, just as important as self-concept is the way nonwhites are perceived by the dominant society. She has studied the concept of "ecological legitimacy" in the Ganados de Valle case in New Mexico, in which working-class Hispanos were told by the government that they could no longer graze their sheep on an open field, even after the community came up with a plan to manage the land and maintain their animals. To the New Mexican government land managers, the group's plan lacked "legitimacy" because the people were poor, Hispanos, lacked formal education, and because they had an economic interest in using the land. The New Mexico Department of Fish and Game responded to the interests of hunters and mainstream environmentalists but not those of Ganados de Valle. Pulido's book on this subject, *Environmentalism and Economic Justice: Two Chicano Struggles in the Southwest*, was published by the University of Arizona Press in 1996. The second struggle described in this book is of the United Farm Workers (UFW), led by César Chávez and Dolores Huerta, who began their fight to improve health and sanitation conditions for migrant farmworkers in the early 1960s. The book discusses UFW's pesticide campaign from 1967 to 1975.

Pulido's latest work studies the radical political movements of African Americans, Asian Americans, and Chicanos in Los Angeles from 1968 to 1978. "The nation as a whole was in political turmoil," she says. However, history seems to give more attention to the "White Left" movement, neglecting the issues that distinguish the "Left of Color." Black, Yellow, and Brown Power radical movements worked to define and empower people beyond the limits set by White society. The "left of color" groups sought empowering solutions through class-based and materialist politics.

Pulido currently is planning a book on left political consciousness among women of color. She resides in Los Angeles.

BIBLIOGRAPHY

Pulido, Laura, "A Critical Review of the Methodology of Environmental Racism Research" *Antipode*, 1996; Pulido, Laura, "Development of the 'People of Color' Identity in the Environmental Justice Movement of the Southwestern U.S.," *Socialist Review*, 1998; Pulido, Laura, "Ecological Legitimacy and Cultural Essentialism: Hispano Grazing in the Southwest," *Capitalism, Nature, Socialism*, 1996.

Raven, Peter

(June 13, 1936–)
Botanist, Director of the Missouri Botanical Garden

A passionate advocate of preserving biodiversity, Peter Raven is a scientist and a highly public activist: in addition to his botanical research and teaching, he spends time making speeches and raising public awareness about the current dangerous rates of plant extinction. His belief that all the fundamental elements of human life, from growing food to building shelter, have their basis in plant life has led him on a quest to explore and maintain the diversity of the plant world. As director of the Missouri Botanical Garden, he transformed a small regional botanical garden suffering from neglect into one of the world's leading research centers specializing in tropical plants. He has become a widely recognized botanist and is credited as the cofounder of the field of coevolution. Among his extensive list of publications is *The Biology of Plants* (1969), which has become a standard textbook at colleges and universities across the country.

Peter Hamilton Raven was born in Shanghai, China, on June 13, 1936, the only child of Walter and Isabelle (Breen) Raven. Not long after he was born, his family moved to San Francisco, where Raven grew up. Precocious from the start, Raven's interest in the natural world had him searching for insects in Golden Gate Park at the age of five. By eight years of age he became a student member of the California Academy of Sciences, the youngest member ever. His scope widened to include plants, and at 12 he joined the Sierra Club and went on plant-collecting expeditions with the club. He attended Catholic school, where he learned Latin and Greek—and continued to find rare and interesting plants. When he was 14 his first scientific paper was published. At 15 he discovered a type of manzanita that botanists had presumed to be extinct. This subspecies, Ravenii, was later named for him.

After high school, Raven entered the University of California at Berkeley, where he received his bachelor's degree in botany in 1957. Recruited by botanists at the University of California at Los Angeles (UCLA) who were impressed with his work, Raven enrolled at UCLA and earned his Ph.D. in plant biology in 1960. During his years as a graduate student, Raven made his first trip to the tropics to conduct research in Colombia. After graduate school he took a postdoctoral fellowship at the Natural History Museum in London, then worked for a year at the Rancho Santa Ana Botanic Garden in Claremont, California. Following that he accepted a job as assistant professor at Stanford University, where he taught biology until 1971. While at Stanford, Raven collaborated with PAUL EHRLICH, a population biologist who wrote *The Population Bomb* (1968). Together they worked on unraveling the complexities of plant-insect interactions and described a phenomenon they called coevolution—a groundbreaking idea that has since gained widespread recognition. In addition to his responsibilities at Stanford, Raven wrote several highly popular textbooks. He had noticed that fewer and fewer botany courses were being taught during the 1960s and wanted to halt this decline and bring the study of plants up to date. His *Biology of Plants* (1969) textbook, cowritten with two other biologists, went on to become the best-selling botany textbook in the country.

At the age of 35, Raven was appointed director of the Missouri Botanical Garden in St. Louis, Missouri, in 1971. The oldest botanical garden in the United States, it was little more than a modest ten acres when Raven took over, even though the garden actually controlled a 79-acre estate. Under Raven's directorship, the Missouri Botanical Garden has grown to fill the entire 79 acres with a wide variety of plant life in many magnificent gardens. When Raven came to the Missouri Botanical Garden, the facility had only 85 people on staff, including three scientists with doctorates. It now employs 350 people—with 55 Ph.D. staff scientists, some of whom live overseas and conduct field research. Raven and his staff have formed one of the world's most active centers of botanical research where they work toward a vitally important scientific goal: to catalog as many of the world's remaining tropical plants as possible. The scientists there identify about 200 new plant species every year in a race against the rapid extinction rates that could erase one of every four plants within the next three decades. Raven is motivated by his belief that as species disappear, the world becomes less interesting and less prosperous—and he hopes to spread the word that plants have value beyond just beautifying our gardens. For example, the future of agriculture lies in the genetic material in the food plants currently in existence. Also, certain plants hold potential as antidisease agents. Over one-fourth of the world's prescription medicines contain a plant-derived ingredient, and preserving biodiversity is the only way to protect millions of other plants that may also have healing properties. Working in the equatorial rain forests of Africa, Asia, and Latin America, the garden's researchers study and collect plants previously unknown to science. They send specimens back to the garden's herbarium, which now contains over five million specimens. In addition to tropical research, the garden also hosts the Center for Plant Conservation, a consortium of botanical gardens dedicated to reintroducing native plant species in the United States.

Raven's success in his growing endeavors has come in part from his talent at networking. Interacting with numerous institutions, giving speeches, lobbying members of Congress—all of his outreach and advocacy efforts have led some to call him an "eco-administrator" or a "biopolitician." Because of his high profile, he has raised public awareness of the importance of preserving tropical rain forests and biodiversity. And he does not stop at discussing the biological consequences of environmental degradation. Believing that rampant consumerism in the Western world, widespread poverty, and Third World debt preclude any easy solutions for preserving the environment, he urges governments and policy makers to take into account the planet's limited resources. The Missouri Botanical Garden has been actively helping poorer countries that contain some of the most critically threatened areas of diversity, providing training and support in making plans for sustainable use of resources.

Peter Raven is a member of many professional organizations, including the national academies of science in at least ten countries. He has written over 500 scientific papers and 19 books. He received the John D. and Catherine T. MacArthur Foundation Fellowship in 1985; the International Prize for Biology, awarded by Japan in 1986; the Volvo Environment Prize in 1992; and a National Wildlife Federation National Conservation Achievement Award. Raven is also a professor of botany at Washington University in St. Louis. He lives with his wife, Kate, in St. Louis, Missouri.

BIBLIOGRAPHY

Beardsley, Tim, "Defender of the Plant Kingdom," *Scientific American*, 1999; Lawren, Bill, "Six Scientists Who May Save the World," *Omni*, 1987; McClintock, Jack, "Peter the Great," *Discover*, 1999; "Missouri Botanical Garden," http://www.mobot.org.

Reilly, William K.

(January 26, 1940–)
Administrator of the Environmental Protection Agency, President of the Conservation Foundation, President of the World Wildlife Fund

Director of the Environmental Protection Agency (EPA) under Pres. George Bush, William Reilly advocated for environmental issues in an administration never fully committed to protecting the environment. Reilly was the first "professional environmentalist" to serve as the agency's head and helped strengthen the Clean Air Act. He served as a check on others in the administration pushing to weaken environmental regulations, including those who would have stripped away wetlands protection. Before his tenure as EPA head, Reilly worked to promote conservation efforts through a variety of groups, including the President's Council on Environmental Quality and the World Wildlife Fund.

William Kane Reilly was born on January 26, 1940, in Decatur, Illinois. He graduated from Yale University in 1962 with a B.A. in history and from Harvard Law School with a J.D. in 1965. He served in the United States Army from 1965 to 1966. In 1971 he received an M.S. in urban planning from Columbia University. He served as the associate director of the Urban Policy Center from 1968 to 1970 and as executive director of the

William K. Reilly (Reuters/Jamil Ismail/Archive Photos)

Rockefeller Task Force on Land Use and Urban Growth from 1972 to 1973.

Reilly's involvement in environmental policy began in 1970, when he joined the President's Council on Environmental Quality, serving as a senior staff member until 1972. In 1973 he became president of the Conservation Foundation, and in 1985, president of the World Wildlife Fund. One of his most important contributions during this period was to facilitate an alliance of the two groups. World Wildlife Fund had particular expertise in the areas of fieldwork and the natural sciences, while the Conservation Foundation was strong in the areas of social science and policy advocacy; joining the groups made it possible to collaborate on pressing issues of international conservation. The groups at first maintained separate operating structures, but by 1991 they had consolidated as a single organization, under the name World Wildlife Fund, Inc.

In 1988, George Bush was elected president of the United States, with promises to be the "environmental president." Conservationists cheered when he appointed Reilly to head the Environmental Protection Agency, because Reilly was an active and respected member of the environmentalist community. In the early days of the administration, it looked as though Bush might uphold his campaign commitment. One of Reilly's first acts was to halt development of the Denver Water Board's Two Forks dam project. Environmentalists had fought hard to stop the large project, proposed for the South Platte River west of the Denver metropolitan area, because of its impact on the surrounding canyon and the wildlife in and around the river, but the EPA under President Reagan had approved the dam. In early 1989, Reilly vetoed the project, angering western Republicans, who formed one of Bush's key constituencies. Reilly also helped craft 1989's Clean Air Act, which required tough new standards on sulfur dioxide emissions to reduce acid rain. Reilly's position, though, was difficult. He described himself as "the man in the middle," caught between the environmental movement and corporate and other groups demanding environmental deregulation. Environmentalists felt Bush and Reilly were not tough enough on polluters and criticized weaknesses in the Clean Air Act. Business interests felt the Republican president should be more aggressive in following the footsteps of President Reagan's deregulatory path. Bush felt he was not getting credit for his environmental efforts and gradually moved away from his campaign promises. Though Reilly maintained a close personal relationship with the president, he clashed with other administration officials, including Bush's powerful chief of staff, John Sununu. Reilly's balancing act came apart visibly during the Earth Summit in Rio de Janeiro, Brazil, in 1992. Reilly led the U.S. delegation but was unable to persuade the administration to sign the biodiversity treaty. More than 100 other countries, including U.S. allies Germany, Canada, and Britain, approved the treaty, but Bush refused to sign on the grounds that the standards could cost U.S. jobs. Reilly's conflict with the administration during the summit made national headlines.

Reilly was also at odds with vice president Dan Quayle, who headed the Council on Competitiveness, a group of wealthy corporate leaders pushing for relaxed pollution standards and blocking new EPA regulations. Two key areas of disagreement between Quayle's group and Reilly were relaxed emissions standards in relatively unpolluted areas and weakening wetlands protection in Alaska. During the election of 1992, it looked as though Reilly would lose, as the administration pushed the EPA to approve the council's recommendations. But when President Bush lost the election, Reilly stalled the proposals, citing the likelihood that the Clinton administration would simply reverse any new regulations put into place in the closing days of 1992. Reilly's own assessment of the successes of his tenure at EPA includes efforts to restore the health of Chesapeake Bay, the Great Lakes, and the Gulf of Mexico; strengthening the role of science in the agency; and integration of environmental and economic concerns.

Since leaving the EPA in January 1993, Reilly has been active in a number of foundations, corporations, and environmental organizations. He serves on the board of directors of the DuPont Corporation, where he has urged the company to support green initiatives. He is the president and chief executive officer of Aqua International partners, a group that finances water projects in developing nations. He serves on the boards of several conservation organizations, including the World Wildlife Fund, the American Farmland Trust, the David and Lucile Packard Foundation, and the Presidio Trust. Reilly resides in San Francisco.

BIBLIOGRAPHY

Dunne, Nancy, "Business and the Environment: Complacency Breeds Contempt," *Financial Times of London*, 1992; "The Presidio Trust," www.presidiotrust.com; Raeburn, Paul, "Bush's EPA Chief Tells of Last-Minute Efforts to Achieve Green Agenda," *Los Angeles Times*, 1993.

Reisner, Marc

(September 14, 1948–July 21, 2000)
Writer, Consultant

Marc Reisner is known for his nonfiction, environmentally focused books. His most famous, *Cadillac Desert* (1986), dramatizes the environmental degradation brought about in the dry American West by federal dam-building policies and by short-sighted water management. He was also an environmental consultant, advocating practical solutions to managing western water resources.

Marc P. Reisner was born on September 14, 1948, to Konrad and Else Reisner in Minneapolis, Minnesota. His father was a family-service executive and a lawyer, his mother a scriptwriter. He attended Earlham College in Indiana, graduating with a B.A. in 1970. He began his career in 1970 as a scriptwriter for environmental telethons and as an environmental lobbyist. In 1972, he took a position with the Natural Resources Defense

Council (NRDC) in New York City as a staff writer. He continued working for the NRDC until 1979, when he became a full-time writer, lecturer, and environmental consultant.

Reisner authored several important, environmentally focused works of nonfiction. He was best known for his first book, *Cadillac Desert: The American West and Its Disappearing Water* (1986), in which he examined the environmental, social, and economic consequences of federal dam-building practices in the American West. The book, which Reisner began researching in 1979 with an Alicia Patterson Journalism Fellowship, describes the competition and dam-building one-upmanship that existed between the Bureau of Reclamation and the Army Corps of Engineers, and the resultant dam building frenzy. Water management in the West, Reisner concluded, has been politically motivated and short sighted and was ecologically unsustainable. *Cadillac Desert* points out the reckless overuse of water resources by city residents and farmers and the polluted waterways and soils that have resulted. It also contains a chapter that recounts the crooked dealings undertaken by Los Angeles County in wresting the water of Owens Lake away from the residents of the Owens Valley and in pumping that water over 200 miles of desert to quench the thirst of Los Angeles's growing suburbs. *Cadillac Desert* was ranked 61st in Modern Library's list of the 100 most notable nonfiction works in English published during the 20th century. It was made into a four-hour Public Broadcasting Service documentary that first aired in 1997.

Reisner's second book also deals with the subject of water in the West. Cowritten with Sarah Bates, *Overtapped Oasis* (1990) is an in-depth analysis of the water allocation network in the American West. The problem, according to Reisner and Bates, is not so much one of a shortage of water as it is of the misallocation and inefficient use of water. This inefficient use is the result of the perplexing array of outdated irrigation practices, federal subsidy programs, state water codes, and resistance to reform. Reisner's third book, *Game Wars: The Undercover Pursuit of Wildlife Poachers*, is also focused on an environmental topic, a U.S. Fish and Wildlife Service agent and his efforts to protect such animals as alligators, walruses, and waterfowl.

After publishing *Cadillac Desert* in 1986, Reisner's views on the culpability of agriculture in the degradation of the western lands changed. Whereas he originally believed that much of the environmental degradation was the fault of farmers and was their responsibility to fix, he later came to believe that agriculture was a more preferable alternative than development (which, he felt, would be the inevitable result if farmers were kicked off of the land). Reisner preferred western farms to western suburbs. To restore balance, he said, three distinct areas need to be reformed: fishery restoration, underground water storage, and decision making. Reisner became an advocate for the free market water trading policies of water banking and the trading of water rights, holding that these policies could be utilized to create additional water supplies without building new dams.

Reisner was consulted by the Institute for Fisheries Resources, assisting in its efforts to remove antiquated and marginally useful dams in California. He was also a consultant to America's Farmland Trust, an organization campaigning to protect California farmland from encroaching development. Reisner originated the Rice Lands Habitat Partnership, which provides incentives to enhance waterfowl habitat on private land. In 1998 he became director of the Vidler Water Company, responsible for furthering the company's interests in wetlands restoration. And in the fall of 1999, he was Distinguished Visiting Professor in the Department of Geology at the University of California at Davis.

Reisner considered himself a writer, a conservationist, a consensus builder, and a deal maker. He received numerous awards for his literary and conservation accomplishments. He received a National Book Critics Circle nomination for *Cadillac Desert* in 1986, and in 1993 he was given the San Francisco Bay Institute annual conservation award. Reisner married Lawrie Mott in 1985, with whom he had two children, Ruthie and Margot. He died of cancer at his home in Marin County, California, on July 21, 2000.

BIBLIOGRAPHY

Associated Press, "'Cadillac Desert' author dies," *Denver Post*, 2000; Lesniak, James G., ed., *Contemporary Authors*, 1994; "Natural Resources Defense Council Profile," http://www.igc.apc.org/nrdc/nrdc/sitings/lookup/proreis.html; *Cadillac Desert, Water and the Transformation of Nature*, http://www.pbs.org/kteh/cadillac-desert/home.html.

Richards, Ellen Swallow

(December 3, 1842–March 30, 1911)
Sanitary Chemist

Ellen Swallow Richards is remembered as the first woman to graduate with a science degree from a university in the United States, for her help opening the heavy doors of science to other women, and for her groundbreaking innovations in the fields of urban and industrial sanitation, food safety, nutrition, and water purity. Many branches of the environmental sciences, including ecology, limnology, oceanography, marine biology, and sanitary chemistry, can be traced back to Richards's pioneering work.

Ellen Henrietta Swallow Richards was born on December 3, 1842, on a farm near Dunstable, Massachusetts. As both of her parents were trained as teachers, she studied at home until she was 17 years old, at which time her family moved to Westford, Massachusetts, so she could study at Westford Academy there. After several years of working at her family's small store and caring for her infirm mother, Richards attended the brand-new all-women's Vassar College in New York, from which she graduated in 1870 with a bachelor's degree. Her thirst for learning still strong, she applied to study at the all-male Massachusetts Institute of Technology (MIT). MIT accepted her as a "special student" but would not accept tuition

Ellen Swallow Richards (Library of Congress)

payment from her in order to reserve the option of expelling her should its experiment with a female student not work out. There she studied sanitary chemistry and nutrition, earning a master's degree in 1873. She had actually completed sufficient work for a Ph.D. but was not awarded one because hers would have been the first Ph.D. in chemistry to be granted by MIT, and MIT did not want a woman to earn one before a man.

Despite the ingrained sexism at MIT, Richards found a niche there. She made herself useful to professors, mending broken suspenders when necessary and astounding them with her brilliant observations in almost any field. Together with her professor, William Ripley Nichols, she designed the nation's first water purity tests, some of which are still used today. In the laboratory of her future husband, mineralogy professor Robert Richards,

she isolated the rare metal vanadium, a near-impossible feat even for mineralogists. She became an instructor of sanitary chemistry and established a Women's Lab where women trained for science-related careers. With the financial support of the Women's Education Association, she also established the marine biology laboratory in Hyannis, Massachusetts that eventually became Woods Hole.

Late-nineteenth-century cities were dirty and disease ridden, lacking clean air and water, sufficient sewage treatment, and garbage collection. Richards's daily walks to school through the filth of Boston led her to important realizations about transmission of disease and the importance of sanitation and hygiene. Richards urged families and city governments to take responsibility for their environment and pushed especially for greater education for women, who were usually those responsible for household sanitary conditions. At that time, it was widely held that women were not meant to study and that those women who did would get sick and become sterile. Richards felt the opposite, that if women did not learn at least enough to improve hygiene in their homes, their families would get sick. Much of her work from 1880 on was devoted to women and the environment and was supported by two major women's organizations of the time: the Women's Education Association and the Association of Collegiate Alumnae. Richards lectured widely to women's clubs, she taught science through a correspondence program whose students were mostly housewives, and she wrote 18 books, among them *The Chemistry of Cooking and Cleaning: A Manual for Housekeepers* (1882); *Food Materials and Their Adulturations* (1886); and *Sanitation in Daily Life* (1907).

Richards became a player in the turn-of-the-century debates on eugenics (the improvement of humanity through controlled breeding) by promoting a counterapproach: "euthenics," which she defined as the improvement of human intelligence and health through an improved environment. The science of euthenics metamorphosed until it became home economics, and today Richards is remembered as the mother of that field.

For several years in the 1880s, Richards worked as chief industrial chemist for the Boston Mutual Manufacturers Fire Insurance Combine, which insured factories. Factories at this time were tinderboxes, and safety precautions were rare. With the support of her friend and employer, the distinguished public figure Edward Atkinson, she drew up new requirements with which factories had to comply in order to take out insurance policies.

Richards was reluctant to marry, since marriage at that time usually entailed the woman's sacrificing her own interests so that she could take care of the household and help her husband. But her suitor, Professor Robert Richards, promised that she could continue her work. She did marry Richards in 1875, and the couple embarked on a plan to make their home into a model showcasing their ideas on healthy environments. The windows were full of green plants to produce oxygen, vents were installed to promote air circulation, and their stove was topped with a hood with a fan—a unique feature in those times.

Why is Ellen Swallow Richards not a household name, given the contributions she made toward public health? Her New England character endowed her with extreme modesty. She almost never allowed herself to be photographed and did not seek press coverage, except to further her projects. Her biographer, Robert Clarke, also points out that her work drew together many fields into a holistic view of the environment at a time when science was fragmenting into many separate microfields.

If Richards felt frustrated at the lack of acceptance of her revolutionary yet commonsense ideas, it never slowed her down. She lectured continuously throughout the United States and Europe during the last third of her life. Richards died on March 30, 1911, at her home in Jamaica Plain, Massachusetts, of heart failure.

BIBLIOGRAPHY

Breton, Mary Joy, *Women Pioneers for the Environment*, 1998; Clarke, Robert, *Ellen Swallow: The Woman Who Founded Ecology*, 1973; Cravens, Hamilton, "Establishing the Science of Nutrition at the U.S.D.A.: Ellen Swallow Richards and Her Allies," *Agricultural History*, 1990; Hunt, Caroline, *The Life of Ellen H. Richards*, 1912; Richards, Ellen Swallow, *The Art of Right Living*, 1904; Richards, Ellen Swallow, *Euthenics—The Science of Controllable Environment*, 1910.

Rifkin, Jeremy

(January 26, 1945–)
Cofounder and Director of the Foundation on Economic Trends, Author

Through his books, lectures, and television appearances, Jeremy Rifkin has emerged as one of the most conspicuous vocal dissenters of the scientific effort to manipulate and alter the genetic makeup of living organisms through biotechnology. Citing possible deleterious consequences to the environment, Rifkin has raised public awareness of the ecological and moral implications of biotechnology and has mobilized support in a campaign against genetic engineering. As cofounder and director of the Foundation on Economic Trends in Washington, D.C., Rifkin has formed coalitions, organized demonstrations, and initiated lawsuits against institutions that neglect to undertake environmental assessments of the risks involved in their biotechnology research. Criticized by some as "antiprogress," Rifkin believes it is time for society to reshape its ideas about the nature of scientific inquiry and to develop a new philosophy of science and technology that is sophisticated, intelligent, and above all, respectful of living things.

Jeremy Rifkin was born on January 26, 1945, in Denver, Colorado, to Milton and Vivette Rifkin and was raised in a middle-class setting in Chicago. He attended the Wharton School of Finance at the University of Pennsylvania, where he was elected president of his class and became known as a fluent and talented public speaker. He earned his bachelor's degree in 1967 in economics and then went on to graduate studies at the Fletcher School of Law and Diplomacy at Tufts University in Medford, Massachusetts, where he re-ceived his master's degree in international affairs in 1968. Through the late 1960s, Rifkin immersed himself in anti–Vietnam War activism, and in 1967 he helped organize the first national rally against the war. Antiwar activism and working in New York City for the Volunteers in Service to America occupied his time until he moved to Washington, D.C., in 1971. At that point he founded the People's Bicentennial Commission to promote alternative bicentennial ceremonies and activities that he hoped would reacquaint the country with revolutionary principles. Rifkin and his organization led protests against large corporations, called for a redistribution of wealth in the United States, and promoted social action.

When the bicentennial festivities died down in 1976, Rifkin redirected his activism efforts toward the business world. He cofounded with Randy Barber the People's Business Commission (renamed the Foundation on Economic Trends in 1977) in Washington, D.C., an organization committed to confronting corporate exploitation and urging reforms that would give workers more autonomy and a larger share of corporate profits. Rifkin's book *Own Your Own Job: Economic Democracy for Working Americans* (1977) expands on the threats posed by multinational corporations to fair labor practices.

Rifkin began gaining a reputation as a skilled popularizer of radical ideas, and he continued to fight for labor unions and union pension funds and for alternatives to the current economic system. He also

Jeremy Rifkin (Bettmann/Corbis)

began examining the implications of scientific research in genetic engineering and manipulation of recombinant deoxyribonucleic acid (DNA), becoming more and more alarmed at the exploitation of life forms through techniques such as cloning and selective breeding. With the discovery of technologies that allowed for gene splicing, chemical companies had begun to isolate genes from one organism and inject them directly into the genetic blueprint of another organism. Proponents claimed it was a quicker, cleaner way to develop new plants with special characteristics such as frost resistance or pesticide-tolerance. But Rifkin worried that biotechnology was outpacing society's ability to assess deleterious consequences to the environment or to identify negative moral or eco-

nomic impacts, and he began voicing his concerns in his characteristically activist approach, appearing on television programs and speaking in public. With the publication of his best-seller book, *Who Should Play God? The Artificial Creation of Life and What It Means for the Future of the Human Race* (1977), cowritten with Ted Howard, Rifkin emerged as one of the most important opponents of genetic engineering.

Unwilling to accept the biotechnology industry's assurance that it was not technology that was inherently bad, it was how it was used, Rifkin began broadcasting his view that technology is not neutral—it reflects the values of the culture. In the early 1980s, in a challenge to what he believed to be dangerous research in genetic manipulation, Rifkin began a series of lawsuits against the National Institutes of Health (NIH) and the University of California, Berkeley. Concerned over potential environmental hazards, Rifkin opposed the NIH's decision to allow university scientists to conduct experiments that would allow the release of genetically modified bacteria (designed to deter frost formation on potatoes) into the environment. The ensuing legal battles brought unprecedented attention to Rifkin, who continued to take to task biotech industries that neglected to undertake environmental assessments of risks involved in their research. Books that he wrote during this time also reflect his concerns. *Entropy* (1980), cowritten with Ted Howard, rejects the contemporary gospel of progress and places modern world problems such as resource depletion and conservation in the context of entropy—a concept holding that disorder increases to a maximum and leads to an ultimate state of inertia. Rifkin argues that unless modern society realizes that its resources are being unsustainably consumed and adjusts its definition of progress to include conserving resources and protecting the environment, a disastrous decline is bound to take over.

His next book, *Algeny* (1983), cowritten with Nicanor Perlas, became one of his most controversial. The term *algeny* refers to the process by which future scientists will be able to alter every aspect of living things. In this futuristic scenario, Rifkin envisions a time when parents will design their own offspring's genetic composition, a scientific paradigm of control that would replace the natural process of evolution. Sounding the alarm about the unforeseeable dangers inher-

ent in this new technological relationship with nature, *Algeny* pleads for a reevaluation of priorities and a rejection of the notion that "if it can be done, it should be done." The book was well-received by the popular press but drew criticism from many in the scientific community, who felt that Rifkin was "antiprogress" and was ignoring the potential benefits of biotechnology.

Through the latter half of the 1980s, Rifkin's work involved protesting military biological testing labs and establishing a coalition to oppose the U.S. government's decision to issue patents for new animal species created by gene splicing. He also formed the National Coalition Against Surrogacy in opposition of surrogate parenthood, arguing against increasing interference with natural biological reproduction processes. In 1989 he published *Time Wars: The Primary Conflict in Human History*, in which he links environmental devastation to the contemporary preoccupation with speed and competition. With society's attention span narrowed to the moment, there is no framework for thinking about sustainability, says Rifkin, and the frantic pace of modern life consumes resources at such an accelerated rate that nature has no chance to replenish itself.

His next effort to draw attention to unsustainable resource consumption came in his 1992 book *Beyond Beef: The Rise and Fall of the Cattle Culture*, an argument against beef production, which he demonstrates to be one of the world's most flagrant sources of ecosystem degradation. He points out that more grain is grown in the United States to feed cattle than to feed people, that beef production wastes water and energy, and that South American rain forests are destroyed to make room for cattle. In 1992 he coauthored *Voting Green: Your Complete Guide to Making Political Choices in the 90s* with his wife, Carol Grunewald Rifkin. The book publicizes politician's voting records on issues of environmental concern.

An engaging public speaker and prolific writer, Rifkin continues to advance his ideas publicly, hoping to engage society in a dialogue on issues such as genetic engineering. He and his wife, also a writer and activist, live near Washington, D.C.

BIBLIOGRAPHY

Otchet, Amy, "Jeremy Rifkin: Fears of a Brave New World," *UNESCO Courier,* 1998; Rifkin, Jeremy, *Biosphere Politics: A New Consciousness for a New Century,* 1991; Stix, Gary, "Dark Prophet of Biogenetics," *Scientific American,* 1997; Windstar Foundation's Choices for a Healthy Environment Speaker Video Series, *Jeremy Rifkin—Global Environmental Security: The Greenhouse Crisis* (video recording), 1989.

Robbins, John

(October 26, 1947–)
Founder of EarthSave International, Author

Best-selling author and lecturer John Robbins promotes a reduction in animal product consumption as a solution to a daunting number of environmental and human health problems and has become a leading advocate for healthy living through a plant-based diet. His meticulously researched *Diet for a New America* (1987) depicts some of the abuses of the meat and dairy industries—from environmental problems such as heavy use of pesticides and unsustainable consumption of water and energy, to the abuse of the animals themselves in their factory living conditions. Robbins also founded EarthSave International, a nonprofit organization that supports healthy food choices, preservation of the environment, and a more compassionate world. An eloquent public speaker, Robbins has received standing ovations at thousands of conferences and speaking engagements worldwide as he continues to explore the link between society's food habits and the health of the planet.

John Robbins was born on October 26, 1947, in Glendale, California, to Irma and Irvine Robbins. His father and his uncle, Burton Baskins, founded Baskin-Robbins, which would one day become the largest ice cream company in the world. John grew up with his two sisters in a mansion in Encino, California, with an ice-cream-cone-shaped swimming pool in the backyard. He spent his summers working at the ice cream plant, and for many years he was groomed for the task of taking over and running the family company. But at the age of 21 Robbins realized that he did not want to make money through a product that contributes to bad health, and he stopped working for Baskin-Robbins. He walked away from his family's fortune and turned down scholarships at Harvard, Stanford, and Yale because he did not want to remain in the company of a privileged few. Instead he enrolled at the University of California at Berkeley in 1965, where he immediately got involved in social activism—joining the free speech, civil-rights, and antiwar movements. He graduated from Berkeley with an individual honors major in the history of political consciousness in 1969. He earned his master's degree in humanistic psychology from Antioch College West in 1976.

Eager to live a life closer to the land, Robbins and his wife, Deo (known as Annette when they met at Berkeley), relocated to a remote island in British Columbia in 1975, where they lived a simple vegetarian lifestyle, growing their own organic vegetables and offering yoga classes. In 1984 Robbins and his wife and son Ocean moved to Santa Cruz, California, where Robbins began working on a book. In 1987 *Diet for New America* was published and became a highly influential best-seller. In the book, Robbins warns that the meat-centered American diet has become profligate, wastefully directing resources toward beef production that could be much more efficiently used on plant crops. For example, production of an average pound of beef in the United States requires 2,500 gallons of water (most of which goes to irrigate the crops that the animal eats). Up to a hundred times more water is necessary to produce a pound of meat than a pound of wheat. If the people in the United States ate just 10 percent less beef, writes Robbins, the amount of grain saved would feed 60 million people. Likewise, there is a direct relationship between livestock production and deforestation; of the 260 million acres of forest in the United States that have been razed to grow food for livestock, well over 200 million acres could be returned to forest if the food crops went directly to people instead. Livestock farms also use large quantities of pesticides, artificial hormones, antibiotics, and other toxic chemicals, and the residues invariably end up in the meat, eggs, or dairy products being produced. Robbins cites studies that indicate that very high percentages of dioxin and other toxic chemical residues in the diets of Americans come from animal-based foods. Other health benefits of a plant-based diet are extolled in the book, such as lower cholesterol, lower blood pressure, and reduced cancer rates. The leading cause of death in the United States is heart disease, warns Robbins, and people with the standard meat-based diet have a 50 percent chance of dying from heart disease, while vegetarians have a 15 percent chance.

Robbins's book attracted a large following and garnered a Pulitzer Prize nomination. Not surprisingly, the National Cattlemen's Association did not agree with the accolades and went on the attack. According to EarthSave, the cattlemen paid $50,000 to the animal science department at Texas A&M University to write a dissenting report that the cattlemen then publicized widely, sending it to television, radio, and print media representatives in advance of Robbins's appearances and interviews. Nonetheless, the book's influence sparked a national reexamination of the impact of food production. In 1992 the WorldWatch Institute issued a landmark report addressing the impact of livestock production on the environment.

As a result of the overwhelming reader response to the book, Robbins founded EarthSave International, a nonprofit organization to support the educational work that *Diet for a New America* began. The group provides education and leadership for a shift toward more healthful and environmentally sound food choices, nonpolluting energy supplies, and sustainable use of natural resources. Robbins served as president of the organization for several years and now continues his involvement as chairman emeritus of its board of directors. In 1994 he received the Rachel Carson Award for his work. In 1996 Robbins's next book, *Reclaiming Our Health*, came out; it examines conventional medicine and alternatives to it and demonstrates some of the many problems with health care in the United States.

Since the publication and popularity of *Diet for a New America*, Robbins has become a widely sought public speaker and has lectured at major conferences sponsored by groups such as the Sierra Club, the Humane Society of the United States, Physicians for Social Responsibility, Oxfam, the United Nations Environmental Program, the United Nations Children's Fund (UNICEF), and other public interest organizations. There is evidence that his views are being accepted—in 1999 the Union of Concerned Scientists reported that eating meat is one of the two most environmentally damaging actions Americans perform (the other is driving cars). Later that year *Time* magazine, usually known for its conventional views, ran a two-page feature article depicting the damage to the environment and human health caused by modern meat production and predicted a collapse of the current meat-centered diet. Robbins has even convinced his own father, who suffered from high cholesterol, high blood pressure, and diabetes, to take some of his advice and eat less ice cream.

Robbins currently works with Youth for Environmental Sanity, helping to educate, inspire, and empower youth to become leaders in the movement to create a just and sustainable society. He lives with his wife in Santa Cruz, California.

BIBLIOGRAPHY

Carlin, Peter Ames, "Quitting Cold: Ice Cream Heir John Robbins Gave up Fortune to Follow a Rocky—But More Satisfying—Road," *People*, 1997; Colodny, Mark M., "Just Say 'No' to Ice Cream," *Fortune*, 1990; "EarthSave International," http://www.earthsave.org; Robbins, John, *Diet for a New America*, 1987; Robbins, John, *Reclaiming Our Health*, 1996.

Robin, Vicki

(July 6, 1945–)
President of the New Road Map Foundation

Vicki Robin is president of the New Road Map Foundation, an all-volunteer educational and charitable organization based in Seattle that champions conscious and practical, satisfying and sustainable alternatives to unabated consumerism—or, as she puts it, "tools for shifting to low-consumption, high-fulfillment lifestyles." She is the coauthor, with Joe Dominguez, of the 1992 book *Your Money or Your Life: Transforming Your Relationship with Money and Achieving Financial Independence*, a longtime best seller that details a nine-step program derived largely from the priceless fiscal education Dominguez received while working as a Wall Street stock analyst. In the Dominguez-Robin lexicon, financial independence is not a fanciful pot of gold at the end of the rainbow, but rather an enrichment of life resulting from conscious consumer choices. Although reducing the impact on the environment of U.S. mainstream society was not the initial motivation for Robin's advocacy of voluntary simplicity, the lifestyle's positive environmental implications became clear to her in the late 1980s, and many of those who choose a simpler lifestyle do so as a way of living in greater harmony with the earth.

Victoria Marie Robin was born on July 6, 1945, in Okmulgee, Oklahoma, to a family that knew both ends of the economic scale. Her mother was frugal from having grown up in a family hit hard by the Depression. But her father was a doctor, so the Robins did not want for money or its spoils. The family moved to Long Island, New York, and Vicki grew up always having what she needed but nonetheless understanding the importance of careful shopping. "Somehow I developed early a passion for experience over stuff," she said. "I learned that the less I spent, the more adventures I could have." For example, during her college years she lived in Europe for a year and a half on money her father had left her to finance her education—attending school in Madrid but taking every opportunity to travel as far north as Norway, as far east as Moscow, as far south as Turkey, and as far west as Portugal. She traveled on third-class train tickets, eating bread and oranges, staying in convents—making her money stretch literally across a continent.

After earning a degree in Spanish from Brown University in 1967, Robin pursued an acting career in New York for a while but became discouraged and took to the road with a $20,000 inheritance. While traveling in Mexico she met Dominguez, whose story intrigued her. Disillusioned by the earn-and-spend treadmill, he had retired from Wall Street at age 31 with a nest egg of $70,000. He had been living on the interest ever since. Although that interest only amounted to about $6,000 a year, Dominguez had determined through many years of tracking and evaluating his expenses that this was just enough for him to lead the life he wanted. Robin recognized the value in Dominguez's approach, and before long she, too, was living according to his definition of financial independence. As she and Dominguez traveled together and eventually settled in the Pacific Northwest, others became interested in Dominguez's strategy, and in 1976 he began giving living room talks, out of which grew the New Road Map Foundation. By 1980, he and Robin were traveling around the country giving seminars. Within five years, there was an audiocasette course, 300,000 copies of which were sold by mail. In 1992 came *Your Money or Your Life*, which to date has sold more than 700,000 copies and been translated into five languages. All proceeds from the book and other New Road Map educational programs are donated to nonprofit groups working toward a sustainable world.

Through the 27 years that Robin lived and worked with Dominguez, who died of cancer in 1997 at age 57, her thinking evolved beyond frugality for its own sake. In 1989 she attended the Globescope Pacific Assembly, the first public hearing in the United States on the World Commission on Environment and Development, and heard speaker after speaker say that the single biggest problem facing the world was the level and pattern of consumption in North America. Yet she heard no solutions, no suggestions of hope. "And there I sat, knowing from a decade of public education and two decades of living that, at the individual level, mere consciousness about aligning money and values tended to adjust consumption down by 20 percent and often far more," said Robin. "That's when I linked my chosen way of life with sustainability."

She has made spreading the word her vocation. She is a founding member and trustee of Sustainable Seattle, a voluntary civic network and forum concerned with the long-term cultural, economic, and environmental health and vitality of the Puget Sound region. She is also a founding board member of the Center for the New American Dream, a national organization with the goal of changing the pattern and overall quantity of consumption in the United States, without sacrificing quality of life. In addition, she also has served on a Task Force on Population and Consumption for the President's Council on Sustainable Development, and in 1994 she delivered testimony on overconsumption to the International Conference on Population and Development PrepCom at the United Nations.

Robin has not simply been preaching to the converted. Having a best-selling book has been her ticket into mainstream America—she has made appearances on *Oprah!* and *Good Morning America* and has been featured in *People* and *Woman's Day*. For those who want to delve deeper, Robin and members of the New Road Map Foundation Outreach Network, Financial Integrity Associates, facilitate Your Money or Your Life study groups. Robin also was instrumental in launching an annual Buy Nothing Day moratorium on the busiest day of commerce each year, the day after Thanksgiving, the kickoff to the holiday shopping season. Her goals in the work may be to help effect deep transformations in U.S. culture, but Robin is not heavy-handed. At one Buy Nothing Day celebration in Seattle, for instance, she dressed as a doctor and dispensed medical advice on a spiritually debilitating malady called "affluenza."

Media coverage of the New Road Map Foundation never fails to note, with all due amazement, that Robin has scaled down her monetary needs to the point where she can live on about $8,500 a year. Robin resides in Seattle.

BIBLIOGRAPHY

Berner, Robert, "A Holiday Greeting Networks Won't Air," *Wall Street Journal*, 1997; Dominguez, Joe, and Vicki Robin, *Your Money or Your Life*, Viking, 1992; "Financial Integrity Associates," http://www.fiassociates.org; "New Road Map Foundation," http://www.newroadmap.org; Oldenburg, Don, "No-Shop Option," *Washington Post*, 1996; Wagenheim, Jeff, "If Money Were No Object," *New Age Journal*, 1990.

Rockefeller, John D., Jr.

(January 29, 1874–May 11, 1960)
Philanthropist

A generous benefactor for the preservation of historic sites and the conservation of areas of natural beauty, John Davison Rockefeller Jr. took a particular interest in the improvement of the nation's national and state parks. His legacy includes the building of the national park on Mount Desert Island in Maine and substantial contributions to parks and natural areas throughout the United States.

John Davison Rockefeller Jr., the son of John D. Rockefeller, the founder of Standard Oil combine, and Laura Celestia Spelman, was born in Cleveland, Ohio, on January 29, 1874. In spite of the family's enormous wealth, Rockefeller and his three sisters were brought up to expect a life of hard work, religious observance, great financial responsibility, and social service. Both parents were devout Baptists, the elder Rockefeller giving both time and money to his church. Rockefeller's father modeled efficiency, rational organization, and hard work, all qualities that had made him one of the richest men in the world.

The Rockefellers lived in Cleveland and spent summers at Forest Hill, outside Cleveland. It was in Forest Hill and away from the city that the family was happiest. Rockefeller particularly enjoyed his first visit to Yellowstone National Park, when he was 12 years old. He kept a diary of the visit, describing walks and sights. This early experience, no doubt, foreshadowed his future efforts as a friend to the nation's parks.

The family moved to New York City while "JDR Jr.," as he later liked to be called, was still a child. After several years in residential hotels, the Rockefellers finally settled down in a spacious brownstone. Rockefeller attended private schools in New York City and received private tutelage during the family's long visits to Cleveland. He entered Brown University in 1893, where he was popular with his classmates, who called him "Johnny Rock." Having had only his sisters and the son of their caretaker to play with when he was a child, Rockefeller would later confess that his time in college was the happiest of his life. While he was not a great scholar, he worked hard, was elected to Phi Beta Kappa, and graduated in 1897.

Unlike his father, Rockefeller never developed an astuteness for the business world, nor did his father ever press him

John D. Rockefeller Jr. (Archive Photos)

to do so. After college, Rockefeller performed odd jobs for his father in the New York office of Standard Oil and married Abigail (Abby) Aldrich, daughter of Sen. Nelson W. Aldrich of Rhode Island. Together they had six children, all of whom went on to notable careers.

During the first years of the twentieth century, a portion of the ever-increasing Rockefeller fortune was transferred from petroleum into philanthropic projects and other industries requiring a boost in capital. Pleased with Rockefeller's handling of the sale of Mesabe Range iron ore properties to J. P. Morgan in 1901, his father gave him the responsibility of overseeing the management of the Colorado Fuel and Iron Company, in which he owned 40 percent. The labor crisis that resulted in the "Ludlow massacre" represented the most devastating publicity of Rockefeller's life and indicated that the days of unrestricted free enterprise and the industry owners' free hand with their labor force, which had made his father rich, were drawing to a close.

For some time, the management of the Colorado Fuel and Iron Company had attempted at every turn to prevent the unionization of its miners. Nevertheless, the western miners were fully determined to achieve company recognition of their right to bargain collectively. The resulting strike in 1914 led management to close the mines and evict the miners' families from their company-owned homes. Tensions rose and the area around Ludlow, Colorado, quickly became a battleground, into which the state militia were sent in an attempt to preserve the last remnants of a tenuous peace. Skirmishes broke out, and a total of 40 people were killed. One famous, though inaccurate, story involves com-

pany thugs and militiamen shooting and killing defenseless women and children, when in reality two women and 11 children crawled into a cave to escape the gunfire and died of suffocation. At the center of the crisis was Rockefeller, who stubbornly supported the management as they tried to treat modern labor problems with brutal, antiquated remedies.

The congressional investigations that followed blamed Rockefeller for the tragedy, and he quickly adopted a more modern policy. Following the recommendations made in a thorough study by Canada's future prime minister and labor relations expert, William Lyon Mackenzie King, Rockefeller went to the mining camps with a plan for employee representation. His willingness to listen and his many informal speeches, in which he emphasized the responsibilities and rights of both labor and management and the necessity for the two to work in harmony, helped to heal the wounds and construct a modified company-union arrangement that lasted until the Great Depression.

After World War I, Rockefeller removed himself completely from the business world and concentrated almost exclusively on public service and philanthropy. In the following years, he would contribute $56 million toward the restoration of Colonial Williamsburg, more than $86 million to schools and colleges, $85 million toward religious causes, and $81 million to medical, charitable, and relief organizations. But it was donations to the nation's national and state parks that captivated his most intense personal interest.

Rockefeller "discovered" Mount Desert Island in Maine when he summered in Bar Harbor with his family in 1908. In 1910, he purchased a summer home in

Seal Harbor and began contributing to the associations that sought to preserve Mount Desert Island and surrounding areas. With the help of HORACE ALBRIGHT, field assistant to the director of the National Parks Service, and landscape architect FREDERICK L. OLMSTED, Rockefeller began a campaign to create a national park on Mount Desert Island. During his lifetime he gave Acadia National Park, which encompasses Mount Desert Island, a total of $3.5 million in money, lands, and roads.

By 1924, Rockefeller's contributions to Acadia National Park were well known throughout the National Park Service. He was regarded as a potential benefactor for other parks and was courted by Park Service management. In 1924, Rockefeller and his three eldest sons visited Yellowstone National Park. They were met by Horace Albright, who at that time was the Yellowstone Park Superintendant. Rockefeller and Albright, who would later become director of the National Park Service, established a close relationship. More than 1,300 letters were exchanged between the two from 1924 to 1960 on various issues relating to the improvement and conservation of national and state parks. During that period, Rockefeller donated some $40 million to national and state parks, and funded private protection of such spectacular areas as the Grand Tetons. Rockefeller's son, LAURANCE ROCKEFELLER, would go on to become a well-known funder of conservation projects as well.

John D. Rockefeller Jr.'s business interests provided thousands of jobs during the Great Depression. Having entered into a scheme to redevelop the midtown section of New York at the end of the 1920s, Rockefeller found himself with a long-term lease on a considerable amount of heavily taxed real estate. In spite of the nation's economic woes, Rockefeller forged onward with plans to build skyscrapers that would house several enterprises, including the emerging network radio industry. The grand result of this project was the Rockefeller Center, the construction of which provided employment for thousands and was not completed until after World War II. That the Rockefeller Center would become a major tourist attraction also suggested that the Rockefeller talent for making money had not completely disappeared.

As his talented sons began to take over and Rockefeller slipped into the shadows, he made one last generous gesture with a donation of a $9 million plot of land on the East River of New York in 1946 to the United Nations for its home.

Rockefeller's father died in 1937, but for the rest of his life, he insisted upon being called John D. Rockefeller Jr., because, he said, there would always be only one John D. Rockefeller. He died in Tucson, Arizona, on May 11, 1960.

BIBLIOGRAPHY

Ernst, Joseph W., *Worthwhile Places: Correspondence of John D. Rockefeller, Jr. and Horace M. Albright*, 1991; Fosdick, Raymond B., *John D. Rockefeller Jr.: A Portrait*, 1956; Newhall, Nancy W., *A Contribution to the Heritage of Every American: The Conservation Activities of John D. Rockefeller, Jr.*, 1957.

Rockefeller, Laurance

(May 26, 1910–)
Philanthropist

Grandson of John D. Rockefeller of Standard Oil fame, Laurance Rockefeller has spent a lifetime sharing his passion for all places wild. He has given millions of dollars to environmental causes through the Rockefeller Brothers and LSR Funds and has been the driving force behind at least two national parks, one in Vermont and one in the U.S. Virgin Islands. Serving on the national Outdoor Recreation Resources Review Commission (ORRRC) in the 1950s, long before green politics were fashionable, Rockefeller helped shape national policy that protected federal lands in the following decades. He received a Congressional Gold Medal for Conservation in 1991.

Laurance Spelman Rockefeller was born on May 26, 1910, in New York City, the fourth child of JOHN D. ROCKEFELLER JR. and Abby (Aldrich) Rockefeller. He was one of six children in the latest generation of a family known as industrialists and bankers. His patriarchal grandfather John D. Rockefeller was among the barons of the oil industry; his matriarchal grandfather Nelson Aldrich was a senator from Rhode Island. The Rockefellers were encouraged to find work of value, and Laurance chose conservation as his specialty.

Rockefeller's father, John D. Rockefeller Jr., may have been the most generous philanthropist in the history of conservation to that time. The developer of New York's Rockefeller Center, he was deeply involved in the establishment of national parks. Instrumental in the founding and expansion of Grand Teton National Park, he also established Colonial Williamsburg and donated land for the United Nations headquarters. From an early age, Laurance shared his father's interest in conservation and wildlife protection. His love of nature was cultivated through family travels to such places as the Rocky Mountains and the Maine seacoast. In 1924 and 1926, Laurance spent summers in the West, visiting Jackson Hole, Wyoming; Taos, New Mexico; and the Grand Canyon in Arizona. The trips shaped his worldview and sparked a lifelong interest in natural places and Native American traditions.

Like other Rockefellers, Laurance attended Ivy League colleges. He graduated from Princeton University in 1932 with concentrations in economics and philosophy and moved on to Harvard Law School. While there he met Mary Billings French, granddaughter of early Vermont conservationist Frederick Billings. With an eye on marriage and the business world, Laurance decided to leave law school. He and Mary French were married in her hometown of Woodstock, Vermont, in August 1934. In time they would have four children: Laura, Marion, Lucy, and Larry.

Stepping into real-world ventures, Rockefeller assumed his grandfather's seat on the New York Stock Exchange in 1937. Ably forecasting business trends, he participated in the founding of Eastern Airlines. At the same time, he assumed some of his father's conservationist roles. In 1940, he stepped in as president of the Jackson Hole Preserve, a corporation the family had formed to protect scenic values and natural habitats of land it held at the base of Wyoming's Teton Range. He and FAIRFIELD OSBORN, president of the New York Zoological Society, established a biological station on the property to facilitate scientific study of the area's wildlife.

Rockefeller soon found a way to combine his twin interests in nature and business. In 1951, he began to construct Jackson Lake Lodge in what would become part of Grand Teton National Park. The first of Rockefeller's grand resorts, it ignited a storm of controversy. Though Grand Teton was not intended as a wilderness park, neither was it considered an appropriate site for major tourist development. Rockefeller argued that a planned resort would preserve the openness of surrounding land. He answered critics by building the lodge to standards of aesthetic beauty tempered by natural interaction. The resort opened in 1955, giving Rockefeller his first success as a developer. Two years later, in 1957, he received the Horace Marden Albright Scenic Preservation Medal from the American Scenic and Historic Preservation Society for his efforts.

After Grand Teton, Rockefeller moved to the Caribbean and bought a large portion of St. John Island in the U.S. Virgin Islands. There he developed the Caneel Bay Plantation Resort, meant to enhance natural surroundings of coral, beaches, and jungle, for guests to engage in what we now call ecotourism. To maintain the lush habitat that surrounded his property, he donated most of his holdings to the National Park system. In all, he gave the U.S. government more than 5,000 acres on St. John, leading to the establishment of Virgin Islands National Park in 1956. The hotel and the park were dedicated on the same day, December 1, 1956. Rockefeller soon built other hotels with similar environmental aesthetics, including Dorado Beach Hotel in Puerto Rico (1958) and the Mauna Kea Resort in Hawaii (1961).

At the same time he was building grand hotels in the 1950s, Rockefeller served as chairman of the Outdoor Recreation Resources Review Commission, the first sweeping federal review of the American outdoors. It was a time when many of the programs that now protect against pollution and designate federally protected wilderness areas were first proposed. Serving under presidents Eisenhower and Kennedy, Rockefeller orchestrated an assessment of the recreation and conservation needs and desires of the American people, outlining the policies and programs required to meet those needs. Reports issued by the commission laid the framework for nearly all significant environmental legislation of the following three decades.

Today, Rockefeller owns the luxurious Woodstock Inn and Resort in Woodstock, Vermont, listed by *Conde Nast Traveler* as one of the top North American resorts. Together with his wife, he has been instrumental in the historic and environmental preservation of Woodstock. Frederick Billings had established a progressive dairy farm and managed forest on the GEORGE PERKINS MARSH farm there. The Rockefellers sustained Billings's mindful forestry practices on the property, which served as their home, and in 1983 established the Billings Farm and Museum to continue the farm's work and to interpret rural Vermont life and history. The farm is now a private nonprofit educational institution within Marsh-Billings-Rockefeller National Historical Park, created in

1992 when the Rockefellers gave the estate lands to the federal government.

Through investments and business ventures, Rockefeller has continued to add to the family fortune. In 1969, the Rockefeller-controlled Venrock Associates was established as a venture capital firm. It had noteworthy successes, including seed money investments in Intel Corporation and Apple Computer, among other companies. The money, along with that remaining in family trusts and philanthropic foundations, has allowed Rockefeller to give generously to environmental interests. Among his giftees are the Laurance S. Rockefeller Library at the California Institute of Integral Studies; the PBS *Religion and Ethics* series; the Library of Congress *American Memory* series, "The Evolution of the Conservation Movement, 1850–1920"; the state of New York, which was granted funds for the acquisition of the 88-acre Rockwood Hall property on the Hudson River near Mt. Pleasant in November 1998; and the Euan P. McFarlane Environmental Leadership Awards for conservation efforts in the Caribbean.

Rockefeller has played a heavy role in the establishment of such conservation organizations as the Conservation Foundation and the American Conservation Association, Inc. He has also served on the board of the Citizens Advisory Committee on Environmental Quality, among hundreds of other board commitments. Through the years he has participated on the boards of the National Geographic Society, the National Recreation and Parks Association, regional conservation societies, and the New York Zoological Society, to name just a few.

Premier among all his awards is the Congressional Gold Medal he received in 1991. That year, the medal was given for the first time in honor of contributions to conservation and historic preservation. It rewarded Rockefeller's "driving passion" for nature, recognizing his service as environmental adviser to five consecutive presidents and the millions of dollars he has donated to numerous conservation organizations. Rockefeller lives in New York's Hudson River Valley.

BIBLIOGRAPHY

Harr, John Ensor, and Peter J. Johnson, *The Rockefeller Conscience: An American Family in Public and in Private*, 1991; Rockefeller, Mary, and Laurance Rockefeller, "Parks, Plans, and People," *National Geographic*, 1967; Winks, Robin W., *Laurance S. Rockefeller: Catalyst for Conservation*, 1997.

Rodale, Robert

(March 27, 1930–September 20, 1990)
Organic Agriculture Expert, Publisher, Writer

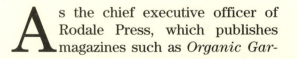

As the chief executive officer of Rodale Press, which publishes magazines such as *Organic Gardening*, *Prevention*, and *Runner's World*, Robert Rodale came to be known as an innovative promoter of healthy active lifestyles. But his influence went beyond that—he spent a lifetime studying and advocating organic farming, focusing especially on its role in regenerating food production in developing countries. His belief in the importance of organic agriculture led him to write *The Basic Book of Organic Gardening* (1971), a syndicated column titled "Organic Living," and *Save Three Lives* (1991), a book about preventing famine by reestablishing traditional farming techniques. Through the Rodale Institute, a nonprofit educational research organization, and its 300-acre research farm in Pennsylvania, Rodale studied ways to improve devastated agricultural systems in Third World countries; he spent the last few years of his life sharing his knowledge around the globe.

Robert David Rodale was born in New York City on March 27, 1930, to Jerome Irving and Anna (Andrews) Rodale. That same year, his father moved his electrical business from New York to Emmaus, Pennsylvania, and set up the family's residence in Allentown. J. I. Rodale then went on to establish a publishing business, which would eventually become Rodale Press. Increasingly interested in organic farming, J. I. Rodale purchased a 60-acre demonstration farm just outside Emmaus, where he pioneered research and demonstration of organic farming practices.

In 1947 Robert enrolled in Lehigh University in Bethlehem, Pennsylvania, to study English and journalism. Two years later, his father pressed him to lighten his coursework so that he could help out with the family business. Robert agreed, on the condition that he be allowed to go on a trip somewhere first. His father gave him a couple hundred dollars, his mother loaned him her new car, and he spent six weeks driving around rural Mexico by himself. This was his first contact with people in a developing country, and their generosity and the relative simplicity of their lifestyles made a great impact on him. In 1951 he married Ardath Harter, a school counselor, with whom he would eventually have five children: Heather, Heidi, David, Maria, and Anthony. Rodale continued attending Lehigh until 1952, but did not receive a degree until he was awarded an honorary doctorate in 1989.

Rodale was named president of Rodale Press in 1951 and helped his father carry on with his commitment to teach people how to grow better food by cultivating healthier soil and eliminating synthetic chemicals. By 1953, Robert Rodale had taken over as editor and publisher of *Organic Gardening* and *Prevention* magazines. Rodale was always a physical fitness enthusiast and was himself a champion skeet shooter. At the age of 38 he returned to Mexico, this time as a member of the U.S. Olympic skeet shooting team in the 1968 Pan-American games, where he won a gold medal.

In 1971, Ballantine published his book *The Basic Book of Organic Gardening*. That year his father died, and Robert Rodale took over the leadership of Rodale Press entirely. Under his guidance, the family-owned Rodale Press grew into a strong positive force, actively seeking to improve the health of its readers and its workers. Rodale's company employs over a thousand people, and all benefit from the nutritious food offered by its cafeteria and from the well-equipped employee fitness center. In 1983, Rodale Press became one of the first large U.S. companies to implement a no-smoking policy.

Perhaps Rodale's greatest impact has come from his role in promoting environmentally sound farming practices worldwide. *Organic Gardening* has become the largest gardening magazine in the world—providing its readers with the tools, ideas, sources, and information they need to successfully grow anything they choose without chemicals. Rodale was also instrumental in the success of the Rodale Institute: a nonprofit research and service organization (separate from the publishing company) that works with people around the globe to develop profitable food systems that preserve environmental and human health. Its headquarters is a 333-acre experimental farm in Berks County, Pennsylvania. He continually raised concerns over the damaging effects of pesticide use, both to humans and the environment. He once wrote an editorial in *Organic Gardening*, called "Tiptoe Through the Toxic Tulips" (1976), in which he issued a prophetic warning about the harm from combinations of pollutants in the environment. Over 20 years later, researchers are finding exactly that—the toxicity of certain chemicals in combination can be 1,000 times more than any of the combined chemicals alone.

Another of Rodale's passions was preventing famine in Third World countries by developing agricultural systems that can sustain people while replenishing their misused land. In his book *Save Three Lives: A Plan for Famine Prevention*, Rodale points out many problems with established famine relief programs, such as the fact that much of the food sent to famine-stricken countries comes from the surplus crops of rich nations like the United States—and is often inappropriate for people who are starving. The worst problem lies in misguided attempts to help poorer countries "improve" their agricultural practices by bringing in heavy machinery and synthetic chemicals. Methods like these were developed in temperate climates and quickly deplete fragile tropical soils. Also, developing countries are urged to plant cash crops such as corn, cotton, or rice—water-intensive crops that are out of place in arid climates and that displace valuable native crops. Rodale praised many of the indigenous food plants he saw on his travels, plants that are drought resistant, highly nutritious, and resistant to diseases and pests native to the land they evolved in. He proposed a system of home gardens and small farms, which return to the use of native food plants and traditional, organic farming methods. He traveled extensively, particularly in Africa, but also in Mexico and Russia to help put his theories into practice. Before his death, China and Eastern Europe had asked for his help in finding solutions to their food crises.

In 1990, shortly after he finished *Save Three Lives*, Rodale was traveling in the Soviet Union. He was there to develop plans for Russian-language editions of magazines and books on sustainable agriculture. On September 20, 1990, while en route to the Moscow airport, he was killed when the car in which he was riding collided with a bus.

BIBLIOGRAPHY

Long, Cheryl, "Chemical Combinations 1,000 Times as Deadly!" *Organic Gardening*, 1997; Rodale, Robert, *The Basic Book of Organic Gardening*, 1971; Rodale, Robert, *Save Three Lives: A Plan for Famine Prevention*, 1991; Rodale, Robert, "Tiptoe Through the Toxic Tulips," *Organic Gardening*, 1976; Simson, Maria, "Bob Rodale and Sierra Club Fight World Hunger," *Publishers Weekly*, 1991.

Rolston, Holmes, III

(November 19, 1932–)
Environmental Philosopher, Theologian, Founder of *Environmental Ethics*

Holmes Rolston III is widely recognized as the father of environmental ethics as a modern academic discipline. He has devoted his career to the development of a philosophical interpretation of the natural world and is regarded as one of the world's leading scholars on the philosophical, scientific, and religious conceptions of nature. His body of work and his role as a founder of the influential academic journal *Environmental Ethics* have been instrumental in establishing, shaping, and defining the modern discipline of environmental philosophy.

Holmes Rolston III was born in Staunton, Virginia, on November 19, 1932, to Holmes and Mary Winifred (Long) Rolston. He grew up in the Shenandoah Valley, where his father was a rural pastor. Their house had no electricity, and water came from a cistern pump outside and another cistern on a hill behind the house from which water flowed by gravity to the kitchen inside. Rolston's mother had been raised on a farm in Alabama, which Rolston would visit for a month each summer and where he explored the woods and swamps. From a very early age he was deeply immersed in nature.

Rolston studied physics as an undergraduate at Davidson College in Davidson, North Carolina, because physics seemed to him to be the science that would best help him understand nature. During his studies, Rolston became interested in biology after taking an entomology class at Davidson; he began to recognize that physics alone could not explain nature the way he had hoped. While physics sought to address order and universal laws of nature, biology explored the more alluring wild nature. Upon completing his undergraduate degree in 1953, Rolston turned to theology and decided to attend Union Theological Seminary. Once he completed his seminary studies in 1956, Rolston pursued theology and religious studies at the University of Edinburgh, completing his Ph.D. in 1958.

Rolston spent the next ten years as a Presbyterian pastor in rural southwest Virginia. Taking two days off each week, Rolston would spend one hiking the southern Appalachian Mountains and the other sitting in on biology classes at nearby East Tennessee State University. After botany and zoology came geology, mineralogy, and paleontology. During this time, Rolston learned the ecology of the mountain woods and discovered that trees and country places had as much to teach him as the scholars.

While he was developing a passion for the wonders of nature, Rolston became alarmed by how quickly the natural world was being lost to development. The sense of wonder he felt for nature turned to horror as he discovered his favorite forests scarred by clear-cuts, mountains stripped for coal, soils eroded away, and wildlife populations decimated. Rolston worked to preserve Mount Rogers, in southwest Virginia, and Roan Mountain, in northeast Tennessee, and to maintain and relocate the Appalachian Trail so that it passed mainly through undeveloped areas.

In his search for a philosophy of nature to complement his biology, Rolston received his first formal training in philosophy when he entered the philosophy program at the University of Pittsburgh. Struggling for respect in a program that considered the philosophy of nature disreputable, Rolston received a master's degree in the philosophy of science in 1968 and accepted a post at Colorado State University in Fort Collins, where he has remained, and currently holds the prestigious position of University Distinguished Professor.

In 1975, Rolston's essay, "Is There an Ecological Ethic?," published in *Ethics*, provocatively probed questions and issues regarding the potential of nature to have an intrinsic value. While debates on this issue had existed since ALDO LEOPOLD's *Sand County Almanac*, published in 1949, Rolston's essay initiated renewed interest in the debate, which ultimately led to the creation of the refereed journal, *Environmental Ethics*, in 1979. Rolston was a founder of that journal and still serves as the associate editor.

Rolston's 1988 book, *Environmental Ethics: Values in and Duties to the Natural World*, is generally recognized as the best available work in its field. Rolston presents a strong argument for a value-centered ecological ethic. He claims that intrinsic values objectively exist at the species, biotic community, and individual levels in nature and that these values impose on humans certain obligations to species and their ecosystems. This intrinsic value of nature is separate from its instrumental value, the latter motivating humans to conserve the environment for their own benefit. Sometimes the two kinds of values complement each other; sometimes they conflict.

Rolston is a prolific writer, having written six books acclaimed in critical notice in professional journals and the national press, chapters in 50 books, and over 100 articles. His work is published and read the world over and has been translated into several languages. He has also served on the Board of Governors of the Society for Conservation Biology. During the 1997–1998 academic year, Rolston delivered the Gifford Lectures at the University of Edinburgh, which resulted in his book, *Genes, Genesis, and God: Values and Their Origins in Natural and Human History*. Because of his prominent status in environmental philosophy, Rolston is often sought after by conservation and policy groups. He has served as a consultant for more than two dozen such groups, including the U.S. Congress and a presidential commission. He is also a member of the Working Group on Ethics of the World Conservation Union.

In early 2000, Rolston visited Antarctica and while there became the only environmental philosopher to have lectured on all seven continents. Avocationally, he is a backpacker, an accomplished field naturalist, and a respected bryologist.

BIBLIOGRAPHY

Rolston, Holmes, III, "A Philosopher Gone Wild," in Karnos, David D., and Robert G. Shoemaker, eds., *Falling in Love with Wisdom: American Philosophers Talk about Their Calling*, 1993; Rolston, Holmes, III, *Philosophy Gone Wild: Environmental Ethics*, 1989; Rolston, Holmes, III, *Science and Religion: A Critical Survey*, 1987; Rolston, Holmes, III, "Values Deep in the Woods," *American Forests*, 1988.

Roosevelt, Franklin D.

(January 30, 1882–April 12, 1945)
U.S. President

Franklin Delano Roosevelt (FDR), state senator and governor of New York and 32nd president of the United States, is known, in addition for his accomplishments in combating the Great Depression, as a dedicated nature lover and conservationist. During his 12 years as president, he appointed such conservationists to his cabinet as Secretary of the Interior Harold Ickes and Soil Conservation Service Director Hugh Hammond Bennett. He established the Civilian Conservation Corps, which gave jobs to two and a half million men in soil erosion prevention, reforestation, and national parks facilities construction.

Franklin Delano Roosevelt was born on January 30, 1882, at his family's Hyde Park estate on the Hudson River in upstate New York. His father, James Roosevelt, was vice president of the Delaware and Hudson Railroad and was 52 years old when Franklin Delano was born. His mother, half the age of her husband, was from a wealthy family that owned the estate across the Hudson River. Roosevelt's patrician heritage allowed for home schooling with tutors and governesses until the age of 14 and frequent trips with his parents to Europe. Isolated from his peers, he spent his free time roaming the estate, tracking rabbits, and shooting birds with a rifle given to him by his grandfather. He amassed a collection of 300 specimens, and as a reward for his interest, his grandfather gave him a life membership in the American Museum of Natural History.

Inspired by his fifth cousin, President Theodore Roosevelt, Franklin Delano Roosevelt became interested in politics while studying at Harvard University. He was editor of the *Harvard Crimson* student newspaper and was elected chair of his class. He graduated in 1903 and went on to study law at Columbia University. He married a distant cousin, Anna Eleanor Roosevelt, in 1905. Although FDR had an affair early in the marriage that deeply hurt his wife, Eleanor remained loyal and did much to advance his political career. She also helped cultivate his concern for the plight of the nation's poorest.

Roosevelt's political career commenced in 1911 when he was elected to the New York legislature. As chairman of the Forest, Fish, and Game Committee, he introduced eight bills to regulate fishing and hunting. Another of his bills proposed that tree cutting be restricted even on private land, if it would cause environmental damage. To testify for that bill, Roosevelt called in U.S. Forest Service founder Gifford Pinchot, whose argument in favor of the bill included some visual evidence, two slides of a Chinese valley. The first was a painting, circa 1500, of the verdant valley; the second was a recent photograph showing the same area denuded and desertlike after four centuries of indiscriminate cutting. Roosevelt was quite affected by these images, and historian Stephen Fox cites them as inspiration for his politics. In 1912 he declared that unregulated competition in the environmental arena could lead to total deforestation, which would in turn cause floods, erosion, and more problems. Cooperation within a community could assure that greed would not lead to such abuse. This, Fox states, was the ideological foundation upon which Roosevelt's progressive politics would be built.

Roosevelt did not consider himself a conservationist as such, since most conservationists in those days were hearty big game hunters like his cousin Theodore. In 1921, Roosevelt had been stricken with polio and lost the use of his legs, which restricted him to his specially designed roadster, a wheel chair, or uncomfortable leg braces and crutches. Roosevelt nonetheless had a deep appreciation of the aesthetics of nature. He loved Hyde Park and closely supervised his gardeners. He was a proponent of the managed forestry that Forestry Division chief Gifford Pinchot had introduced during Theodore Roosevelt's presidency and purchased a 1,200-acre degraded farm next to Hyde Park that he had his gardeners reforest in an attempt to regenerate the land.

The conservation policies that Roosevelt implemented when he was elected to his first term as president in 1932 reflected a belief that he shared with Pinchot: that natural resources should provide "the greatest good for the greatest number of people." When he took office, Roosevelt was faced with millions of unemployed people suffering in a severely depressed economy and an increasingly degraded environment. Several top officials in his administration, including two men of his cabinet, helped design projects that would provide jobs and further his idea of conservation. Secretary of Agriculture Henry Wallace, worried about the environmental effects of drought and the dust bowl it was causing, successfully lobbied for legislation that would pay farmers to leave degraded land fallow. Director of the new Soil Conservation Service, Hugh Bennett, set up demonstration projects and instructed tens of thousands of farmers in soil conservation methods. Secretary of the Interior Harold Ickes oversaw Roosevelt's pet project, the Civilian Conservation Corps. It put two and one-half million unemployed men to work at such tasks as reforestation, prevention of soil erosion, flood control, and improving accessibility in national parks and forests. In addition to providing employment, the CCC fostered a love for the outdoors that many of its employees had never before had the opportunity to develop. Roosevelt also initiated the Tennessee Valley Authority (TVA) hydroelectric project, which was seen as a cure-all for that poor and neglected region of the country. The series of dams would create jobs and electric power. Decades later, environmentalists would criticize its environmental impact, but at the time, few found fault with the TVA.

Conservation organizations found Roosevelt and Secretary Ickes eminently open to their concerns. They listened to conservation leaders' arguments for preserving wildlife and acted on those leaders' recommendations to close government lands to duck hunting, to refuse to extend hunting season, and to protect trumpeter swans from an artillery range in Utah. Ickes in particular sympathized with supporters of unspoiled wilderness. He and National Forest supervisor ROBERT MARSHALL worked hard to keep certain areas roadless despite the complaints of Roosevelt, who when asked to keep Olympic National Park free of roads, asked "How would I get in?"

Roosevelt, an immensely popular president, was elected president four times. He died shortly after he commenced his fourth term, on April 12, 1945, in Warm Springs, Georgia, of a cerebral hemorrhage.

BIBLIOGRAPHY

Fox, Stephen, *The American Conservation Movement: John Muir and His Legacy*, 1981; Nixon, Edgard B., ed., *Franklin Roosevelt and Conservation, 1911–1945*, 2 volumes, 1957; Riesch-Owen, Anna L., *Conservation under FDR*, 1983.

Roosevelt, Theodore

(October 27, 1858–January 6, 1919)
U.S. President

While remembered for his prolific accomplishments during his 30-year political career, conservationists honor Theodore Roosevelt as the first U.S. president with a deep commitment to conserving the country's wild lands. During his tenure as president, Roosevelt established over 50 wildlife refuges, 18 national monuments, and five national parks. Roosevelt was the first president who considered conservation an issue of national import, and he left a legacy that has encouraged the presidents following him to contribute to the national treasury of resources.

Theodore Roosevelt Jr. was born to a prominent New York City family on October 27, 1858. Like many conservationists, his intense interest in nature became apparent early in his childhood. At the age of nine, he opened the Roosevelt Museum of Natural History in an upstairs closet of his family's New York City home. Inspired by his father's role in founding the American Museum of Natural History and the Metropolitan Museum of Art, Roosevelt and his cousins eventually amassed a collection of sev-

eral hundred animal specimens, most of them shot and stuffed by young Roosevelt himself. A fan of the living as well, the boy packed notebooks with his observations of the fauna he studied at his family's summer home in the country and the animals he saw during extended family voyages to Europe and the Middle East. The notebook he kept during his 11th summer was entitled "About Insects and Fishes, Natural History" and was 40 pages long.

Roosevelt intended to become a professional naturalist, but once he began studying at Harvard University in 1876, he quickly learned that natural history studies there were conducted in laboratories, not outdoors as he had hoped. That disappointment, as well as his romance with Alice Lee, a high-society young woman who was not interested in natural history, led him to switch his major to political economy. But although science would not become his profession, he remained a devoted promoter and protector of the natural world throughout his life.

After graduation from Harvard in 1880 and his marriage later that year to Alice

Theodore Roosevelt (Library of Congress)

Lee, Roosevelt attended law school at Columbia University and became involved in politics. He joined the Republican Club and in 1881 was elected to the New York State Assembly. During the 1882 Assembly recess he traveled to the Dakota Badlands on a buffalo hunt. The vastness of the West enchanted him, and at the conclusion of his visit he invested in land and a herd of cattle and hired his buffalo hunt guides to tend the ranch for him. After two tragic deaths within two weeks of each other in 1884, that of his mother, preceded by that of his wife in childbirth, Roosevelt retreated to his new ranch, where he spent more than half of the next eight years.

Living in the Dakotas, Roosevelt was an unhappy witness to the rapid disappearance of the buffalo and other large game animals of the West. He shared his concern with GEORGE BIRD GRINNELL, and

together the two founded the Boone and Crockett Club in 1887. Others could join by invitation only; one of the prerequisites for membership was to have killed at least one adult male of three separate large game species. This elite club had both social goals (to promote a gentlemanly approach to hunting and provide a community network for themselves) and conservationist ends (to work for the preservation of large game and promote research on the habits of wild game). They were successful in their first major battle, one on behalf of Yellowstone Park, in which they convinced Congress to halt the proposed construction of a railroad spur into the park, to limit tourism concessions there, to protect the forest bordering the park, and to enforce the laws against hunting within the park boundaries.

Roosevelt's boundless energy, powerful network of friends, and impressive intellect combined to make him a national figure by the late 1890s. In 1897 he had been named assistant secretary to the navy, a post he stepped down from in 1898 in order to form the Rough Rider cavalry squad. He was recognized as a war hero after his triumph in Cuba during the Spanish-American War and was elected governor of New York. He was also the father of five children, who were born after he married Edith Kermit Carow in 1886. Roosevelt ran for vice president of the country with William McKinley on the Republican ticket and was elected in 1900. Six months after the inauguration, McKinley was assassinated, and Roosevelt assumed the presidency.

The first speech that Roosevelt gave to Congress as president included a strong conservationist message. Roosevelt—who by this time had renounced hunting—urged Congress to see conservation as an insurance policy for future generations of Americans. Natural resources such as forests, he told Congress, should be "set apart forever, for the use and benefit of our people as a whole and not sacrificed to the shortsighted greed of a few." By the turn of the century, half of the nation's forests had been cut, thousands of tons of topsoil had washed away owing to careless farming methods, and many visible species—including the buffalo and the formerly prolific passenger pigeon—were disappearing.

Roosevelt's aggressive conservation program included setting aside forest reserves and establishing wildlife refuges. Roosevelt worked with forester GIFFORD PINCHOT to establish the U.S. Forest Service and add over 150 million acres to its reserves. This tripled the acreage that all previous presidents together had set aside as reserves. In 1903, at the urging of bird-watchers and ornithologists, he established the first federal wildlife refuge in the nation, Pelican Island in Florida. At this time in history, birds with beautiful feathers were in great danger of extinction because plumes were widely used to adorn women's hats.

While president, Roosevelt personally surveyed the country for potential parks and refuges, exploring natural areas new to him. He toured western wilderness areas extensively and begged JOHN MUIR to guide him through Yosemite in 1903. Muir complied, and the two spent four days together in the park, hiking, camping, and sharing stories and visions for the future of wilderness conservation. Back in Washington, Roosevelt issued a standing invitation to his favorite conservationists and naturalists, many of whom would visit him for long evenings of strategizing and early morning bird-watching expeditions on the White House lawn or at the president's country retreat.

In the ten years he was to live after stepping down from the presidency, Roosevelt continued promoting conservation and indulging in his naturalist avocations. He wrote prolifically and corresponded with his naturalist friends, reviewing their books and discussing the fine points of speciation, a topic he was particularly interested in. On January 6, 1919, at his family's home in Sagamore Hill, New York, Theodore Roosevelt died in his sleep of an arterial thrombosis.

BIBLIOGRAPHY

Cutright, Paul Russell, *Theodore Roosevelt: The Making of a Conservationist*, 1985; Fox, Stephen *The American Conservation Movement: John Muir and His Legacy*, 1986; Roosevelt, Theodore, *Diaries of Boyhood and Youth*, 1928; Roosevelt, Theodore, *Hunting Trips of a Ranchman*, 1886; Roosevelt, Theodore, *Theodore Roosevelt: An Autobiography*, 1913; Roosevelt, Theodore, *The Wilderness Hunter*, 1893.

Roszak, Theodore

(1933–)
Writer, Professor

Theodore Roszak is an author of nonfiction and fiction whose works explore Western society, its fixation with technology, and the effects this fixation has on human relationships with the natural world. He is critical of science and technology and believes that they have led humans to lose touch with nature and spirituality. His best-known work is *The Making of a Counter Culture*, published in 1969, a book that examines the student protest movement of the 1960s in an attempt to reveal it as a backlash against the technocratic, spiritually lacking society of the 1960s.

Theodore Roszak reveals little about his early life beyond that he was born in 1933 in Chicago, Illinois, to Anton and Blanche Roszak and that his father was a cabinetmaker. Roszak attended the University of California, Los Angeles, receiving a B.A. degree in 1955. He then went to Princeton University and earned a Ph.D. in 1958. After graduating with his doctorate, Roszak became a history instructor at Stanford University, where he remained until 1963. In 1964, he moved to London, England, where he had accepted a position as editor for the pacifist journal, *Peace News*. He returned to the United States in 1965, accepting a history professorship at California State University, Hayward, where he still teaches.

Roszak began publishing in 1966. He is a prolific author, and he has published in journals such as *Atlantic, Harper's, Nation, Utne Reader,* and *New Scientist*. His first book, *The Dissenting Academy*, which he edited, appeared in 1968. It was a collection of critical and controversial essays written by prominent, socially conscious educators, and it established Roszak as an agitator for social and political activism. Roszak is probably best known for his second book, *The Making of a Counter Culture: Reflections on the Technocratic Society and Its Youthful Opposition*, which was published in 1969. It was this book that first used the term *counter culture*. In it, Roszak outlined the various influences of such intellectual and social subversives as Timothy Leary, Allan Ginsberg, Herbert Marcuse, and PAUL GOODMAN on the student protest movement. Roszak was critical of science and of scientific ways of viewing the world that do not acknowledge subjective ways of knowing. He stated that the "youthful opposition" was calling for a new culture "in which the non-intellective capacities of the personality—those capacities that take fire from the visionary splendor and the experience of human communion—become the arbiters of the good, the true, and the beautiful." The book was nominated for a National Book Award.

Roszak next compiled two anthologies, editing *Masculine Feminine: Readings in Sexual Mythology and the Liberation of Women* (1969) and *An Anthology of Contemporary Material Useful for Preserving Personal Sanity While Braving the Great Technological Wilderness* (1972). Also in 1972, and as a result of work done on a Guggenheim Fellowship throughout 1971 and 1972, he published *Where the Wasteland Ends: Politics and Transcendence in Post-Industrial Society*, a book in which

Roszak criticized not only the technological society of the 1970s, but also the human "mindscape" that could result in the type of society that he found himself inhabiting. He called for a transformation of this "mindscape" and for a return to the lost synergy between humans and nature. In the book's introduction, he explained that he was endeavoring to discover the processes that have caused humans to abandon their religious impulses and to become separated from nature and from a mystical view of the universe. This book also was nominated for a National Book Award.

Roszak has written many other works of nonfiction, exploring the relationships between nature, humans, and the afflicted societies in which they live, including *Unfinished Animal* (1975), *Person/Planet* (1978), *The Cult of Information* (1986), and *The Voice of the Earth: An Exploration in Ecopsychology* (1992). He has also written several novels. Among these are *Pontifex* (1974), *Bugs* (1981), *Flicker* (1993), and most recently, *The Memoirs of Elizabeth Frankenstein* (1995). In an interview with Catherine Maclay, Roszak commented that all of his books treat the common theme of "the moral and spiritual dilemmas of living in a society that worships technology." He also likened all of industrial civilization to a monster of the Frankenstein variety, stating "All of these monsters are connected with one great thing, and that is our screwed-up relations with the natural world. Science tells us that we can steal the earth's resources, and we can remake Mother Earth into anything we want her to be." In 1996, *Elizabeth Frankenstein* received the James Tiptee Jr. award for literature that reflects an understanding of gender issues.

Although Roszak is an advocate for wilderness protection, he is also a self-professed city slicker. He is not, as he wrote in an article for *New Scientist*, "a hiker, camper, climber, or outdoorsman." He is, he says, unable to meet the wilderness on its own terms. Therefore, he writes, "I simply leave the wilderness alone. It is enough for me to simply know it is there." Roszak lives in Berkeley, California, with his wife of 35 years, Betty. They have a grown daughter, Kathryn.

BIBLIOGRAPHY

Locher, Frances Carol, ed., *Contemporary Authors*, 1979; Maclay, Catherine, "Theodore Roszak: The Monster in the Laboratory," *Publishers Weekly*, 1995; Roszak, Theodore, "Leave the Wilderness Alone," *New Scientist*, 1988.

Ruether, Rosemary Radford

(November 2, 1936–)
Theologian

Georgia Harkness Professor of Applied Theology at Garrett Evangelical Theological Seminary in Evanston, Illinois, Rosemary Radford Ruether is a leading feminist and radical theologian who has written extensively

Rosemary Radford Ruether (Courtesy of Anne Ronan Picture Library)

about ecofeminism. Her 1992 book *Gaia and God: An Ecofeminist Theology of Earth Healing* discusses strains of Christian thought that contribute to the destruction of the planet as well as those that offer possibilities for forming a "biophilic" relationship to the environment. Ruether's work shows connections between the patriarchal privileging of men over women and Man over Nature and argues for a new ethic of equality and mutual care.

Rosemary Radford Ruether was born in St. Paul, Minnesota, on November 2, 1936, to a Protestant father and a Catholic mother. The youngest of three daughters, Ruether's early life was spent in the Georgetown area of Washington, D.C. Her father died when she was 12,

and the family moved to La Jolla, California, her mother's childhood home. In *Disputed Questions*, an account of her "intellectual and personal journey of faith and action," Ruether describes the Catholicism of her youth as "free-spirited and humanistic," connected to the Jesuit tradition and formed in private rather than parochial Catholic school. Ruether attributes her embrace of radical ideas to her mother's influence, particularly to growing up amid her mother's friends, in a community of mothers and daughters. Having grown up in California and Mexico, Ruether's mother had a respect for people from diverse backgrounds and a belief that she could do or be anything she wanted—and she bequeathed the same to her daughter.

Ruether attended Scripps College in Claremont, California, where she majored in philosophy, studying classics with Prof. Robert Palmer. In 1958 she graduated from Scripps College and married Herman Ruether, a political scientist and cultural historian. She went on to earn an M.A. in ancient history (1960) and a Ph.D. in classics and patristics (1965), both from Claremont Graduate School. Her background in classical history led Ruether to reject religious exclusivism in favor of respect for all religions, seeing connections between all such historically and culturally specific quests for meaning. Ruether's work has been particularly attentive to the Christian roots of anti-Semitism and anti-Islamic prejudice.

After completing her Ph.D., Ruether became increasingly active in the struggle for civil rights, first during the summer of 1965 when she went to Mississippi to work with the Delta Ministry, and later as part of the faculty of the School of Religion at Howard University, where she taught for ten years, beginning in 1966. Ruether lived in Washington, D.C., during the height of the movement against the war in Vietnam. She was arrested numerous times, often with other Catholic radicals. Her participation in the peace movement gave her an awareness of the global structure of social inequality, of the ways in which racism in the United States was linked to and mirrored in global imperialism and capitalism. Ruether developed a particular interest in the politics and theology of Latin America, her mother's ancestral home, and center of the flowering of liberation theology in the 1970s and 1980s. Liberation theology stresses the importance of social justice on earth, rather than fulfillment in heaven, and thus serves as one way for Christians to reimagine the human relationship to the natural world.

Reenvisioning our relationship to nature is linked in Ruether's thought to the need to challenge the dualistic thinking that underpins structures of social dominance. She traces the connections between the Christian valuing of God over man, and man over woman, to the dominance in much Christian thought of humans over animals and heaven over earth. Ruether argues that instead of opposition between these terms, we need to see connection. Rather than placing humankind in dominion over nature, we need to see that we are part of nature, joined to other living things through community and compassion. In *Gaia and God*, Ruether traces the damaging effects of Christian apocalyptic thinking, the belief that the world is destined for destruction. Apocalyptic Christians are prepared to sacrifice the earth as part of the vanquishing of evil, and she cautions environmentalists not to let their more dire predictions reinforce the belief in a coming apocalypse.

Ruether's global consciousness is evident throughout her writings on ecofeminism. In 1996 she edited *Women Healing Earth: Third World Women on Feminism, Religion and Ecology*. She has contributed articles to several collections on ecofeminism, including the 1994 book *Ecotheology: Voices from South and North*, edited by David Hallman. Ruether stresses the importance of linking economic justice to environmental reform and the need to build coalitions between working people throughout the world.

Rosemary Radford Ruether is still active as a scholar and speaker. She has written or edited more than 30 books and numerous articles. She has three grown children and contributes regularly to journals such as *The National Catholic Reporter* and *Sojourners*. She resides in Illinois.

BIBLIOGRAPHY

Bouma-Prediger, Steven, *The Greening of Theology: The Ecological Models of Rosemary Radford Ruether, Joseph Sittler, and Jürgen Moltmann*, 1995; Ruether, Rosemary Radford, *Gaia and God: An Ecofeminist Theology of Earth Healing*, 1992; Tardiff, Mary, *At Home in the World: The Letters of Thomas Merton & Rosemary Radford Ruether*, 1995.

Safina, Carl

(May 23, 1955–)
Marine Ecologist, Director of the Living Oceans Program for the National Audubon Society, Writer

One of the first marine ecologists to study seabird behavior out at sea, Carl Safina has contributed to a greater understanding of the complex interactions among sea birds, prey fish, and competing predatory fish. But he is as much a compelling writer and advocate as a research scientist, and he uses his other talents to bring marine conservation problems to public attention, often employing his favorite motto, Fish are wildlife, too. He established and directs the Living Oceans Program for the National Audubon Society, where he spearheads such major campaigns as rewriting federal fisheries law in the United States, attempting to gain international agreements to achieve sustainable fish catches, and pushing legislation that will protect and restore depleted and endangered marine life and marine ecosystems. After years of working with the science and public policy sides of marine conservation, Safina also began looking at the human side of the fishing culture and incorporated this aspect, along with in-depth ecological issues, into his landmark book *Song for the Blue Ocean: Encounters along the World's Coasts and beneath the Seas* (1997).

Carl Safina was born on May 23, 1955, in Brooklyn, New York, and grew up near the ocean on Long Island. His father often took him fishing for striped bass, and he acquired a general fascination with nature. When the woods near his house were bulldozed to make room for more houses, he spent more and more time along the coast, where he could still find a sense of nature. His affinity for the ocean stayed with him, and when he entered the State University of New York at Purchase in 1974, he enrolled in biology courses and realized that he could create a career out of the things that interested him. He graduated from Purchase in 1977 with a major in environmental science and went on to graduate school, where he chose to study seabirds so that he could continue to explore coastal habitats. While conducting his research, it became evident to him that fish populations were diminishing, and he started questioning why the conservation movement had made no efforts to protect marine and coastal species. He received his master's degree from Rutgers University in ecology in 1982. And despite having professors tell him that conservation was for people who were not smart enough to get Ph.D.'s, Safina became determined to make marine conservation a legitimate and well-known issue.

Safina began working for the National Audubon Society in 1980 as a research ecologist, formulating innovative field research projects on the ecology of seabirds and fish populations. While studying the breeding biology of endangered terns in their coastal colonies, he noticed gaps in the knowledge of the birds' biology that could only be filled by observing them at sea. He followed the birds offshore and used sonar to survey prey fish schools; he became one of the first scientists ever to study seabirds as marine animals. His work on the interactions among seabirds, prey fish, and competing predatory fish led up to his doctoral dissertation in ecology, which he completed in 1987 at Rutgers University.

Safina's doctoral research and the work he conducted for the National Audubon Society had him spending many hours at sea, and again he observed what appeared to be marked local declines in sea turtles, sharks, and several fish species. When he looked into the problem—researching the records of fishery management agencies and other literature—he saw that the declines he found in New England waters were just part of a larger trend that extended around the globe. Motivated by an increasing sense of urgency over dwindling marine life and the lack of public awareness of the issue, Safina dropped his research projects and devoted his time to getting marine fish issues onto the wildlife conservation movement's agenda. In 1990, he founded and became director of the Living Oceans Program at the National Audubon Society, a marine conservation program that uses science-based policy analysis, education, and grassroots advocacy to campaign on behalf of marine fish and ocean ecosystems, focusing on fisheries depletion and restoring fish abundance through sustainable fishing. In 1991 Safina accepted an appointment by the U.S. secretary of commerce to the Mid-Atlantic Fisheries Management Council and broke new ground by becoming the only professional conservationist to serve on any of the eight federal fishery management councils. That same year he received the Pew Charitable Trust's prestigious Scholar's Award in Conservation and the Environment. Also in 1991, Safina served as a U.S. delegation member

to the International Commission for the Conservation of Atlantic Tunas (ICCAT) and found out from Japanese and U.S. representatives that despite the dangerous declines reported to them by their own scientists, they had no intention of lowering bluefin tuna catch quotas. Spurred to action, Safina wrote a petition to the U.S. government to list the bluefin under the Convention on International Trade in Endangered Species (CITES), an action that would effectively suspend commercial fishing for this species. His efforts increased international awareness of the need for conservation of the bluefin tuna: by 1993, tuna quotas were cut for the first time in a decade, and ICCAT made a historic commitment to cut catch quotas in half by 1995. Safina was appointed deputy chair of the World Conservation Union's Shark Specialist Group in 1993 and brought attention to declining populations of sharks. By 1997, the Living Oceans Program had convinced the U.S. government to heed severe declines in Atlantic sharks and cut commercial catches in half. The program also engaged in major campaigns to ban high seas drift nets and to achieve passage of a new high seas fisheries treaty through the United Nations.

Meanwhile, Safina had also been working on a book that he had always wanted to write—one that summed up his experiences as a fisherman and an ecologist and that also addressed some of the cultural issues of fishing. In 1997 his book, *Song for the Blue Ocean: Encounters along the World's Coasts and beneath the Sea*, was published. In researching the book, he traveled to ten countries on four continents over five years, conducting investigations from small aircraft and commercial fishing vessels and engaging in conversations everywhere he went. He talked to families who depend on the fishing industry for their living, economists, officials from various international maritime commissions, and environmentalists. He also collected sobering evidence regarding the status of underwater wildlife. In the northeastern United States the Atlantic bluefin tuna had declined nearly 90 percent over the previous two decades as a result of overfishing, while hydroelectric dams and logging in the Northwest had almost wiped out the salmon. In the South Pacific, coral reefs are destroyed through fishing practices that use dynamite or cyanide-poisoning and by increasing tourism and development that leads to excess silt and sewage runoff. Though Safina protests unsustainable fishing practices, he admits to eating seafood himself and enjoys marine fishing. And unlike many environmentalists, he places blame not only on the fishers or the loggers but also on the excesses of consumers in developed nations. There is no doubt that humans will continue to depend on the oceans as a source of food, writes Safina, and the answer is not to halt fishing altogether. But he does stipulate that for regional fisheries to survive, new regulations will have to be implemented to allow for recovery of fish populations.

In addition to his book, Safina has published over 100 scientific and popular articles on ecology and marine conservation. He has testified at numerous congressional and state hearings and has served on the Smithsonian Institution's Ocean Planet Advisory Board. In 1998 Safina was selected by *Audubon* magazine as one of the top 100 most influential conservationists of the century. He continues his work at the Living Oceans Program and is also a lecturer at Yale University. He lives in Islip, New York.

BIBLIOGRAPHY

Lee, M., and Carl Safina, "Effects of Overfishing on Marine Biodiversity," *Current: The Journal of Marine Education*, 1995; McKibben, Bill, "The Next Waves," *Interview Magazine*, 1998; Safina, Carl, *Song for the Blue Ocean: Encounters along the World's Coasts and beneath the Seas*, 1997; Safina, Carl, "Song for the Swordfish," *Audubon*, 1998; Safina, Carl, "To Save the Earth, Scientists Should Join Policy Debates," *Chronicle of Higher Education*, 1998; Safina, Carl, "Where Have All the Fishes Gone?" *Issues in Science and Technology*, 1994; Safina, Carl, "The World's Imperiled Fish," *Scientific American*, 1995.

Sagan, Carl

(November 9, 1934–December 20, 1996)
Astronomer, Television Show Host

Carl Sagan was one of the best-known scientists among the public in the United States and perhaps in the world, recognized for his contributions to our knowledge of Venus and Mars, for his extensive work on Mariner missions of the National Aeronautics and Space Administration (NASA), and for his search

Carl Sagan (Joseph Sohm; ChromoSohm Inc./Corbis)

for extraterrestrial life. Sagan became a household name to U.S. audiences, starting with regular guest appearances on the popular television talk show, *The Tonight Show,* and on the public television series, *Cosmos,* that he hosted, and continuing with his novel, *Contact* (1985), which was made into a major Hollywood film at the end of his life. Additionally, Sagan was one of the authors of a scientific paper predicting that a "nuclear winter," or severe climate disruption that would threaten all life on the planet, would result from nuclear war. This paper influenced national and international nuclear weapons policy, helping to convince the American public of the necessity to end the cold war and the nuclear arms race.

Carl Sagan was born in Brooklyn, New York, on November 9, 1934, to Rachel Gruber Sagan and Samuel Sagan, a Russian immigrant who worked in a clothing factory. As a young child, Sagan was captivated with stars. He also became a devoted science fiction reader, especially enjoying novels about Mars by Edgar Rice Burroughs. Sagan began studying simple mathematical relationships involving stars and planets at an early age. When his family moved to New Jersey, Sagan entered Rahway High School, where he maintained a strong interest in science, and decided to become a professional astronomer. When he graduated in 1951, he was voted the student most likely to succeed. Sagan attended the University of Chicago, receiving a B.S. in physics in 1955, and then, supported by a National Science Foundation grant, he earned an M.S. in physics in 1956. He was awarded his Ph.D. in astronomy and astrophysics in 1960, also from the University of Chicago.

Beginning in 1960, Sagan became increasingly involved with the National Aeronautics and Space Administration, studying Jupiter, and he was asked to advise the NASA astronauts who would be landing on the moon. Sagan won NASA's Apollo Achievement Award in 1970 and worked on NASA's Pioneer 10 and 11 explorations of Saturn and Jupiter and the Mariner 9 and Viking expeditions to Mars.

Sagan did postdoctoral work at the University of California at Berkeley and in 1962 joined the Smithsonian Astrophysical Observatory in Cambridge, Massachusetts, as an astrophysicist. He taught at Harvard University as an assistant professor. Over the next few years, Sagan conducted significant studies of Mars, which culminated with an article for *National Geographic* magazine. In 1968, Sagan moved to Cornell University in Ithaca, New York, where he would live for the rest of his life. At Cornell, he first served as an associate professor of astronomy at the Center for Radiophysics and Space Research, then was promoted to professor and associate director of the center. In 1977 Sagan became the David Duncan Professor of Astronomy and Space Science at Cornell. By this point in his career, Sagan had contributed more than 200 articles to scientific journals.

Sagan began producing a television series for the Public Broadcasting Service (PBS) entitled *Cosmos* in 1970. Making scientific exploration comprehensible and accessible to television viewers around the world, the series became one of the most popular in PBS history. Sagan became a regular guest on Johnny Carson's *Tonight Show*, bringing astronomy and science into the living rooms of millions of American viewers. In 1973 he published his first book, *The Cosmic Connection*, which was very successful, and he continued writing books that brought science home to the average reader. In 1978 he won a Pulitzer Prize for *The Dragons of Eden* (1977), a book about the evolution of the human brain. Sagan then wrote *Cosmos* (1980), a companion book to the television series, and *Murmurs of Earth: The Voyager Interstellar Records* (1978). Sagan's first novel, *Contact*, was published in 1985 and was made into a major Hollywood film in 1997.

Sagan continued his involvement in space exploration throughout the 1980s and 1990s. He believed it was possible that Jupiter's moon might also have some form of life. Sagan also helped NASA to establish a groundbreaking radio astronomy search program for extraterrestrial life called Communication with Extra-Terrestrial Intelligence (CETI).

Actively involved in politics since graduate school, Sagan participated in student protests against the Vietnam War and was an avid supporter of the Democratic Party. In December 1983 Sagan cowrote an article for *Science* warning against the harmful consequences of nuclear war on the global climate. The article stated that even a limited number of nuclear explosions could severely alter the world's climate, causing a "nuclear winter." Strongly debated among respected scientists worldwide, this article was followed with a number of studies by scientists around the world on the effects of nuclear war as well as other human interference in the world's climate, including global warming. Their findings influenced many countries, institutions, and individuals to avoid the potential for nuclear war and its devastating global environmental consequences. As a result of his antinuclear research, Sagan won the National Committee for a Sane Nuclear Policy (SANE) National Peace Award in 1984, the Kennan Peace Award from SANE/Freeze in 1988, and the Helen Caldicott Leadership Award, presented by Women's Action for Nuclear Disarmament in 1988.

Sagan felt strongly that space exploration was misunderstood by many environmentalists, who opposed space missions such as the Cassini probe to Saturn. In Sagan's view, the space sciences and the life sciences were interdependent. In addition to providing unusual access to weather and meteorological studies that can monitor long-term trends in global warming, space exploration and technology offer unparalleled views of the effects of ozone layer depletion, nuclear winter, deforestation, strip mining, ocean dumping, and other environmental hazards. Sagan felt this was extremely important to furthering our understanding of how to protect the environment on earth.

Sagan was married three times and had three sons and two daughters. In 1994, he was diagnosed with myelodysplasia, a serious bone-marrow disease. Despite his illness, Sagan continued to work on numerous projects. His last book, *The Demon-Haunted World: Science as a Candle in the Dark*, was published in 1995. At the time of his death, Sagan was coproducing a film version of his novel, *Contact*, with his wife, Ann Druyan. Released in 1997, the film received popular and critical acclaim as a testimony to Sagan's enthusiasm for the search for extraterrestrial life. On December 20, 1996, Sagan died at the Fred Hutchinson Cancer Research Center in Seattle, Washington.

BIBLIOGRAPHY

"The Planetary Society," http://www.planetary. org/; RP Turco et al. "Nuclear Winter: Global Consequences of Multiple Nuclear Explosions," *Science*, 1983; Sagan, Carl, *Billions and Billions: Thoughts on Life and Death at the Brink of the Millennium*, 1997; Sagan, Carl, *The Demon-Haunted World: Science as a Candle in the Dark*, 1996; Sagan, Carl, *Pale Blue Dot: A Vision of the Human Future in Space*, 1994; Sagan, Carl, and Ann Druyan, *Shadows of Forgotten Ancestors: A Search for Who We Are*, 1992; "Scientific American," http://www.sciam. com/explorations/010697sagan/010697explrations.htm; "SETI Institute Online," http://www. seti_inst.edu/.

Sale, Kirkpatrick

(June 27, 1937–)
Cofounder of the E. F. Schumacher Society, Writer, Bioregionalist

Kirkpatrick Sale has advocated social change throughout his long career as a writer. A cofounder of the E. F. Schumacher Society, he has often voiced discontent with the environmental deterioration caused by large-scale commerce. "Something is deeply wrong with America," he wrote in *Human Scale*, his

1980 book on the problems of modern society, and pointed to our decaying environment, the nuclear threat, urban violence, promiscuity, and increased feelings of alienation as markers of societal illness. A former editor of *The Nation* and a freelance journalist, he is author of eight books on subjects ranging from African culture to the neo-Luddites.

Born in Ithaca, New York, on June 27, 1937, to William M. and Helen (Stearns) Sale, John Kirkpatrick Sale grew up in middle-class postwar prosperity. After graduating from high school, he enrolled at Swarthmore College in 1954, but stayed only one year before moving on to Cornell University, where his father was an English professor. Sale edited the student newspaper, often writing scathing editorials. He graduated in 1958 with a B.A. in history, but journalism was his true calling.

A persuasive writer who was bored with convention, he found work in 1959 as editor of the leftist paper, *The New Leader*. It launched him into the publishing world, where he debated social issues for the next three years. He longed for life experience, however, and in 1961 journeyed to Africa, serving as foreign correspondent to the *Chicago Tribune* and *San Francisco Chronicle*. He went back to New York to marry fellow editor Faith Apfelbaum in mid-1962 and in 1963 returned to Africa, taking a post as lecturer in history at the University of Ghana. He absorbed as much local culture as he could, gathering material that would eventually lead to his book *The Land and People of Ghana* (1972).

Returning to New York in 1965, where he has lived since then, Sale became editor of the *New York Times Magazine*. The late 1960s were heady political times, and he was caught in the fervor of

Vietnam protest. The student movement at Cornell drew his attention, and in 1968 he decided to turn freelance in order to write about the new activism. He devoted the next two years of his life to studying the Students for a Democratic Society. The experience, in his own words, "further radicalized" him. Sale's book about the movement, *SDS*, came out in 1973.

Established as an independent thinker and writer, Sale published another political volume, *Power Shift* (1975), then turned his attention to another crisis: the damaged environment. He joined the Small is Beautiful movement, inspired by E. F. Schumacher. Schumacher was a Swiss decentralist who advocated small, sustainable communities in the face of massive industrialism. Sale embraced such ideas and expounded on them in his book *Human Scale* (1980). In 1981 he helped form the E. F. Schumacher Society, which is dedicated to Schumacher's vision. Sale still serves on its board.

For the next year Sale took on the prestigious role of editor at *The Nation* magazine, but he continued to study bioregionalism, convinced that humans would benefit from a return to the land. He spoke up for geographic understanding and small-scale agriculture as cures for society's excesses. His ideas resonated among environmentalists. *Dwellers in the Land*, his next opus, was published by the Sierra Club in 1985. Sale was intent on documenting the environmental movement in much the same way that he had studied political change. He spent much of the next several years writing a history of the environmental movement in the United States, which culminated in *The Green Revolution*, published in 1993.

Sale has written widely on the damage technology does to modern society. For a time in the mid-1990s, he was even suspected of being the Unabomber. When Ted Kaczynski's manifesto appeared anonymously, Sale gained media attention for suggesting that its author was "rational" and "reasonable." Although he disagreed with violent action against individuals, Sale said the Unabomber was not all wrong. The alienation of modern people and the resulting ill effects on the land were what bioregionalism also sought to cure. His most recent book, *Rebels against the Future* (1995), explored the link between the Luddites, who protested the Industrial Revolution, and the neo-Luddites, who speak out against a technology- and profit-driven society.

Sale was appointed codirector of the Hudson Bioregional Council in 1985. He has also served on boards for the School for Living, Project Work, the PEN American Center (as vice president), and The Learning Alliance. A contributing editor for *The Nation*, he continues to send articles to various publications, including the *Evergreen Review, Mother Jones, Green Revolution*, and *New Roots*. He lives in New York City. He and his wife, Faith, have two children, Rebekah Zoe and Calista Jennings.

BIBLIOGRAPHY

Sale, Kirkpatrick, *Human Scale*, 1980; Sale, Kirkpatrick, *Dwellers in the Land: The Bioregional Vision*, 1985; Sale, Kirkpatrick, *The Conquest of Paradise: Christopher Columbus and the Columbian Legacy*, 1990; Sale, Kirkpatrick, *Rebels against the Future: The Luddites and Their War on the Industrial Revolution*, 1995.

Sanjour, William

(February 19, 1933–)
Hazardous Waste Expert at the Environmental Protection Agency, Whistleblower

An administrator and consultant for the Environmental Protection Agency (EPA) for 30 years, William Sanjour has played a key role in exposing failures and corruption at the agency. Sanjour helped craft the landmark 1976 Resource Conservation and Recovery Act (RCRA) and called attention to the Carter administration's efforts to weaken it. Sanjour then continued to "blow the whistle" on agency waste, fraud, and abuse. He has fought a number of battles to maintain his position within the EPA, winning important grievances and Department of Labor civil actions. With the help of the National Whistleblower Center, Sanjour won a federal court case blocking a regulation by the Office of Government Ethics that would have limited the ability of whistleblowers to speak to citizens groups on their own time. Sanjour has been an important resource for legislators, journalists, and environmental groups by providing an insider's view of the workings of the EPA.

William Sanjour was born in New York City on February 19, 1933. He served in the United States Army from 1952 to 1954, then studied at the City College of

New York, receiving a B.S. in physics in 1958. He earned an M.A. in physics from Columbia University in 1960 and worked for the United States Navy's Center for Naval Analysis as an operations analyst until 1964. From 1964 to 1966 he worked as an operations research analyst for American Cyanamid Corporation, and from 1966 to 1967 he worked for the Chemical Construction Corporation. In 1967 Sanjour joined Ernst & Ernst, a consulting firm in Washington, D.C., where he first began consulting for the EPA. He conducted computer simulations of air pollution emissions, designing a model to achieve maximum pollution reduction with minimum costs. From 1972 to 1974 he continued his work for the EPA as a private consultant.

Sanjour joined the EPA full time in 1974, as a branch chief within the newly created Hazardous Waste Management Division. His branch conducted studies to assess the scope of the hazardous waste problem in the United States. Its studies detailed treatment methods, analyzed health and safety impacts, and documented more than 600 cases of damage caused by hazardous waste. These studies were instrumental in the passage of the Resource Conservation and Recovery Act, the first national hazardous waste management law. In 1978, with the country gripped by inflation spurred by the energy crisis, the Carter administration sought to protect industry from the costs of complying with RCRA. The EPA was neither implementing nor enforcing hazardous waste regulations, and after first attempting to deal with the matter internally, Sanjour went public with claims that the EPA was failing to live up to the letter and spirit of RCRA. In response, the agency trans-

ferred Sanjour in 1979 to a position with essentially no duties. Sanjour fought the transfer and won appointment as head of the Hazardous Waste Implementation Branch in 1980, a position he held until 1983. This branch helped regional EPA offices implement RCRA and drafted regulations for the transportation of hazardous waste.

During this period Sanjour continued to speak out on important hazardous waste issues, testifying before Congress and speaking and writing for grassroots environmental groups. During the Reagan administration, Sanjour and fellow EPA whistleblower HUGH KAUFMAN were targeted for their actions, and the EPA tried to silence them. Rita Lavelle, the political appointee in charge of Sanjour's office, was responsible for fabricating an unsatisfactory performance review of Sanjour. He fought the evaluation and won, under the agency's own grievance procedures. Lavelle was later jailed for lying to Congress. In 1984, weary of agency harassment, Sanjour went on loan to the Congressional Office of Technology Assessment (OTA), where he wrote a report critical of EPA hazardous waste regulations. The report appeared in OTA's 1985 study, *Superfund Strategy*.

When Sanjour returned to the EPA, he had been removed from his position as head of the hazardous waste implementation branch and reassigned as a policy analyst. In this capacity he wrote regulations for government procurement of recycled materials, including lubricating oil, retread tires, and insulating materials. Sanjour received an outstanding performance award for these achievements. In 1989 Sanjour and Kaufman garnered national press attention for their criticism of plans to clean up Boston Harbor, which they charged did not address adequately the problem of contaminated sludge. In 1995 Sanjour was assigned to assist the Superfund ombudsman, investigating citizen complaints against regional EPA offices in their implementation of the Superfund program. He eventually became frustrated in the position by the lack of cooperation from the agency and was reassigned to a position as policy analyst for the Technology Innovation Office at the EPA.

Sanjour has held this position since 1997. His duties include developing nationwide programs for public-private partnerships to promote new technologies for hazardous waste cleanup. Sanjour lives in Arlington, Virginia.

BIBLIOGRAPHY

Sanjour, William, "In Name Only," *Sierra*, 1992; Tye, Larry, "US Judge Lashes out at EPA Pair," *Boston Globe*, 1989; "William Sanjour Homepage," http://pwp.lincs.net/sanjour/.

Sargent, Charles Sprague

(April 24, 1841–March 22,1927)
Dendrologist

Dendrologist Charles Sprague Sargent was the country's foremost expert on trees during the late 1800s, the author of a 14-volume illustrated encyclopedia of the trees growing in North America (*The Silva of North America*, 1891–1902). Sargent made the first official surveys of the nation's forests and became an influential advocate of the establishment of state and national forest preserves.

Charles Sprague Sargent was born into a wealthy Boston family on April 24, 1841. His father was a merchant who traded goods from the East Indies. Sargent was sent to private preparatory schools and then to Harvard University, where he studied under the eminent botanist ASA GRAY. Upon graduation from Harvard in 1862, Sargent enlisted in the Union Army, rising to the rank of major during the Civil War. When the war

ended, he traveled to Europe, where he spent three years studying horticulture. He returned to Harvard in 1872, where he was appointed director of Harvard's Botanic Garden, which Gray had developed. In 1873 Sargent became director of the university's new Arnold Arboretum, a 260-acre estate in Jamaica Plain, Massachusetts, a recent bequest to Harvard. Sargent worked with landscape architect FREDERICK LAW OLMSTED SR. to design the plantings throughout the property's meadows, rolling hills, and gentle valleys. Over the course of his 54 years as director of Arnold Arboretum, Sargent oversaw its transformation into the nation's most important study center for dendrology, the study of trees and shrubs. In addition to the groves of hemlocks, cedars of Lebanon, and rare oaks, Sargent assembled a well-stocked herbarium and a grand botanical library on the property.

Students traveled from the world afar to study at Sargent's Arnold Arboretum.

In 1880, Sargent was asked to coordinate a survey of the nation's forest resources for the tenth census. He worked with a large team of field botanists and then prepared his *Report on the Forests of North America* for publication in 1884. At the same time he engaged in a collection project for the American Museum of Natural History in New York City. He assembled specimens of American trees, and his wife, Mary Allen Robeson, made more than 400 watercolor paintings of flowers and fruits for a large exhibit. This work led to his appointment to the Northern Pacific Transcontinental Survey, which traveled westward in 1882 and 1883 to survey the wild lands of the West. Impressed by the botanical value of the lands he passed through, Sargent began to lobby for the preservation of certain tracts, particularly the area that later became Glacier National Park in northern Montana. Following that, in 1884 and 1885, Sargent was named chairman of the New York State Commission on Forestry, which successfully recommended the preservation of the Adirondack and Catskill forests. During the 1890s, the weekly *Garden and Forest* magazine that Sargent edited sounded a steady call for the systematic preservation of the valuable forests of the United States.

Sargent and the young GIFFORD PINCHOT, who was at the dawn of his career in forestry, acted together to lobby the federal government for a national forest policy. After garnering endorsements from such influential institutions as the New York Board of Trade and New York Chamber of Commerce, the National Academy of Sciences appropriated $25,000 to pay for a survey and

Charles Sprague Sargent (Library of Congress)

a study of the forests of the United States. A six-man commission, headed by Sargent with Pinchot serving as secretary, set out for a two-year survey of the forests of the West, from Montana to Arizona. The forested areas the group was visiting were obviously in trouble: unregulated lumber cutting and livestock grazing were seriously degrading fragile forests. But the committee was split on what type of policy to recommend. Sargent and committee adviser JOHN MUIR believed their task was to recommend which areas of forest were most in need of preservation; Pinchot and committee member Arnold Hague of the U.S. Geological Survey surveyed the forests with an eye to which ones could be exploited for timber. Sargent and Muir were in favor of immediate instatement of army patrols to protect the forests. Pinchot favored control of the forests by trained forest managers, who would supervise limited cutting and grazing. The report that the committee drew up leaned toward the Sargent-Muir position, but Pinchot signed it anyway. It recommended the establishment of 13 forest preserves to protect an area totaling 21.4 million acres. Pres. Grover Cleveland approved the creation of these new preserves during his last days in office, but soon after he left office, Congress—swayed by the western logging and livestock lobbies—voted to undo 11 of these 13 reserves and to allow managed use of the remaining two. Pinchot soon became director of the country's new Division of Forestry and implemented a forest management policy that preser-

vation-oriented Sargent and Muir bitterly disagreed with.

The rest of Sargent's career was dedicated to travel and collection of specimens, further study of dendrology, and development of the Arnold Arboretum. He received a medal of honor for his services to horticulture from the Garden Club of America in 1920, and the American Genetics Association presented him with the Frank N. Meyer Horticultural Medal in 1923 for his work in plant introduction. Sargent died on March 22, 1927, in Brookline, Massachusetts.

BIBLIOGRAPHY

Fox, Stephen, *The American Conservation Movement: John Muir and His Legacy*, 1981; Nash, Roderick, *Wilderness and the American Mind*, rev. ed., 1973; Rehder, Alfred, "Charles Sprague Sargent," *Journal of the Arnold Arboretum*, 1927.

Sawhill, John

(June 12, 1936–May 18, 2000)
President of The Nature Conservancy

John Sawhill began his environmental career in 1974 when he was named director of the Federal Energy Office, or "energy czar," for the White House. In 1990, he became president of The Nature Conservancy (TNC), one of the largest conservation organizations active in the United States today. During his tenure at TNC, which ended with his death in 2000, TNC was able to protect more than seven million acres of land in the U.S.

John Crittenden Sawhill, the oldest of four children, was born to James Mumford Sawhill and Mary Munroe Gipe Sawhill on June 12, 1936, in Cleveland, Ohio. Soon after his birth his family moved to Baltimore, where Sawhill attended private schools. He went to college at Princeton, where he studied in the Woodrow Wilson School of Public and International Affairs. During this time, Sawhill was active in many extracurricular activities. He wrestled, played intramural hockey, and was a member of the prestigious Colonial dining club. In 1958, Sawhill married Isabel Van Devanter, with whom he had one child, James Winslow Sawhill. He graduated in 1959 with a B.A. degree and spent the next two years working in the underwriting and research departments of the investment brokerage firm of Merrill, Lynch, Pierce, Fenner, and Smith.

In 1960, Sawhill enrolled in New York University's Graduate School of Business Administration, earning a Ph.D. in economics, finance, and statistics by 1963. During his time there, he was employed as assistant to the dean and instructor in 1960 and 1961 and later as assistant dean and assistant professor. He also worked as senior consulting economist for the House Committee on Banking and Currency of the U.S. Congress. From 1963 to 1965 Sawhill was director of credit research and planning for the Commercial Credit Company in Baltimore, and from 1965 to 1968 he was a senior associate for McKinsey and Company in Washington. In 1968, he returned to the Commercial Credit Company, where he advanced to senior vice president in the following year.

In April 1973, Sawhill gave up this $100,000 a year job and accepted a $38,000 per year position as one of the four associate directors of the Office of Management and Budget for the Nixon administration. In this capacity Sawhill was responsible for programs relating to energy, science, and natural resources. During the fall of 1973, energy became an increased concern in the United States when the Arab oil-producing nations imposed a trade embargo on the West. The newly created Federal Energy Office (FEO), headed by John A. Love, failed to relieve the pressure of the intensifying fuel shortage crisis, and in early December Richard Nixon announced the resignation of Love. Nixon then signed an executive order that officially established the FEO within the White House and named Sawhill as the deputy director, subordinate only to William E. Simon, the newly named administrator. The FEO's most important task was preventing fuel shortages, through price regulation of domestic crude oil and a mandatory allocation program.

In April 1974, Simon was named Treasury Secretary, and Sawhill was chosen to replace him as FEO chief, the so-called energy czar. This post took on additional significance in May when Nixon signed a bill converting the FEO into the Federal Energy Administration. This bill turned the FEO into a temporary independent executive agency and gave it authority to impose conservation measures, collect data from industry, and otherwise support the administration's efforts to reduce U.S. reliance on foreign energy sources. Sawhill remained in this post until the beginning of 1975, pushing the development of domestic energy supplies and encouraging energy conservation. And even after the embargo ended in March 1974, Sawhill's agency continued to support conservation by offering energy-saving advice to consumers, encouraging the production of fuel-efficient automobiles, and asking big businesses to cut down on energy use. In late 1974, having frustrated the administration of Gerald Ford through his continued advocacy of mandatory conservation (a position not shared by Ford or the newly appointed head of the Energy Resources

Council, Rogers C. Morton), Sawhill was forced to resign.

Sawhill became the 12th president of New York University on April 12, 1975, serving until 1979, when he briefly returned to public service. He was appointed secretary of the Department of Energy until 1980, when he returned to the private sector as chairman of the U.S. Synthetic Fuels Corporation. He stayed in this position for one year, joining the board of directors of McKinsey and Co. in Washington in 1980, where he remained for the next ten years. In 1990 Sawhill became president of The Nature Conservancy. At this time, The Nature Conservancy was feeling the effects of the rapid growth that had occurred throughout 1980s and was, in many ways, outgrowing its own infrastructure. When Sawhill arrived, many basic organizational systems had been unable to keep up with the growth: the financial system was unable to produce reports on time, and the marketing system was not able to provide accurate up-to-date information on members. Sawhill brought new leadership and organization to The Nature Conservancy, correcting these problems and encouraging a shift in conservation strategy away from simple land acquisition and toward the science-based, compatible economic development strategies being utilized today. During Sawhill's 10 years as president, The Nature Conservancy membership doubled, its revenues from land sales increased fivefold, and the size of the staff tripled.

The Nature Conservancy, founded in 1951, has been referred to as "the titan of the green groups." The Conservancy has nearly a billion dollars in assets and is constantly in the process of obtaining cash donations from individual members and corporate sponsors such as Arco, DuPont, British Petroleum, and many others. Known as "nature's real estate agent," it is a nonpolitical organization, focusing on the purchase of private lands for the purpose of conserving habitats and wildlife.

As president of The Nature Conservancy, Sawhill's main focus shifted from energy conservation to habitat conservation. In a talk given at Harvard University in February 1998, he spoke about what he saw as the most pressing of the environmental issues humans face; the extinction of species. He said that a normal extinction rate is about one or two species per year, while today, largely because of human activities, we face the extinction of up to one species per hour. This rate is similar to that of the massive extinction event at the end of the Cretaceous that led to the end of the dinosaurs.

According to Sawhill, the most significant culprit leading to these extinctions is the fragmentation of natural habitats due to development. In his talk, he outlined the tactics advocated by The Nature Conservancy for dealing with this problem. The first step is to identify successful, environmentally responsible economic strategies. An example of this type of strategy can be found in The Nature Conservancy's attempts to work with Malaysian companies in identifying ecologically compatible forest management practices that generate profits and meet the needs of local people. The next step in preventing extinction is to build strong public-private partnerships, a practice that combines private and corporate funds to purchase private land in the public's interest. This type of cooperation can lead to "acceptable conservation strategies to seemingly intractable problems." Finally, and most important, people need to develop a strong conservation ethic. From Sawhill's perspective, only when people understand and embrace the importance of conservation will the problem of extinction be effectively and meaningfully challenged.

Sawhill died of diabetes on May 18, 2000 in Richmond, Virginia.

BIBLIOGRAPHY

Howard, Alice, and Joan Magretta, "Surviving Success: An Interview with The Nature Conservancy's John Sawhill," *Harvard Business Review*, 1995; Levy, Claudia, "Environmentalist John Sawhill, 63, Dies," *Washington Post*, 2000; "The Massachusetts Chapter of the Nature Conservancy," http://www.tnc.org/infield/State/Massachusetts/article8.htm; Moritz, Charles, ed., *Current Biography Yearbook*, 1979; "The Nature Conservancy," http://www.tnc.org/.

Schlickeisen, Rodger

(January 24, 1941–)
President of Defenders of Wildlife

Rodger Schlickeisen has been the president and chief executive officer (CEO) of Defenders of Wildlife since 1991. Defenders of Wildlife is a national organization dedicated to the protection of native plant and animal species and their natural habitats. Through advocacy, education, litigation, and research, the group strives to conserve species, habitat, and international biodiversity. Its efforts include helping to return wolves and bears to their natural habitats, creating and maintaining wildlife refuges, conserving wild birds and marine life, and promoting wildlife education and appreciation through their quarterly publication, *Defenders* magazine.

Born to parents Oscar Schlickeisen and Alice Rennemo on January 24, 1941, Rodger Schlickeisen grew up in Seattle, Washington, and southern Oregon, spending much of his childhood exploring the Cascade Mountains. In 1963 he earned a B.A. in economics from the University of Washington. He went on to the Harvard Business School, earning an M.B.A. in 1965, and to George Washington University, where he received his doctorate in business administration in 1978.

Schlickeisen began his professional life with a string of government jobs. From 1966 to 1968 he was a captain in the United States Army. From 1968 to 1970 he served as a finance loan officer for Latin America with the United States Export-Import bank. From 1970 until 1974 Schlickeisen volunteered with the Virginia group of Common Cause, where he helped to put in place a number of open government reforms, making government meetings and decisions more freely accessible to the voting public. In 1974 Schlickeisen joined the U.S. Senate Committee on the Budget, working with that group until 1979. In 1979 Schlickeisen became the associate director for economics and government in the Office of Management and Budget for the

Rodger Schlickeisen (Courtesy of Anne Ronan Picture Library)

Carter administration, serving in that office until 1981.

Schlickeisen's environmental career was launched when he joined Craver, Mathews, Smith & Company as its CEO in 1982. The Washington, D.C., and California firm consults with a number of national environmental and other organizations on such issues as fundraising and organizational development. Schlickeisen directed operations at Craver, Mathews, Smith & Company until 1987. During this time the company consulted for various congressional campaigns as well as organizations including Greenpeace, the Sierra Club, and the Natural Resources Defense Council, improving their fundraising through direct mail marketing techniques.

In 1987 Schlickeisen became the chief of staff for Sen. Max Baucus, a Democrat from Montana and one of the only proenvironment senators from the northern Rockies states. Baucus is a ranking member of the Senate Environment and Public Works Committee. Schlickeisen worked with Baucus until 1991, helping him to win reelection to a third term in the Senate.

Defenders of Wildlife hired Schlickeisen as president and CEO in 1991. Since joining Defenders of Wildlife, Schlickeisen has increased membership in the organization from 62,000 to nearly 400,000. He is a staunch supporter of the Endangered Species Act (ESA), claiming that it is the "single most effective tool for keeping our lands biologically healthy for future generations." Defenders of Wildlife continues to help enforce the ESA and is a member of the Endangered Species Coalition, which is housed in the Defenders of Wildlife's offices. Schlickeisen has also been a champion of the efforts to return wolves, the top natural predator, to Yellowstone National Park and central Idaho and hopes to see the animal reintroduced to Washington and New York states as well as the Southwest. Schlickeisen believes that the wolf represents "the very embodiment of wild nature" to Americans. For this reason he sees the success of the reintroduction efforts as pivotal to other similar conservation endeavors.

In addition to remaining president of Defenders of Wildlife, Schlickeisen is currently a member of the board of the League of Conservation Voters and the Natural Resources Council of America. He resides in Alexandria, Virginia, with his wife, Susan, and their son, Derek.

BIBLIOGRAPHY

"Defenders of Wildlife Homepage," http://www.defenders.org; Rembert, Tracey C., "Rodger Schlickeisen: Defending America's Wilder Ways," *E Magazine*, 1998; Schlickeisen, Rodger, "After 25 Years, Still Protecting Endangered Species," *Christian Science Monitor*, 1998; Schlickeisen, Rodger, "Connection with Wolves," *Defenders*, 1994/1995.

Schneider, Stephen

(February 11, 1945–)
Climatologist

Climatologist Stephen Schneider was one of the first prominent U.S. scientists to warn of the impending dangers of global warming in the late 1970s. Throughout his career, Schneider has attempted to encourage long-term, scientifically grounded policy decisions, by means of sharing his findings with the public and with policy makers through books, magazine articles, and television and radio interviews. He believes in the strength of interdisciplinary research and has collaborated on research teams and government committees with experts and scientists from a variety of disciplines.

Stephen Henry Schneider was born in New York City on February 11, 1945, to Samuel and Doris (Swarte) Schneider. He loved science, having built a five-foot-long telescope before he grew that tall, and used it with wide-eyed excitement to see the craters of the moon, the rings of Saturn, or the phases of Venus. By high school that enthusiasm turned to the drag racing track, where he and his brother won trophies (Schneider, not yet 16, was the mechanic; his older brother Peter was the driver).

Schneider entered Columbia University in 1962, where he studied mechanical engineering and received his B.S. in 1966. He continued his studies at Columbia, earning an M.S. in 1967 and a Ph.D. in mechanical engineering and plasma physics in 1971. Schneider was working in plasma physics as a graduate student in 1970 at Columbia University when the first Earth Day came along. He was shocked to learn that pollution could affect the global climate and even more surprised to discover that very little research had been done on the subject. As a result, he changed his field of research and became a postdoctoral fellow in 1971 at the Goddard Institute for Space Studies (GISS) of the National Aeronautics and Space Administration (NASA) in New York City as a National Research Council research associate, working on the role of greenhouse gases and suspended particulate material on climate. He remained at GISS until 1972, when he was awarded a postdoctoral fellowship at the National Center for Atmospheric Research (NCAR), which facilitated his move to Boulder, Colorado, for the year-long fellowship. He remained with NCAR until 1996, working in various capacities on climate issues.

In 1975, Schneider founded the journal, *Climatic Change*, an interdisciplinary, international journal devoted to the description, causes, and implications of climatic change. He continues to serve as its editor. As a research scientist, Schneider has done pioneering modeling work in the fields of atmospheric science and global climatology and in particular has explored the relationship of biological systems to global climate change. He is known for coupling models of the atmosphere to models of other climate subsystems such as oceans, ice, or biosphere, as well as to social systems (via economic models, and so on). He has initiated new research and policy directions in environmental issues, in part by crossing disciplinary boundaries and combining disciplinary research contributions, original interdisciplinary syntheses, popular publications, legislative testimony, media appearances, television documentary productions, and the organization of scientific and public policy meetings and seminars.

With support from numerous federal and private foundation grants, Schneider has conducted research on the earth's climate, exploring such issues as the climatic effects of nuclear war, climatic impacts on society, the ecological implications of climate change, and the application of climate modeling to policy making. Of particular interest to Schneider is the greenhouse effect, the atmospheric process whereby carbon dioxide and water vapor in the earth's atmosphere trap solar radiation, creating temperatures capable of sustaining life as we know it on the surface of the planet. These "greenhouse gases" are necessary for sustaining life on earth. However, the greater the amount of greenhouse gases there is in the atmosphere, the more solar radiation gets trapped. Thus, when we emit unnatural amounts of carbon dioxide into the atmosphere, we are creating a situation where more heat will be trapped and the surface temperature of earth will rise. As early as the mid 1970s, Schneider was issuing warnings about the global warming dangers brought about by the massive amounts of carbon dioxide humans were pumping into the atmosphere through the burning of fossil fuels. Nearly 25 years later, and after a decade of slowly rising temperatures, the question among scientists of whether or not global warming will take place is largely moot. Nowadays it is more a question of how much the climate will change and what type of impact these rising temperatures will actually have. It is believed that they will lead to a rise in sea level, to more variable and extreme weather, to changes in local precipitation patterns, and to the expansion of deserts into grazing and agricultural lands. Schneider's suggestions for dealing with global warming include the reduction of emissions, the elimination of the use of chlorofluorocarbons, the halting of deforestation, as well as the implementation of reforestation efforts.

While at NCAR in Boulder, Schneider also examined the theory of nuclear winter, a theory that attempts to explain the possible climatic and atmospheric ramifications of a nuclear war. He is also interested in advancing public understanding of science and in improving formal environmental education in primary and secondary schools. Schneider firmly believes in the public airing of scientific issues in language understood by the majority of people (including and especially lawmakers). This belief has made Schneider unpopular with some of his colleagues, but he has made a continuous practice of "going public" with his scientific findings. Ultimately, he believes, science can only provide probabilities, and it is up to the citizens to decide how much risk to take.

Schneider is the author of *The Genesis Strategy: Climate and Global Survival* (1976), *The Co-Evolution of Climate and Life* (1984), *Global Warming: Are We Entering the Greenhouse Century?* (1989), and *Laboratory Earth: The Planetary Gamble We Can't Afford to Lose* (1997). And he has edited several other important books, including the *Encyclopedia of Climate and Weather* (1996). He has authored or coauthored over 200 scientific papers, proceedings, legislative testimonies, edited books, and book chapters as well as some 120 book reviews, editorials, and published newspaper and magazine interviews. He is a frequent contributor to commercial and noncommercial print and broadcast media on climate and environmental issues: *Nova, Planet Earth, Nightline, Today Show, Tonight Show, Dateline,* and programs on the Discovery Channel and British, Canadian, and Australian Broadcasting Corporations, among many others.

Schneider has received many awards for his scientific achievements. In 1992, he received a MacArthur Foundation Prize Fellowship; in 1991, he was awarded the American Association for the Advancement of Science/Westinghouse Award for Public Understanding of Science and Technology; and in 1979, he was selected as one of 35 Top Americans Under 35 by *U.S. Magazine.* Schneider is a member of the American Meteorological Society, the Society for Conservation Biology, and the Ecological Society of America, among other organizations. Schneider is a frequent witness at congressional and parliamentary hearings. He has been a member of the Defense Science Board Task Force on Atmospheric Obscuration; was a consultant to the White House in the Carter, Nixon, Reagan, Bush, and Clinton administrations; and met with British prime minister Margaret Thatcher to discuss global warming issues during her visit to Colorado in 1990. Today, Schneider continues his research into global change and teaches courses on climate and environmental policy at Stanford University in California.

Since 1992, Schneider has lived in California. He has two grown children, Rebecca and Adam, from a previous marriage, and is married to Prof. Terry Root of the University of Michigan.

BIBLIOGRAPHY

Athanasiou, Tom, "Greenhouse Blues," *Socialist Review*, 1991; Johnson, Dan, "Earth's Changing Climate," *The Futurist*, 1997; Schneider, Stephen H., *Global Warming: Are We Entering the Greenhouse Century?*, 1989; Schneider, Stephen H., "The Greenhouse Effect: Science and Policy," *Science*, 1989.

Schultes, Richard Evans

(January 12, 1915–)
Ethnobotanist

Widely considered the father of ethnobotany and the last of the great explorers, Richard Evans Schultes has been deeply influential in forwarding both conservation and a respect for indigenous cultures and knowledge in his work and research. The foremost botanist of the Amazon of the twentieth century, Schultes's career has significantly advanced conceptions of the plant world in both academic and public spheres.

Richard Evans Schultes was born January 12, 1915, in Boston, Massachusetts. Though he would often describe himself as a fourth- or fifth-generation Bostonian, he was also the grandson of German immigrants on his father's side. Schultes was quite ill as a child and was advised by his family doctor to stay close to home instead of commuting by trolley through the damp and congested tunnel that connected East Boston, where he lived, to the mainland to attend Boston Latin, the most prestigious public school in the city and the most appropriate for a student of Schultes's caliber. So he attended East Boston High School, where he excelled in all disciplines, especially Greek, Latin, chemistry, and foreign languages. In his spare time he read, raised rabbits, worked in the family garden, and ran errands, working for a nickel a day, which he put toward future college expenses. He applied only to Harvard, did well on his entrance exams, and became the first member of his family to attend university. His plan was to become a doctor. Rather than staying in residence at Harvard, Schultes saved money by commuting from home.

At the end of his first year, Schultes received financial support in the form of the Cudworth Scholarship, a small award given by the Unitarian Church of East Boston. The scholarship was endowed to help Harvard students of good moral standing who hailed either from East Boston or Lowell. Looking for work during his second year to supplement his scholarship, Schultes took a job at the Botanical Museum, filing cards and stacking books in the Economic Botany Library. There, he fell under the influence of economic botany professor Oakes Ames, the museum's director and a millionaire many times over.

Intrigued by the material he was organizing, Schultes decided to enroll in Ames's biology course—"Plants and Human Affairs"—in the spring. As part of the course requirements, Schultes had to prepare a report on a monograph entitled "Mescal: The Divine Plant and Its Psychological Effects," by the German psychiatrist, Heinrich Klüver. Published in 1928, this monograph was the only study in English that described the pharmacological effects of peyote. Inspired by Klüver's descriptions of peyote's uses and effects, Schultes asked Ames whether he might write about peyote for his undergraduate thesis. Ames agreed on one condition: that research of the literature would not be sufficient; Schultes would have to travel west to Oklahoma and see the plant in use. Ames told Schultes that little was known about peyote, and a history of the plant's use would be a valuable study. He even promised his eager pupil some grant money; Schultes discovered some years later that the money had come directly from Ames's own pocket. So began a long and distinguished career of field study into the cultural and scientific mysteries of plants with psychoactive powers. Schultes completed his A.B., *cum laude*, in 1937 and immediately went to work on an A.M., completing his Ph.D. in biology in 1941.

Torn between accepting a one-year research fellowship in South America and staying in the United States in case the war in Europe should spread, Schultes was persuaded by Ames, who believed that biologists should work in the field as well as the laboratory, to opt for the research fellowship. Shultes went to Bogotá, Colombia. From Bogotá, he took buses as far as the roads would go, before transferring himself and his equipment to horses to cross the Andes. Across the Andes, he changed to dugout canoe to paddle down uncharted tributaries of the Amazon River. The area that Schultes had chosen to study remains today one of the wildest and least-explored areas on earth. His only guide was the diary of his boyhood hero Richard Spruce, an English botanist who had explored the same area some 90 years earlier. For a botanist who grew up in New England where there are 1,900 species of plants, the Amazon jungle with an estimated 80,000 plant species was overwhelming.

Despite the remote location of his study site, Schultes learned of the Japanese attack on Pearl Harbor on a trader's radio within a few weeks of the event. He decided to return to Bogotá, to the American embassy, to enlist in the U.S. military. The trip took a month and a half. Embassy officials decided that Schultes could be of most use by continuing his botanical studies. The Japanese had captured all the rubber plantations in Southeast Asia, and the Allies were in desperate need of a new source of natural rubber for the war effort. Accordingly, in early 1942, Schultes was hired as a field agent for a government entity known as the Rubber Development Corporation. His task was to locate rubber trees in Amazonia and to teach local natives how to extract the latex.

When the war ended, the Colombian government and the U.S. Department of Agriculture asked Schultes to stay on and help develop a living nursery of rubber trees. He gladly accepted, as it would also allow him to continue his research of the medicinal and poisonous plants used in the Amazon jungle. Over the next 13 years, Schultes returned to New England only for occasional two-month Christmas breaks, where he could see his family, skate, and ski.

Schultes's relationships with the indigenous people with whom he lived and from whom he learned is among the most impressive aspects of his career. Desiring to learn from them rather than teach them, Schultes gained virtually full acceptance among many tribes. Though he acknowledged proficiency in understanding only two Amazonian languages, he apparently managed to communicate with little difficulty with tribal peoples who spoke other tongues.

Schultes's one-year fellowship turned into a 13-year sojourn and might have lasted even longer, had not someone in Harvard's administrative offices discovered that an honorary research fellow named Schultes had been on a leave of absence since 1941. Schultes returned to Harvard in 1954. Between annual trips back to the northwest Amazon, he taught economic botany, the oldest science course at Harvard and the one that had captivated him when he was a student. He was eventually appointed director of the Botanical Museum and has since received countless awards, most notably the highest honor offered by the Republic of Colombia, the Cross of Boyaca, in 1983 and, the following year, the World Wildlife Fund's Gold Medal and the Tyler Prize for environmental achievement. In 1986, in recognition of his contributions to conservation in that country, the government of Colombia named a 2.2-million-acre protected tract "Sector Schultes."

Schultes married Dorothy Crawford McNeil, an operatic soprano, in 1959. Together they had three children. Portraying himself as conservative and almost Victorian, Schultes nevertheless was concerned about the need for conservation in the Amazon in the 1940s before the emergence of the general recognition that the region was in danger. Chewing coca leaves and sampling potent hallucinogens and various witch doctors' brews also seems out of character for such a straitlaced Bostonian, who always took tins of baked beans into the jungle along with his botanical supplies.

By the time he retired in 1985 to suburban Massachusetts, Schultes had identified three previously unidentified genera and more than 120 species of Amazonian plants. In addition, a large species of Amazonian cockroach was named for him. Over the course of his long career, Schultes collected 24,000 species, hundreds of which were previously unknown to science. Schultes's scholarship and contributions of specimens to the Botanical Museum at Harvard have significantly increased scientists' understanding of the plant world. Furthermore, his recognition of the importance of conserving ethnobotanical lore superceded the awareness of most biologists of any conservation problems in tropical areas.

BIBLIOGRAPHY

Clark, Tim, "Old Man Amazon," *Yankee*, 1986; Davis, Wade, *One River: Explorations and Discoveries in the Amazon Rain Forest*, 1996; Davis, Wade, *The Serpent and the Rainbow*, 1985; Kahn, E. J., Jr., "Jungle Botanist," *New Yorker*, 1992; Krieg, Margaret B., *Green Medicine*, 1964.

Schurz, Carl

(March 2, 1829–May 14, 1906)
Secretary of the Interior

Journalist, senator, abolitionist, and revolutionary, Carl Schurz was a visionary in several fields, including forestry. As secretary of the interior for the administration of Pres. Rutherford B. Hayes, Schurz proposed conservationist policies that were far ahead of his time and that would be adopted 30 years later.

Carl Schurz was born on March 2, 1829, near the town of Liblar, Germany. He was the oldest of three children. His mother, Marianne, was the daughter of a tenant farmer, and his father, Christian, was a schoolteacher turned small businessman. Schurz attended local schools until his father was thrown into debtors' prison. After completing his studies on his own, Schurz entered the University of Bonn in 1847, with a view toward studying history and entering the professorate. While at the university, Schurz met Prof. Gottfried Kinkel. Kinkel was a leader in the German revolutions of 1848 and 1849, and Schurz got caught up in the movement, becoming a staff aide to Kinkel during the last battles of 1849. Schurz fled the country in late 1849 but returned in 1850 to rescue Kinkel, who had been sentenced to life in prison. After months of planning, Schurz bribed a guard, lowered Kinkel out of the window of the Spandau prison, and fled to France to begin a life of permanent exile.

After being forced to leave France because of his political views, Schurz stayed for a time in England before emigrating to the United States in 1852. In 1856 Schurz bought a farm in Wisconsin. He became a passionate abolitionist and one of the great orators of his day, which attracted the attention of Wisconsin Republicans. He was nominated but never elected as lieutenant governor and served as chairman of the Wisconsin delegation to the Republican convention of 1860, which nominated Abraham Lincoln. During the Civil War Schurz served as a brigadier-general for the Union Army, where his German ancestry made him the scapegoat for several defeats. After the war Schurz began a career in journalism, eventually accepting the editorship of the German-language paper *Westliche Post* in St. Louis, Missouri. Schurz continued to be active in Republi-

Carl Schurz (Bettmann/Corbis)

can politics and was elected to the U.S. Senate from Missouri in 1868. He grew disillusioned with the corruption and imperialism of the Republican administration of Ulysses S. Grant and led a revolt that resulted in the formation of the Liberal Republican Party. The reformers nominated Horace Greeley, but the revolt was unsuccessful, and President Grant was reelected in 1872.

In 1876 Schurz supported the presidential candidacy of Republican Rutherford B. Hayes and was rewarded with the post of secretary of the interior. Schurz was the first German-born citizen to hold a seat in the cabinet. One of Schurz's drives was to eliminate the corruption that political patronage had wrought upon the executive branch. He advocated, somewhat successfully, for civil service reform throughout the Department of the Interior. This principle led to real reform in the treatment of Native Americans, whose treaties had often been negotiated and administered by agents whose first priority was their own pocketbooks. Schurz held hearings to expose unscrupulous agents and calmed the hysteria that had arisen after Custer's defeat at the Battle of the Little Bighorn. Schurz also managed to halt a move in Congress to transfer Indian affairs to the War Department, which almost certainly would have resulted in even bloodier resolution of Indian conflicts.

Schurz brought to his post a German sensibility regarding forest conserva-tion. Germany had long been interested in managing its forests and had developed early principles of reforestation. Schurz saw that the U.S. government's management of natural resources, just as in Indian affairs, was dominated by greed and corruption. Lumber companies stole timber from public lands, and Schurz intervened to try to pass legislation halting the practice. Schurz put forth the first policy for public conservation, though Congress failed to enact the reforms he suggested until the administration of Pres. THEODORE ROOSEVELT set aside the first national forests in the early 1900s, 30 years later. As secretary, Schurz did successfully institute the U.S. Geological Survey, which conducted geological research and made land use policy recommendations.

After leaving the cabinet, Schurz settled in New York City, where he continued to press for civil and conservation reform in writing. He wrote for and edited a number of journals, including the *New York Evening Post, The Nation*, and *Harper's Weekly*. Carl Schurz died on May 14, 1906, of pneumonia. He was buried in Sleepy Hollow Cemetery in Tarrytown, New York.

BIBLIOGRAPHY

Camp, Helen, "Carl Schurz," *American Reformers*, 1985; "Carl Schurz," http://rs6.loc.gov/ammem/today/oct29.html; Trefousse, Hans, *Carl Schurz*, 1998.

Chief Sealth (Seattle)

(mid-1780s–June 7, 1866)
Chief of the Suquamish and Duwamish, Orator

Chief of the Suquamish and Duwamish, two tribes in Puget Sound, Sealth is best known as a powerful orator and a friend of White settlers, who named a small settlement after him. Although Sealth's gift of oratory and friendliness toward Whites are well documented, the authenticity of a famous speech attributed to him, often presented as representative of Indian views of the environment, is open to question.

Sealth was born on what is today known as Bainbridge Island sometime during the mid-1780s. He was the son of Schweabe, chief of the Suquamish, and Scholitza, a Duwamish woman. Sealth was a boy when he encountered his first White men, George Vancouver and the crew of the *Discovery*, who anchored off Bainbridge Island in May 1792. Young Sealth was awed by the White men's technology and later in life often spoke of the favorable impression Vancouver made on him.

When Sealth came of age, he chose a bride from his mother's tribe. His first wife, who was known as La Daila, bore him one child, Kick-is-mo-lo, whom Whites later called Princess Angeline. La Daila died shortly after giving birth, greatly grieving Sealth, who took at least one other wife from his mother's tribe.

When Sealth was in his early twenties, he responded to a call for help from the Duwamish, who were threatened by a large group of warriors from a hostile inland tribe. Sealth developed a unique defensive strategy that resulted in the rapid defeat of the invaders, a feat for which he was elected chief. After his rise to the chieftainship, Sealth became known as a peacemaker and never again participated in armed combat. He converted to Catholicism in the late 1830s after the arrival of "blackrobes" from Quebec and remained a follower of the faith for the rest of his life.

The first White American settlers came to the area presently occupied by the city of Seattle in 1851, three years after Britain had relinquished its claim to the area below the 49th parallel. The settlers named their small settlement after the chief, modifying his Indian name to the present-day spelling Seattle. The chief was upset over the tribute, however, since his people's custom prohibited the use of a deceased person's name. Though the chief was still very much alive, he was worried that after he died, his name would be spoken by thousands of people everyday.

In January 1855, Sealth participated in the Point Elliott Council, signing the resultant treaty with the United States on behalf of the Duwamish and Suquamish. His recorded comments were conciliatory and hopeful, and he pledged his support to the Whites. The years immediately after the treaty council saw many skirmishes between Whites and Indians, including a famous if not particularly successful attack on the city of Seattle led by Owhi of the Yakamas, Quilquilton of the Upper Puyallups, and Leschi of the Nisquallis and Lower Puyallups. Sealth took no part in any of the conflicts and, keeping with his reputation as a man of peace and friend of the Whites, he has been credited with warning the

White settlers of more than one impending attack. Sealth maintained friendly relations with local Whites for the remainder of his life, though he did not refrain from expressing his displeasure with the Whites' less than forthright implementation of the promises made during the 1855 Point Elliott Council. He died on June 7, 1866, and was buried at Suquamish, across Puget Sound from the city of Seattle.

In 1887, Henry Smith, a Seattle doctor and occasional newspaper columnist, wrote what would become one of the most famous documents in American Indian history. Smith claimed that the piece was based on a speech Sealth made at a reception welcoming Isaac Stevens, Indian agent and first governor of Washington Territory, to the region. Smith, who claimed to have witnessed the event, did not assert that the speech was recorded verbatim but rather that it was but a "fragment of his speech," one that lacked "all the charm lent by the grace and earnestness of the sable old orator." Smith's version of the speech, which is not documented anywhere else, is not the environmentalist tract that contemporary renditions have made it out to be. Instead, it is a dirge to the indigenous inhabitants of Puget Sound, who the author of the speech was convinced were headed toward extinction, and an admonition to the Whites that the spirits of the dead Indians would forever stay with the land. "At night, when the streets of your cities and villages shall be silent and you think them deserted, they will throng with the returning hosts that once filled and still love this beautiful land. The white man will never be alone."

Since the 1960s, non-Indians have embellished and manipulated the text so as to make it more amenable to a national, non-Indian audience concerned about environmental degradation. The particularities of place are often removed, reinforcing the notion of a generalized Indian free from a specific tribal affiliation, and anachronisms have been added—the shooting of buffalo from trains, for example. Although the line does not appear in Smith's version, "How can we buy or sell the sky?" has even become a bumpersticker slogan.

Although some scholars, such as anthropologist Crisca Bierwert and Lutshoodseed elder Vi Hilbert, argue that the core of the speech is Sealth's, others, such as historian Albert Furtwangler and archivist Jerry Clark, argue that the version as recorded by Smith is too far removed from its source to be a useful account of Sealth's views of White settlement and the environment. Literature scholar Denise Low asserts that contemporary versions of the speech, based loosely on Smith's rendition, are more representative of a dominant culture that has created a "white man's Indian" of Sealth, a stereotyped "noble savage" who has been used as a pawn to advance a political agenda. Whatever the case may be, the authenticity of every version of the famous speech by Chief Seattle is open to question.

BIBLIOGRAPHY

Bierwert, Crisca, "Remembering Chief Seattle: Reversing Cultural Studies of a Vanishing Native American," *American Indian Quarterly*, 1998; Clark, Jerry L. "Thus Spoke Chief Seattle: The Story of an Undocumented Speech," *Prologue*, 1985; Furtwangler, Albert, *Answering Chief Seattle*, 1997; Hilbert, Vi, "When Chief Seattle Spoke," *A Time of Gathering: Native Heritage in Washington State*, Robin K. Wright, ed.,

1991; Kaiser, Rudolf, "Chief Seattle's Speech(es): American Origins and European Reception," *Recovering the Word: Essays in Native American Literature*, Brian Swann and Arnold Krupat, eds., 1987; Low, Denise, "Contemporary Reinvention of Chief Seattle: Variant Texts of Chief Seattle's 1854 Speech," *American Indian Quarterly*, 1995; Metcalfe, James Vernon, "Chief Seattle," *The Catholic Northwest Progress*, 1964.

Seeger, Pete

(May 3, 1919–)
Singer, Songwriter

A major force in the movement to revive the American folk song, troubador Pete Seeger has composed songs and sung about social justice and the environment since 1940. From his home overlooking the Hudson River, Seeger coordinated the creation of *Clearwater*, a replica of an eighteenth-century Hudson River sloop that has become both a tool and a symbol of the restoration of Seeger's beloved river.

Peter Seeger was born in New York City on May 3, 1919, to Charles and Constance Seeger, a musicologist and violin teacher respectively, both of whom taught at the Juilliard Institute. After forcing his two older brothers to play the violin, the Seegers decided not to impose music lessons on young Pete, yet he always had access to the family's collection of musical instruments and showed an early curiosity in whistles, the squeezebox, marimbas, and the like. At the age of eight he learned to play the ukelele, then at 13 switched to the tenor banjo, and when he traveled to a North Carolina folk music festival with his father in 1936 and discovered the five-string banjo, he immediately fell in love with that instrument. Seeger entered Harvard College with the class of 1940 but dropped out in 1938 to join a troupe of puppeteers, which toured New York State and performed at meetings for a large strike of dairy farmers. He wrote his first songs for the strike, pulling tunes from old folk songs and making up new strike-inspired lyrics for them.

In 1940 Seeger began working for folklorist Alan Lomax at the Library of Congress Archives of Folk Song. At a benefit concert for migrant workers that year, he met WOODY GUTHRIE, who became his inspiration and mentor. They formed the Almanac Singers, together with Lee Hays and Millard Lampell, and performed songs that melded folk traditions with progressive politics, in support of unions, migrant workers, and other social causes. The group disbanded at the outset of World War II. Seeger served in the United States Army during the war. During a period of leave in 1942 he married Toshi Ohta, to whom he gives much credit for the success of all of his later endeavors.

When the war ended, Seeger and Lee Hays, Ronnie Gilbert, and Fred Hellerman founded the Weavers, whose songs decried injustice yet persisted in their

Pete Seeger (Courtesy of Anne Ronan Picture Library)

constructive optimism. Hayes and Seeger collaborated on many songs, including the classic "If I Had a Hammer." The song's inspirational power was recognized immediately by the editors of the folk-song publication *Sing Out*, who printed it on the cover of the magazine's first issue. The Weavers were very popular until Seeger's Federal Bureau of Investigation file was leaked to the press, and it was revealed that he and other members of the Weavers were being investigated for their leftist political leanings. The Weavers' concerts began to be canceled unexpectedly at the last minute, and eventually the group fell apart. Seeger was subpoenaed by Joseph McCarthy's House Un-American Activities Committee and testified in 1956 about his own involvement in radical political organizations, including the Communist Party (he had been a member for about eight years during the 1940s). He refused to give the committee the names of his friends and associates who were also involved, and for this he was indicted for contempt of Congress and sentenced to ten years in prison. He never had to serve time, however, as his case was eventually dismissed on a technicality. Because he was blacklisted during the McCarthy area, Seeger was able to obtain bookings only at college campuses and small-town school auditoriums, scraping together a living for his family. One of the favorite songs of his small audiences was a rearrangement of an old gospel tune that he had learned from union organizer Zilphia Horton, "We Shall Overcome." Highlander school song leader Guy Carawan taught it to civil rights protesters in 1960, and it became the veritable anthem of those doing civil disobedience through the South during the 1960s.

The folk music revival blossomed during the 1960s with such singers as Bob Dylan and Phil Ochs incorporating the words and melodies that Lomax, Guthrie, and Seeger had rediscovered. Some of the songs that Seeger wrote or recorded became better known in their recordings by folk singers Joan Baez, Peter, Paul and Mary, and others; Seeger was never offended or hurt by that, since he himself had done the same with old folk songs or gospel songs.

During the 1960s, Seeger found a new interest—still firmly rooted in social justice—in the environment. He read and was spurred into action by RACHEL CARSON's *Silent Spring* when it was serialized in the *New Yorker*, in 1963. "Up till then," he told Steve Curwood on the public radio program *Living on Earth*, "I'd thought the main job to do is to help the meek inherit the Earth. . . . But I realized if we didn't do something soon, what the meek would inherit would be a pretty poisonous place to live." Seeger began an all-out campaign to educate himself, reading books by PAUL EHRLICH and BARRY COMMONER. He recorded his first environmental-themed album in 1965, *God Bless the Grass*, titled after a song by Malvina Reynolds.

Seeger also started paying more attention to what was happening close to his home. He and Toshi had built a log cabin on a few forested acres overlooking the Hudson River in 1949, and by the mid-1960s, the river was widely acknowledged to be a contaminated mess. In addition to the raw sewage and industrial waste flowing into the river, environmentalists were greatly concerned by a proposal to build a pump storage power plant at perhaps the most famous Hudson River landmark, Storm King Mountain. Local environmental groups were engaged in a massive lawsuit against the Federal Power Commission to stop its construction. Seeger soon dreamed up his own creative response to the Hudson's sorry state. After learning from a friend that the main form of transportation in the Hudson River Valley during the eighteenth century had been sloops—graceful, sleek sailboats of at least 70 feet in length—he came up with an idea that he felt had the power to both symbolize and enable the movement for a cleaner environment. Seeger recounted to Curwood, "I wrote a letter to my friend: wouldn't it be great to build a replica of one of these? Probably cost $100,000. Nobody we know has that money, but if we got 1,000 people together we could all chip in. Maybe we could hire a skilled captain to see it's run safely and the rest of us could volunteer." Seeger and a group of Hudson River neighbors and friends did just that; 3,000 people formed a nonprofit corporation called the Hudson River Sloop Restoration, Inc., and raised funds to build a boat. Seeger held a series of "sloop concerts" that raised about $60,000 for the project. The *Clearwater*, a 106-foot wooden sailing ship, was finally completed in 1969. It has been used as a classroom, laboratory, and stage, sailing from community to community to take nearly 20,000 schoolchildren on educational field trips each year. The boat's motto is "Clearwater's cargo is people. Her work is making them care." The organization, now called Clearwater, with more than 15,000 members, works with other environmental organizations up and down the river to fight for the cleanup of polychlorinated biphenyl

(PCB) contamination, improved water quality, better public access to the river's shores, and more effective local water conservation. Clearwater organizes dozens of festivals each year, from the Great Hudson River Revival, which attracts some 20,000 people, to smaller town shad, strawberry, corn, and pumpkin harvest festivals. At all of these, the sloop serves as a waterborne stage for performers, and guests are invited on board for short river rides.

Seeger has composed numerous songs about the *Clearwater* and the Hudson, including "My Dirty Stream":

> Sailing up my dirty stream
> Still I love it and I'll keep the dream
> That some day, though maybe not this year
> My Hudson River will once again run clear.

Another song is "Sailing Down My Golden River":

> Sailing down my golden river,
> Sun and water all my own
> Yet I was never alone.
> Sun and water, old life-givers
> I'll have them where'er I roam
> And I was not far from home.

Pete and Toshi Seeger, now grandparents in their eighties, continue to live in their cabin overlooking the Hudson and to serve Clearwater as active members.

BIBLIOGRAPHY

"Clearwater: Hudson River Sloop," http://www.clearwater.org/; Dunaway, David King, *How Can I Keep from Singing*, 1981; Hope, Jack, "A Man, A Boat, a River, a Dream," *Audubon*, 1971;

"Living on Earth Transcript January 1, 1999," http://www.livingonearth.org/archives/990101. htm#feature5; Seeger, Pete, *The Incompleat* *Folksinger*, 1972; Seeger, Pete, *Where Have All the Flowers Gone: A Musical Autobiography*, 1996.

Selikoff, Irving

(January 15, 1915–May 20, 1992)
Physician, Asbestos Researcher

A pioneer in environmental and occupational medicine, Irving Selikoff conducted the first studies to demonstrate the damaging effects of asbestos exposure. Selikoff's research resulted in a ban on most asbestos products in the 1980s and successful compensation claims by thousands of workers who had contracted cancer and asbestosis on the job. He was the founding director of the first hospital division in the United States devoted to environmental and occupational medicine; it was at New York's Mount Sinai Medical Center. Selikoff was the author of numerous books and articles in this field, including 1988's *Living in a Chemical World.*

Irving J. Selikoff was born in Brooklyn, New York, on January 15, 1915. He attended Columbia University, graduating with a B.S. in 1935. He received his medical training at the Royal Colleges of Scotland, qualifying in 1941. Selikoff completed an internship in internal medicine at Newark, New Jersey's Beth Israel Hospital in 1944 and his residency at Sea View Hospital in New York in 1947. He joined the staff of Barnert Hospital in Paterson, New Jersey, in 1947. There, in collaboration with Dr. Edward Robitzek, he conducted successful clinical trials of a powerful new drug, isoniazid, for the treatment of tuberculosis. In 1955 they were awarded the Albert Lasker Award in Medicine from the American Public Health Association for this work.

In his work at the chest clinic in Paterson, Selikoff began treating 17 patients who worked at a local asbestos plant, Union Asbestos and Rubber Company (UNARCO). All 17 had unusual illnesses, and within the first seven years, 6 of the men were dead. Eventually, 13 of Selikoff's patients died of cancer or asbestosis, and Selikoff began what he called "shoe-leather epidemiology." He first tried to get personnel files from UNARCO, and when the company refused to cooperate, Selikoff used other means, including union records and Federal Bureau of Investigation files, to track down 877 of the 933 men who had worked at the plant during World War II. Their lung cancer rates were seven times higher than normal, their stomach and colon cancer rates twice as high. In 1964 Selikoff published the results of his research in the *Journal of the American Medical Association* as "Asbestos Exposure and Neoplasia." This study was one of the first to establish the link between cancer and asbestos. In 1983 the article was voted one of 50 landmark essays published in the journal's first 100 years.

In 1966 Selikoff founded the Division of Environmental and Occupational Medicine at Mount Sinai, where he and his colleagues continued to document asbestos damage and the effects of other environmental substances. They completed studies of carpenters, firefighters, painters, plumbers, roofers, and textile workers. The research helped convince the Occupational Safety and Health Administration to impose standards protecting workers against exposure. It also convinced the Environmental Protection Agency (EPA) in 1989 to ban most uses of asbestos. Not content to merely study the effects of asbestos, Selikoff began an active campaign to educate Congress and the public about its dangers. Selikoff was a public health advocate, testifying before Congress, addressing unions and public health groups, and writing on the topic in a wide variety of venues, including the op-ed page of the *New York Times*. He served as a consultant to a number of government agencies, including the World Health Organization. His research convinced him that 95 percent of cancers have environmental causes, and his work helped generate public awareness of the dangers of chemicals in the environment. Selikoff earned particular respect from unions, with which he worked closely; he was named an honorary member of the International Association of Heat and Frost Insulators and Asbestos Workers.

Selikoff has also played an important role in establishing the legitimacy of environmental and occupational medicine. In addition to founding the division at Mt. Sinai, Selikoff founded and played an active editorial role in two important journals, the *American Journal of Industrial Medicine* and *Environmental Research*. In 1982 he founded the Collegium Ramazzini, a group of scientists and public health advocates organized "to advance the study of occupational and environmental health issues around the world." The Collegium defines its purpose as serving "as a bridge between the world of scientific discovery and those social and political centers which must act on these discoveries to conserve life and prevent disease." The Collegium conducts research and political advocacy, in part to counter the efforts of industry to limit regulation of environmental hazards. In 1988, Dr. Ellen Silbergeld, a toxicologist employed by the Environmental Defense Fund, said of Selikoff, "He is one of the great historical figures in medicine and public health. His major contribution has been to move occupational medicine out of a second-class position in the field of medicine and in the opinion of the public."

Selikoff worked at Mt. Sinai until his retirement in 1985. He died of cancer in Ridgewood, New Jersey, on May 20, 1992. In 1997 the Collegium Ramazzini conducted a symposium in Selikoff's honor, published by the New York Academy of Sciences as *Preventive Strategies for Living in a Chemical World*.

BIBLIOGRAPHY

"Collegium Ramazzini," www.collegiumramazzini.org; Goodman, Billy, "Health Science Pioneer," *Environmental Defense Fund Newsletter*, 1988; Hooper, Joseph, "The Asbestos Mess," *New York Times*, 1990; Washburn, Lindy, "Irving J. Selikoff," *Bergen Record*, 1992.

Seo, Danny

(April 22, 1977–)
Entrepreneur, Founder of Earth 2000

Danny Seo founded the youth environmental organization Earth 2000 at the age of 12 in 1989. Within six years there were 25,000 members of Earth 2000, and the group had become internationally famous for its success in protecting undeveloped land from real estate developers, inspiring a massive British boycott of Faroe Island seafood, and passing a law in the Pennsylvania state legislature that gave students the right to refuse to dissect animals in secondary school biology classes. At the age of 18, he retired from his position as chief executive officer (CEO) of Earth 2000, and since then he has dedicated his time to fundraising, writing, and promoting sustainable and vegetarian lifestyles.

Danny J. Seo says that being born on Earth Day, April 22, 1977, was just the first of a series of "accidents" that led him down the path to his environmental commitment and the fame that followed. During Seo's 12th year, in 1988–1989, the revival of Earth Day was receiving a lot of press. At the same time, newspaper headlines about environmental catastrophe—more pessimistic every day—were terrifying him. Seo threw a small birthday party that year and asked his friends not to bring him gifts—for his realization that material goods did not lead to happiness was another key event in his life that year—but instead to join Earth 2000, a youth environmental organization that he had just decided to establish.

Earth 2000's first project was to raise money through recycling cans to sponsor an animal at the local Philadelphia zoo.

The animal was a prairie dog, which escaped from the zoo soon after Earth 2000 sent in the money. The group's next project was to fight a housing development in Seo's neighborhood: developers wanted to raze a 66-acre forest with a pond and build new homes. Seo recalls that this project was another turning point for him because he learned how to be an effective ecosoldier. Seo recruited top environmental lawyers, who agreed to work pro bono for him. Seo learned to appeal to the public's affection for children. He says that everyone, even the real estate developers, thought he was "adorable," until it became evident that he could win. The officials at his private prep school, proud at first to have such an ecohero in the student body, disbanded his organization once Seo began to be served subpoenas during gym class. The clincher in this fight was triggered by another "accident." Seo was hiking through the forest in question, tripped over an unidentified, partially buried piece of rubble, dug it up, and discovered a colonial-era bottle. The forest was an archaeological site! Through some savvy media work, Seo and his group convinced the public that no one would want to ruin an archaeological site by building houses on it. The developers abandoned their plans.

Seo's next fight was against the pilot whale hunters of the Faroe Islands, a territory of Denmark. In March 1993, Earth 2000 organized a demonstration at the Danish embassy in Washington, D.C. Advance press coverage was extensive; Earth 2000 had learned how important media attention was to the success of its events. The Danes alerted the State Department, which arranged for extensive vigilance of the demonstration. News of Earth 2000's demonstration—13 kids and their inflatable dolphin accompanied by seven State Department agents—caused hilarious uproar in Great Britain, which buys 98 percent of Faroe Island seafood exports. A major boycott was organized in Great Britain; the boycott is still in existence because Faroe Island fishermen have not ceased hunting whales. This experience reiterated for Seo a lesson he had learned with the housing project fight: that if he allowed the media to exploit him, his adorableness, and his youth, his projects would enjoy overwhelming success. This was free advertising, reaching many more people and touching them much more deeply than traditional advertising. Over the past ten years, Seo has collected more than 500 newspaper and magazine articles about himself and his projects.

Earth 2000's various projects took a toll on Seo's academic life. Seo, by his own admission, was a terrible student. He graduated from Governor Mifflin High School in Shillington, Pennsylvania, 129th out of a class of 130. He saw the irony of the situation: at the same time that he was working with Pennsylvania state legislators to write, lobby for, and pass the right-not-to-dissect bill, he was earning an F in his civics class. Seo's later degrees have been issued by the school of life; he has not attended university nor does he have any plans to do so. In 1995, Seo stepped down as CEO of Earth 2000 and began writing books (*Generation React*, 1997; *Heaven on Earth*, 1999; *Conscious Style*, 2000) to inspire young people. His message is that young people are capable of tremendous success if they really believe in what they are doing. His book *Heaven on Earth* (1999) lists ten steps to assured success and instructs activists in pain-free fundraising. Seo writes a column for *Vegetarian Times* magazine (he has been vegan for four years) and is developing a television program that will marry his concerns about the environment with lifestyle and altruism. He resides in New York City.

BIBLIOGRAPHY

Czape, Chandra, "The Most Powerful People in Their Twenties," *Swing Magazine*, 1998; "Danny Seo," http://www.dannyseo.com; Seo, Danny, "Activism 101," (speech at University of Colorado–Boulder), 1998.

Sessions, George

(June 10, 1938–)
Ecophilosopher

A leading proponent of deep ecology in the United States, George Sessions is a popular spokesperson for the long-range deep ecology movement, which he sees as the forwarding of the western environmental ethic first set out by ALDO LEOPOLD. His studies of and writings about deep ecology have helped the movement to gain ground by clarifying much of this ecophilosophy's foundation and positing methods for its pragmatic implementation.

George Sessions was born June 10, 1938, in Stockton, California, and was raised in Fresno. He attended Fresno State University, completing a B.A. in philosophy in 1960. After three years in the United States Air Force, Sessions started his graduate studies in philosophy at the University of Chicago in 1964. He completed his M.A. in 1968, specializing in the philosophy of science.

As a youth, Sessions enjoyed outdoor recreation—especially rock climbing—and worked as a garbage collector in Yosemite National Park. He joined the Sierra Club when he was 15 and became influenced by the writings of JOHN MUIR and DAVID BROWER. It was not until 1968, however, after the completion of his studies, that Sessions's interest in ecology began to develop. Sessions started teaching at Humboldt State University and shared an office with sociologist BILL DEVALL. Both were influenced by the growing popularity and impetus of ecology, which culminated with Earth Day in 1970. Together, Sessions and Devall sought some kind of ecophilosophy that would push toward an overriding environmental ethic. In 1973, Arne Naess published his now-famous short paper, "The Shallow and the Deep, Long-Range Ecology Movement," in *Inquiry*. Like Devall, Sessions knew that this work—and the long-range deep ecology movement in general—provided the philosophical framework within which to develop his own analyses of the environmental movement and its progress in North America toward a biocentric environmental ethic. Naess distinguished between the anthropocentric environmental concern of the shallow ecology and the spiritual, biocentric perspective of the deep ecology.

Sessions left Humboldt State University in the fall of 1969 to teach philosophy at Sierra College, in Rocklin, California, where he would remain for the duration of his career. He and Devall remained in contact, and between 1978 and 1981 they further developed the distinction between shallow (or reform) ecology and deep ecology and used it as a basis for classifying and describing the various ecophilosophical positions. In 1984, Sessions worked with Naess to draft a more neutral deep ecological platform designed to appeal to a broader audience, from different philosophical and religious backgrounds. Sessions and Naess raised eight widely quoted points that were central to the long-range deep ecology movement:

1. The well-being and flourishing of human and nonhuman life on earth have a value in themselves. These values are independent of the usefulness of the nonhuman world for human purposes.
2. The richness and diversity of all life forms contribute to the realization of the previously mentioned values and are also values in themselves.
3. Humans have no right to reduce this richness and diversity except to satisfy vital needs.
4. The flourishing of human life and cultures is compatible with a substantial decrease of the human population. The flourishing of nonhuman life requires such a decrease.
5. Present human interference with the nonhuman world is excessive,

and the situation is rapidly worsening.

6. Policies affecting basic economic, technological, and ideological structures must therefore be changed. The resulting state of affairs will be deeply different from the present.

7. The ideological change is mainly that of appreciating life quality (dwelling in situations of inherent value) rather than adhering to an increasingly higher standard of living. There will be a profound difference between big and great.

8. Those who subscribe to the foregoing points have an obligation directly or indirectly to try to implement the necessary changes.

Sessions collaborated with Devall on *Deep Ecology: Living as if Nature Mattered* (1985), a widely read primer on this ecophilosophy, or ecosophy. More recently, Sessions wrote *Deep Ecology for the Twenty-First Century* (1995), arguing in it that "the crucial paradigm shift the Deep Ecology movement envisions as necessary to protect the planet from ecological destruction involves the move from an anthropocentric to a spiritual/ecocentric value orientation. . . . Humanity must drastically scale down its industrial activities on Earth, change its consumption lifestyles, stabilize and then reduce the size of the human population by humane means, and protect and restore wild ecosystems and the remaining wildlife on the planet." Although critics of deep ecology still question the relative validity or feasibility of a nonanthropocentric biocentrism, Sessions has continually defended such a perspective on the grounds that while it may seem to lack any distinct precedent in the history of Western civilization, such a paradigm shift toward an environmental ethic is crucial if we are to salvage what is left of this fragile planet.

Sessions has written extensively and remains active since his retirement from Sierra College in 1998. He has acted as an editorial adviser to *Wild Earth*, the journal associated with the Wildlands Project, and has conducted several interviews for the *New Dimensions* radio show and other interest groups on deep ecology and bioregionalism. He lives in the Sierra foothills, east of Sacramento.

BIBLIOGRAPHY

Devall, Bill, and George Sessions, *Deep Ecology: Living as if Nature Mattered*, 1985; "New Dimensions Presents Deep Ecology for the 21st Century," http://www.newdimensions.org/html/ecology.html; Sessions, George, "The Deep Ecology Movement: A Review," *Environmental Review*, 1987; Sessions, George, ed., *Deep Ecology for the Twenty-First Century*, 1995.

Seton, Ernest Thompson

(August 14, 1860–October 23, 1946)
Writer, Illustrator, Lecturer

Ernest Thompson Seton (also known as Ernest Seton-Thompson and Ernest Evan Thompson) was one of North America's most effective publicists of nature and wild animals during his more than 50-year career as a writer, illustrator, and lecturer. He published scientific guides featuring his own pen-and-ink drawings of birds and animals, including the four-volume *Lives of Game Animals* (1925–1928), but was best known for his prolific animal stories, of which *Wild Animals I Have Known* (1898) was the most famous. Critics—most notably naturalist JOHN BURROUGHS—held that these stories, usually starring an animal of great intelligence and loyalty that lives a good life but dies valiantly at the end, did not portray animals realistically. Nonetheless, Seton's stories, for all the fiction they contained, inspired many young readers of the early 1900s to pursue careers as naturalists and conservationists.

Ernest Evan Thompson was born in South Shields, Durham County, England, on August 14, 1860, the eighth of the ten surviving sons of Alice Snowden and Joseph Logan Thompson. He changed his last name later in life to honor his father's side of the family, the Setons. A member of this side of the family had been forced to change his name to Thompson to escape retribution after being on the losing side of the Stuart Rebellion of 1745. The Thompson family moved to a farm in Lindsay, Ontario, in 1866, after Joseph Thompson's shipping business went bankrupt. Young Ernest fell in love with nature and life on the frontier while living on the farm. He was particularly moved one day when he watched a tiny kingbird act with enough bravery to drive off an eagle. His first literary work, written when he was 16 years old and published when he was 19, would be about this animal, the poem "The King Bird: A Barnyard Legend."

Although the family sold the farm and moved to Toronto in 1870, Ernest returned periodically during his teenage years, to recover from bouts of ill health. He studied birds by shooting and dissecting them, and he developed a key to identifying hawks and owls that would later lead to *A Key to the Birds of Canada*, published in 1895. Seton divided his time during the 1870s and 1880s between studying art in London, New York, and Toronto and exploring the wilds of Manitoba, to refine his sketching skills and knowledge of natural history. He was invited to join the American Ornithologists Union in 1884 and began writing for its publication, *The Auk*. His *The Birds of Manitoba* came out in 1891, and he was contracted by ornithologist FRANK CHAPMAN to illustrate two of his bird books, *Handbook of Birds of Eastern North America* (1895) and *Bird Life* (1897).

Seton studied painting at Julian's Academy in Paris from 1890 to 1892 and was fascinated by news articles about a hunter who was killed in the Pyrenees by wolves, after the hunter had killed several members of the wolf pack. Seton painted a huge canvas of this scene, showing the hunter's clothes scattered on the ground, and wolves gnawing a human skull. Because of its gruesome realism, *The Triumph of the Wolves* was not shown at exhibits he submitted it to in Paris or Toronto, and although the jury of the Chicago World's Fair in 1893 voted not to show it because it would present a bad image of Canada, the painting was controversial enough that the organizers of the fair agreed to exhibit it.

Later that year, Seton was hired for a five-month stint killing wolves on a New Mexico cattle ranch. His experience yielded "Lobo, the King of Currumpaw," published in *Scribner's Magazine* in 1894. The story was about Lobo, a male wolf of extraordinary size, strength, and intelligence, who died courageously in one of Seton's traps. This was followed by *Wild Animals I Have Known* (1898), an immediate best-seller that is still in print more than 100 years later.

In 1896 Seton married Grace Gallatin, a writer and activist for women's rights, who accompanied him on some of the many expeditions he made during the first few years of their marriage and helped him edit and design his books. Seton's popularity grew, and he became a well-known lecturer, earning around $12,000 per year for his presentations. In 1903, naturalist and writer John Burroughs, who believed that writings about natural history should be as scientifically accurate as possible, attacked Seton in an *Atlantic* article entitled "Real and Sham Natural History." Burroughs claimed that Seton's stories portrayed animals unrealisitically, especially when they were shown to possess the same types of intellectual and reasoning processes as humans. He was especially adamant that it was wrong to pretend that fiction was fact, as he claimed Seton had in his stories. Seton arranged to be seated next to Burroughs at a dinner party given by Andrew Carnegie shortly after this critique was published, and Seton reportedly interrogated Burroughs until he found out that Burroughs had no personal experience with wolves. Burroughs accepted an invitation to visit Seton's home, with his thousands of stuffed animal specimens, and Burroughs acknowledged in a 1904 *Atlantic* article that Seton was indeed a knowledgeable naturalist. Yet he did give the caveat in this article that Seton's work was "truly delightful" for those readers "who can separate the fact from the fiction in his animal stories."

In addition to his popular stories, and the highly praised, scientific works *Life Histories of Northern Animals* (1909) and the later four-volume *Lives of Game Animals* (1925–1928), Seton is remembered as the founder of the scouting movement in the United States. He recruited a group of boys who lived near his Greenwich, Connecticut, home to visit his "Indian village," where he led them in athletic competitions and outdoor skills for camping, hunting, and tracking. The organization was called the Woodcraft Indians, and the boys organized themselves into what Seton told them was "Indian society," with an elected chief, medicine man, war chief, and so on. The charter and by-laws of his group were published in *Ladies Home Journal*, and soon similar groups sprouted up throughout the country and abroad. Grace Gallatin Seton founded the Girl Pioneers in 1910, which later changed its name to the Camp Fire Girls. English lieutenant general Sir Robert Baden-Powell adopted Seton's 1906 handbook, *The Birch Bark Roll of the Woodcraft Indians*, but changed the Indian-inspired titles and mission of the organization to military ones. Baden-Powell officially founded the Boy Scouts in 1908, with Seton—despite his misgivings about Baden-Powell's appropriation of the group and its new militaristic model—heading the committee that organized the Boy Scouts of America. Seton wrote *Boy Scouts of America Handbook* in 1910, which has sold more copies than any other book except the Bible. Seton's Woodcraft Indians had emphasized activities based on nature, but on the eve of World War I, the military-oriented leaders were pushing the Boy Scouts of America away from that orientation, and Seton resigned in protest in 1915.

Seton and his wife slowly drifted apart, and Seton became involved with his secretary, Julia M. Buttree Moss, an expert on mysticsm and Indians and author of several books herself. In 1930,

Seton sold his home in Connecticut and moved to New Mexico, where the Seton Castle was built on 2,500 acres near Santa Fe, to house Seton's vast collection of books and artwork. Seton divorced Gallatin and married Moss in 1935. Seton and Moss enjoyed a productive literary and intellectual partnership and adopted a daughter, Beulah, in 1938. Seton died at home on October 23, 1946, of pancreatic cancer.

BIBLIOGRAPHY

Anderson, H. Allen, *The Chief: Ernest Thompson Seton and the Changing West*, 1986; Keller, Betty, *Black Wolf: The Life of Ernest Thompson Seton*, 1984; Seton, Ernest Thompson, *Trail of an Artist-Naturalist, The Autobiography of Ernest Thompson Seton*, 1940; Wadland, John Henry, *Ernest Thompson Seton: Man and Nature and the Progressive Era, 1880–1915*, 1978.

Shabecoff, Philip

(March 5, 1934–)
Journalist

Philip Shabecoff has documented the growing environmental movement in the United States since the 1970s. As chief environmental reporter for the *New York Times*, Shabecoff was the first in his field; he made the environment a topic worthy of regular media attention. When taken off the environmental beat in 1991, he left the *Times* to write *A Fierce Green Fire*, his acclaimed overview of environmentalism in American history. He was founder and publisher of the daily news service *Green-wire* until 1999. Recently, Shabecoff has delved more deeply into environmental history and has produced two more volumes on the topic, *A New Name for Peace* (1996) and *Earth Rising* (2000).

Philip Shabecoff was born on March 5, 1934, in the Bronx, New York City. He received his bachelor's degree in journalism from Hunter College in 1955, then continued his studies to earn a master's degree from the University of Chicago in 1957. After a two-year stint in the United States Army, Shabecoff was hired by the

New York Times in 1959 and soon proved his mettle as a reporter. As foreign correspondent for the paper in the mid-1960s, he reported from Europe and Asia. When he returned to the United States, he satisfied his lifelong love of nature by acquiring property in an unspoiled section of the Berkshire Hills of Massachusetts. With his wife, Alice, and their two children, he built a summer home, clearing much of the heavily wooded site by axe. Their "remote" farmstead was quickly surrounded by development, which chagrined Shabecoff and increased his interest in environmental issues.

In 1970, Shabecoff was transferred to the *Times*'s bureau in Washington, D.C., where he saw firsthand that the environment had become a matter of national concern. Although his first requests to write about environmental subjects were turned down, by 1977 he was able to convince editors to let him cover such issues part-time. The job soon filled all his hours. When the struggle for environmental action came to the forefront during the Reagan administration, Shabecoff was there, on the scene.

As witness to a major social movement, Shabecoff documented the changing national zeitgeist. Americans had begun accepting responsibility for protecting and preserving the natural world, and they wanted to be informed about progress and setbacks. Reporting on industrial pollution and waste of resources, Shabecoff often fought federal agencies and corporations for access to data. For his efforts, he won the James Madison Award of the American Library Association (ALA) in 1990. The ALA credited him with "leadership in expanding freedom of information and the public's right to know."

Shabecoff's balanced, reasoned pieces about environmental efforts and legislation were never radical. Nevertheless, in 1991, Washington bureau chief Howell Raines told Shabecoff that the newspaper wanted to move him to the Internal Revenue Service beat, suggesting he was too "proenvironment" and had ignored the "economic costs" of environmental protection. Shabecoff felt betrayed and was unwilling to take the new assignment. He left the *Times* to become founder and publisher of *Greenwire*, the electronically distributed environmental news daily.

Within two years he had also produced a book that blazed through environmental circles. Published in 1993, *A Fierce Green Fire* was among the first and best overviews of the green movement in the United States. It traced the history of environmentalism, from Henry David Thoreau to Rachel Carson to 1970s activists. Shabecoff's position in Washington had granted him access to governmental officials and environmental leaders, and these contacts informed his work. In the mid-1990s, the book and his past record won him the Sierra Club's David Brower Award, given annually to environmental journalists.

Shabecoff continued to follow environmental issues. His contacts with policy makers expanded, and he kept tapes of his many interviews. In spring 1997 he donated all his taped archives to the environmental journalism program at Michigan State University, sharing his many years' experience at the *New York Times* and *Greenwire* with students and others in his field. He was also among the initial group of journalists who agreed to found the Society of Environmental Journalists.

In 1992, the secretary-general of the United Nations Conference on Environment and Development asked him to cover the Earth Summit in Rio de Janeiro. He was granted full access to diplomatic sessions on acid rain, ozone depletion, and more. The conference helped him formulate theories on the interconnected economy and environment. His 1999 volume, *A New Name for Peace*, explores these ideas. Moving from ancient Babylonian times until the present, Shabecoff outlines the efforts of international organizations, nations, and environmental groups to deal with the economic and ecological ramifications of overconsumption. He argues that poverty, hunger, and disease are environmental issues, since the world will not reform its economic systems until these are conquered.

As a freelancer based in Becket and Newton, Massachusetts, Shabecoff has continued to explore environmental issues and policy. His latest work has taken a global approach. *Earth Rising* (2000) examines issues and problems confronting today's environmentalists. It notes that future activists will have to take on broader roles, protecting the earth by reforming education, taming the global economy, and working for political reform, especially in regard to corporate campaign finance.

Shabecoff has served on several environment-related boards, including the Commission on Environmental Cooperation/NAFTA, the Dow Chemical Company Environmental Advisory Council, the Rachel Carson Council, and the Urban Tree Commission in the city of Newton, Massachusetts.

BIBLIOGRAPHY

Shabecoff, Philip, *Earth Rising: American Environmentalism in the 21st Century*, 2000; Shabecoff, Philip, *A Fierce Green Fire*, 1993; Shabecoff, Philip, *A New Name for Peace*, 1999.

Shuey, Christopher

(January 13, 1955–)
Environmental Health Specialist with the Southwest Research and Information Center

Christopher Shuey started his career as a journalist but found himself frustrated by the limits of a reporter's power. He turned instead to environmental activism, becoming an advocate for communities harmed by the environmental and health impacts of uranium mining and oil exploration. He has spent much of his time working on behalf of the Navajos and Navajo communities in the Four Corners states, battling mining companies that have left groundwater polluted, land permanently contaminated, and people dying of cancer. Since 1981, Shuey has been a member of the senior professional staff of the Southwest Research and Information Center (SRIC) in Albuquerque, New Mexico. As an environmental health specialist at the nonprofit research and advocacy organization, Shuey directs the Community Water, Wastes, and Toxics Program, which offers technical and organizational assistance on pollution and environmental health concerns to communities and citizen groups locally, regionally, nationally, and internationally.

Christopher Lincoln Shuey was born in Springfield, Ohio, on January 13, 1955. He is one of five sons born to farmers Lin and Ruth Ann, who continue to live in Ohio. Growing up and working on the family farm, Shuey developed an appreciation of the land at an early age. Much of this came from his father, who was an early practitioner of soil conservation techniques, using crop rotation and contour cropping to prevent soil erosion and protect water quality. Shuey cites two events that shaped him as a teenager and led him to question government decisions in the 1960s. The first was the Vietnam War; the second, the government's condemnation of half the family farm's acreage in a river bottom for a dam project. Although questions were raised about the necessity of the dam and just compensation for landowners, the Shuey family lost its case in court.

Shuey attended Ohio University from 1973 to 1974, with a major in journalism. He transferred to Arizona State University (ASU) in 1974 and continued there until 1981. While pursuing his education at ASU, he worked as a staff writer for the *Daily Progress* in Scottsdale and freelanced for the Time/Life Inc. magazine chain. While Arizona was, and continues to be, a very conservative state, the publisher of the *Daily Progress*, Jonathan Marshall, was a mainstream liberal Democrat. Shuey believes that because of this the paper was much more willing to cover all sides of an issue, which he saw as a valuable lesson. He wrote about water and environmental and Native American issues; he also covered the police department and the courts. Several of his articles on the Fort McDowell Yavapai Apache tribe's opposition to the Orme Dam and the Central Arizona project won statewide journalism awards. While at the paper, he covered early uranium miner issues and developed a strong interest in the health problems of the miners.

Shuey was enjoying success with his writing but said he became "disillusioned by the institutional constraints of traditional journalism." While covering public debates on the Palo Verde nuclear power plant in Arizona in the mid-1970s, Shuey realized that the views of thoughtful, intelligent people opposed to the plant were systematically censored by the mainstream Phoenix press. At the same time, he said, he felt that the small "liberal" paper he worked for was not being taken seriously. "If that were the case from a journalistic perspective, why not just be an advocate? Journalism did not provide an outlet to be an advocate," he said.

He left the paper in 1978 and became involved with the antinuclear movement, working with like-minded individuals to start Arizonans For a Better Environment in 1979. The project had difficulty in raising funds and was given up after a year. Shuey became acquainted with staffers of the Southwest Research and Information Center in 1978 while tracking down information about the high lung cancer rate among uranium miners in the Four Corners area of Arizona and New Mexico. SRIC is a nonprofit, progressive public education and scientific organization whose mission is "to provide timely, accurate information to the public on matters that affect environment, human health and communities in order to protect natural resources, promote citizen participation, and ensure environmental and social justice now and for future generations."

Shuey started working at SRIC in 1981 after it had raised some money to publish a magazine on mining impacts. When Shuey was offered the job as editor of *Mine Talk*, he packed up his car and made the move to Albuquerque. Funding soon ran out for *Mine Talk*, but SRIC's work with communities impacted by mining continued, and Shuey became a key member of SRIC's team, delivering technical services to communities that requested them. He sees his role with SRIC as a "poor people's consultant" and aims to help people understand the technical aspects of pollution and the science behind the data. That way, he hopes to help people use the law to make changes in their communities. "We want to be able to not just conduct research and give out information, but to take the next step to help people build their own expertise and skills, so it's community people advocating for themselves," said Shuey. Shuey helped launch the Eastern Navajo Diné Against Uranium Mining (ENDAUM); the group is now fighting to keep future uranium mines from opening on Navajo land.

Once in Albuquerque, Shuey finished his undergraduate studies at the University of New Mexico (UNM), earning a B.A. in university studies in 1990. Since then, he has continued his studies and is soon to complete a masters degree in public health with a concentration in environmental epidemiology from UNM. His thesis work involves the study of pollution of groundwater due to uranium mining, the chronic ingestion of uranium in drinking water, and the development of kidney disease as a result. He is also interested in the challenges of measuring and modeling health risks from multiple contaminant sources in areas that produce oil and natural gas. And he wants to help communities within mining and mineral development areas to do their own environmental monitoring.

Much of the uranium mining and oil and gas development takes place on Indian lands, and as a result, Shuey has developed a keen interest in Native American history and culture, particularly that of the Navajo. He is taking courses and has spent considerable time in Navajo communities to better his proficiency in both the written and spoken language of the Diné, "the People."

Shuey has written numerous papers and reports on issues relating to uranium mining and other resource development as it occurs on Navajo Nation land. In one 1979 article, "The Widows of Red Rock," published in the *Saturday Magazine* of the Scottsdale *Daily Progress*, he examined a Navajo community torn apart by uranium mining. By the late 1970s, dozens of Navajo miners had died of lung cancer, leaving families destitute and women to pick up the slack. Years later, Shuey contributed to an SRIC report, "Uranium Mining in Navajo Ground Water: The Risks Outweigh the Benefits," that pointed out deficiencies in the U.S. Nuclear Regulatory Commission's Draft Environmental Impact Statement on the Crownpoint Uranium Solution Mining Project. The deficiencies identified in the report eventually became the core set of issues that ENDAUM used in its legal challenge of new uranium mining.

Groundwater issues on a state and local level continue to occupy much of

Shuey's time. He served on the New Mexico Governor's Ground-water Quality Advisory Committee from 1988 to 1989. On a local level, he served on the Ground Water Protection Advisory Committee for the Albuquerque City Council and Bernalillo County Commission from 1988 to 1994. That work resulted in a Groundwater Protection Act in 1994 and a policy that is now integrated into the area's overall water resources management plan. In addition, he has served on U.S. Environmental Protection Agency–sponsored technical teams that evaluated oil and gas regulatory programs in oil- and gas-producing states.

In 1995, Shuey was awarded the Karl Souder Award by the New Mexico Environmental Law Center. Named for a former state hydrologist, the award honors those who have had significant roles in advocacy and in the protection of groundwater in New Mexico. Shuey and his wife, Laura, live in Albuquerque with their sons, Bryant and Conor. He is active in both sons' school activities and in the coaching of their baseball and basketball teams.

BIBLIOGRAPHY

Shuey, Chris, "Policy and Regulatory Implications of Coal-bed Methane Development in the San Juan Basin, New Mexico and Colorado," *Proceedings of the First International Symposium on Oil and Gas Exploration and Production Waste Management Practices*, New Orleans, Louisiana, sponsored by the U.S. Environmental Protection Agency, 1990; Shuey, Christopher, "The Widows of Red Rock," *Daily Progress*, 1979; Shuey, Christopher, and W. P. Robinson, "Characterization of Ground Water Quality Near a Uranium Mill Tailings Facility and Comparison to New Mexico Standards," *Proceedings of a Symposium on Water Quality and Pollution in New Mexico*, New Mexico Bureau of Mines and Mineral Resources Hydrologic Report 7, 1984; Shuey, Christopher, Paul Robinson, and Lynda Taylor, "The Costs of Uranium: Who's Paying with Lives, Lands and Dollars?," *The Workbook*, 1985; "Southwest Research and Information Center," http//www.sric.org.

Silkwood, Karen

(February 19, 1946–November 13, 1974)
Anti-Nuclear Activist

Karen Silkwood was a nuclear-plant worker and Oil, Chemical, and Atomic Workers International Union (OCAW) activist in life, but she is remembered for her brave, self-sacrificial efforts to expose the plutonium safety violations and dangerous work practices of the Kerr-McGee Nuclear Corporation in Cimarron, Oklahoma.

Karen Gay Silkwood was born in Longview, Texas, on February 19, 1946, and grew up the very lively and outspoken child of house painter Bill Silkwood and bank loan officer Merle Biggs. She was raised in the town of Nederland, Texas, where she played sports and the flute and was an honor student at the local high school. She graduated in 1964

Karen Silkwood (Courtesy of Anne Ronan Picture Library)

and earned a scholarship to study medical technology at Lamar College in nearby Beaumont. Silkwood married Bill Meadows one year later, at the age of 19, and together they had three children. The couple divorced in 1972, and Silkwood gave custody of the children to her ex-husband.

In August of that year, Silkwood began work at the Kerr-McGee Nuclear Corporation. She took an analyst position in the metallurgy laboratory, where she ran quality-control tests on plutonium pellets manufactured at the plant. Shortly after beginning work at Kerr-Mcgee, Silkwood became an active member of the OCAW and participated in a two-month strike against the company that ended in January 1973. Her involvement in the strike gave her the opportunity to learn about the plant's work hazards. Coworkers

taught her about the dangers of working with plutonium and of the company's poor handling of incidents in which workers and town residents risked contamination from the plant. Her outspoken mistrust of the company and increased involvement in the OCAW earned her a position—the first ever to be held by a woman—on the union's negotiating committee.

While serving on the committee, Silkwood began collecting evidence of the company's cover-ups. Beginning in 1974, she gathered documentation of careless plutonium handling, defective plutonium rod production, and safety violations that went unreported to the Atomic Energy Commission (AEC). She was increasingly harassed as her accumulation of information and public speeches on health and safety continued, and she was put in physical danger when her apartment was contaminated with radioactive materials. Silkwood endured threats, jeers, and insults from antagonistic coworkers and company management on a daily basis and was afraid for her safety during most of the fall of 1974, but she felt driven to see the work to completion. On the night of November 13, 1974, very afraid but only 40 minutes away from her meeting with boyfriend Drew Stephens and a reporter from the *New York Times*, her white Honda Civic ran off of the road into a ditch. She died instantly. The side of Silkwood's car had new dents, indicating that she had been run off the road, but local and federal investigators concluded that Silkwood had been driving while on antidepressants and had caused her own death.

Silkwood's family and friends were devastated and outraged and sued the Kerr-McGee Nuclear Corporation for negligence, noncompliance with AEC regulations, civil rights violations, and interference in union activities. On May 18, 1979, Silkwood's work was finally rewarded. The jury in the case found Kerr-McGee guilty of negligence in its plants and noncompliance with federal safety regulations. Silkwood's death was not considered during the case, but her sacrifice fueled the country's antinuclear and occupational safety and health movements and accelerated the passage of state "whistle-blower" laws that protect workers who speak out against their employers.

BIBLIOGRAPHY

Jackson, Kenneth T., Karen E. Markoe, and Arnold Markoe, eds., *Dictionary of American Biography*, 1994; Kohn, Howard, *Who Killed Karen Silkwood?* 1981; *Silkwood*, 1983 (motion picture); "The Silkwood Mystery," *Time*, 1975.

Sive, David

(September 22, 1922–)
Professor of Environmental Law, Environmental Litigator, Cofounder of the Natural Resources Defense Council

Known in legal circles as the "father of environmental law," David Sive has distinguished himself as an outstanding environmental litigator, scholar, and advocate. For 30 years, he has argued in the courts in defense of the environment, lectured in classrooms to future and current lawyers about its importance and how to best defend it, published numerous articles on environmental law, and provided leadership to several significant environmental organizations.

The middle of three children, David Sive was born to Abraham and Rebecca Sive on September 22, 1922, in Brooklyn, New York. Raised in an urban setting, the young Sive had very little exposure to the great outdoors, and the idea of hiking and camping was completely foreign to him. His fondness for snow, however, led him to begin exploring the Adirondack and Catskill Mountains of New York State as a teenager, and it was there, at the age of 15, that he began to discover for himself the wonders and beauty of nature. Sive also enjoyed the classic nature writers, and the combination of reading HENRY DAVID THOREAU, RALPH WALDO EMERSON, WALT WHITMAN, and JOHN MUIR and climbing mountains instilled in the young Sive a deep commitment to nature and its preservation.

After high school, Sive attended Brooklyn College, graduating in 1943 with a bachelor's degree in political science. That same year he enlisted in the United States Army and joined the European troops in World War II, serving in combat in Europe for ten months as part of the Ninth Infantry Division. He was twice wounded before being discharged in 1945; he received both the Purple Heart and the Oak Leaf Cluster military honors. After discharge, Sive, like many of his generation, was able to take advantage of a tuition benefit program for disabled veterans. Sive enrolled at Columbia Law School at the age of 23, and upon graduation three years later he passed the New York bar exam and moved to the New York suburbs with his wife, Mary Robinson.

Sive entered law practice as an associate of Levien, Singer & Neuburger, later moving to Seligson, Morris & Neuburger as an associate and then a partner. During his early years as litigator, Sive focused on commercial litigation. In 1961, he founded Sive, Paget & Riesel, specializing in commercial litigation and transactions.

Sive had been an attorney for about 12 years when the environmental movement began to take shape, and he soon began applying his legal skills to protecting the environment. During the early years of the development of environmental law, Sive was involved in several significant landmark cases, including the founding case of environmental law, the *Scenic Hudson Preservation Conference v. Federal Power Commission* (1965, 1971). At issue was whether the natural beauty of Storm King Mountain and the surrounding area was important enough to preserve from becoming a site for a pump storage power plant. Sive's initial involvement in the case was as a board member of the Scenic Hudson Organization, but by the time the case made its way through the courts, he was on board as counsel. Ultimately, Sive and the other attorneys representing the organization were successful in obtaining a mediated settlement that protected the mountain and the river. Sive has been fighting legal battles to protect the environment ever since.

Sive was lead counsel on several groundbreaking cases that followed the Scenic Hudson case, including *Citizens Committee for the Hudson Valley v. Volpe* (1970), *Concerned About Trident v. Schlesinger* (1975), and *Mohonk Trust v. Board of Assessors of Town of Gardiner* (1979). *Hudson Valley v. Volpe* was the first major environmental lawsuit that permanently prevented a major construction project; in this case, construction was halted on the Hudson River Expressway. *Trident v. Schlesinger* attempted to stop the construction of the Trident Nuclear Submarine Base on the Hood Canal in Washington. Although the injunction was refused, the case established the principle that projects involving the navy, army, and air force are subject to the environmental review and assessment process of the National Environmental Policy Act. The Town of Gardiner case established tax exempt status for nature trusts and preserves in New York State, and similar tax exemption laws have been adopted in many states based on this case.

In the 1970s, Sive also began writing and lecturing about environmental law, contributing significantly to its development as a separate body of legal study and practice. His extensive list of publications includes articles on subjects ranging from examining expert witnesses and legal instruments to protect the environment, to the role of litigation in environmental policy, the use of scientists as expert witnesses, and ethical issues in environmental litigation. His articles have appeared in major law journals such as the *Columbia Law Review*, *Michigan Law Journal*, *Iowa Law Journal*, *National Law Journal*, the *New York Law Journal*, and *Environmental Law Commentary*. For nearly three decades he has guided the organization and development of environmental continuing legal education courses for the American Law Institute–American Bar Association, and his visiting professorships include the law schools at Columbia University, New York University, Cornell University, and the University of Wisconsin. In 1994, he received a Fulbright Scholar Award and traveled to Australia to lecture on environmental law, and in 1995, he joined the faculty of Pace Law School as a professor of environmental law.

Sive has served in leadership roles in numerous environmental organizations throughout his career. In 1971, he helped found the Natural Resources Defense Council, and for more than two decades he served on that organization's executive and legal committees. He chaired the Atlantic Chapter of the Sierra Club from 1965 to 1969, the Friends of the Earth Foundation from 1986 to 1992, and the Environmental Law Institute from 1972 to 1982. He has been a board member of numerous other organizations, including New York Lawyers for the Public Interest, Sierra Club, Association for the Protection of the Adirondacks, National Research Council, and Adirondack Council.

In large degree due to Sive's work, the environment is now one of the two or three most important areas of public law, and environmental law has expanded to become a part of the curriculum, or a distinct area of study, at numerous law schools throughout the United States. Sive and his peers have established a large body of environmental law, and a strong foundation of laws and regulations is in place. The work of future environmental lawyers, Sive believes, will be to utilize this foundation to extend protection to greater areas of land and to more populations threatened by development, contamination, and the like.

Sive has been honored with numerous awards throughout his career from organizations such as the New York State Bar Association (Outstanding Volunteer Service to the Community Which Has Reflected Honor on the Legal Profession, 1977), the American Law Institute–American Bar Association (Award of Merit for Contributions to Continuing Legal Education in the Field of Environmental Law, 1983), and the Environmental Law Institute (Award in Recognition of Outstanding Contributions to the Improvement of Environmental Law and Policy, 1984).

Sive has five children and six grandchildren. He hikes and climbs mountains from his homes in Montclair, New Jersey, and Margaretville, New York. He is professor emeritus at Pace University and continues as a member of his law firm.

BIBLIOGRAPHY

"Environmental Law & Litigation," http://www.sprlaw.com/html/body_david_sive.htm; "Pace University School of Law, Faculty Biographies," http://www.law.pace.edu/pacelaw/facbios/sive.html; Sive, David, and Frank Friedman, *A Practical Guide to Environmental Law*, 1987.

Smith, Rocky

(April 30, 1950–)
Public Lands and Forest Policy Analyst, Activist

Rocky Smith, an expert on environmentally sound forest planning and land use, analyzes management plans and project proposals for public lands for several Colorado environmental organizations, including Colorado Wild, Inc. His effective rhetoric and speaking skills made him an effective member of the team that defeated the controversial Two Forks dam project near Denver, Colorado.

Born on April 30, 1950, in Oak Park, Illinois, Rocky Smith grew up spending as much time as possible outside. With his school buddies, Smith rode his bicycle to parks and undeveloped lots, fished in a neighborhood fishing hole, and vacationed with his family at lakes and forest preserves near Chicago. At the University of Wisconsin–Madison, Smith initially studied meteorology, but he switched and graduated with a B.A. in communication and public address in 1972.

After moving to Colorado in 1975 and experiencing a tremendous awe at the beauty of the Rocky Mountains, Smith spent all the time he could hiking, mountain climbing, and cross-country skiing.

A 1979 Colorado Mountain Club (CMC) trip to the Weminuche Wilderness Area in the San Juan National Forest catalyzed his latent environmentalism. His group shared a meadow camp with a 43-person group from the Chicago Mountaineering Club. The excessive size of that group, the destruction wrought by their pack horses, and—to top it off—the sauna that they had packed in seemed wholly inappropriate for a wilderness area. Smith wrote an article about this encounter in the CMC magazine *Trail and Timberline* in late 1979. The CMC's conservation coordinator called Smith and asked him to write a letter to the San Juan National Forest administration about the problem. The Forest Service response was to invite Smith to participate in a Land and Resource Management Planning Process, which was to determine rules for the use of that national forest. He did take part, and within a couple of years he participated in the design of similar plans for all of Colorado's national forests. In the process Smith became known as the state expert on national forest planning.

Until 1983, Smith did this as a volunteer. To earn a living, he worked at a factory, where he made packing crates and did other carpentry. In 1983, he began earning a $100 per month stipend from the CMC, which soon rose to $400. As his reputation grew, he was asked to analyze individual timber sales and was invited to speak at conferences.

When the Two Forks dam controversy erupted in Colorado in the mid-1980s, Smith studied the project and the alternatives to it and concluded that it should not be built. In 1985, the Colorado Open Space Council hired him part-time to begin organizing against Two Forks. The dam project would have created a very large reservoir near Deckers, Colorado, to supply water to Denver. Environmentalists claimed that it would have violated the Endangered Species Act, since a decrease in water flow in the South Platte River would have threatened three rare birds that inhabit the Platte in central Nebraska: the whooping crane, the piping plover, and the interior least tern. Smith became an effective publicist, convincing anglers, boaters, fiscal conservatives, environmentalists, and hikers that Two Forks was a bad idea and that the environmentalists' alternative for water delivery was much better. He spoke to groups frequently for three years and rallied dam opponents to attend public hearings and to write letters. Smith's efforts paid off; in 1990 Environmental Protection Agency administrator WILLIAM K. REILLY rejected the dam and adopted the alternative plan.

During most of the 1990s, Smith worked for the Colorado Environmental Coalition (CEC; formerly the Colorado Open Space Council), responding to forest management plans and individual timber sales. Smith traveled throughout Colorado's public land system on his mission; he also developed a network of locals familiar with public land sites. He and environmentalist Roz McClellan of the Southern Rockies Ecosystem Project developed a new way of communicating environmentalist objectives to the U.S. Forest Service: "citizens' management alternatives." Instead of merely reacting to a land and resource management plan, they would mobilize local environmentalists to write their own alternative for such a plan, always with the goal of maintaining the biological diversity of the area in question. This effort has paid off; U.S. Forest Service documents now include the goal of biodiversity maintenance in their plans.

Smith left CEC in 1998. He currently works with the Forest Watch committee of Colorado Wild, a group he helped found in 1998, whose objective is to conserve the integrity of Colorado forests. Smith resides in Denver, Colorado.

BIBLIOGRAPHY

Garner, Joe, "Coalition Battles Ski Resort; Foes Organize to Appeal Forest Service Approval of Lake Catamount Area Planned near Steamboat," *Rocky Mountain News*, 1991; Hood, Andrew, "Vail Expansion Fight Isn't Over," *Denver Post*, 1997; Roberts, Chris, "Salvage Logging Impact Debated," *Boulder Daily Camera*, 1995.

Sneed, Cathrine

(1955–)
Founder of the Horticulture Project and the Garden Project of San Francisco

Recognizing the power of tending a garden to transform both individual and community, Cathrine Sneed founded the Horticulture Project, an innovative program in San Francisco's county jail that puts prisoners to work growing organic vegetables and provides them with the gratification of productive work and a therapeutic appreciation for their environment. She later created a similar postrelease program called the Garden Project that provides work experience in planting trees, gardening, and community cleanup and has become one of the most successful community-based crime prevention programs in the country. Besides giving recently released inmates an opportunity to restructure their lives, the Garden Project promotes organic farming and sells its fresh produce to restaurants in the area and at the local farmer's market. Sneed's motivation comes from her belief that it is not possible to restore the earth unless the people are restored first; in a state like California, where more money is spent on prisons than on schools, she has gone to the heart of matter, providing job training, employment, and environmental appreciation to people in need.

Cathrine Sneed was born in 1955 and grew up in urban Newark, New Jersey. The first of fourteen children, she dropped out of high school and moved to California. With two children herself and few job prospects, Sneed soon wound up on welfare. But out of concern for her children's future, she sought a new direction: She passed her high school equivalency test, enrolled in college, and began working in the San Francisco County Jail No. 3 counseling women prisoners, many of whom were drug users or prostitutes. Then in 1982, when she was only 27, she contracted an untreatable kidney disease and nearly lost her life. A friend brought her a copy of John Steinbeck's *The Grapes of Wrath* while she was in the hospital, which gave her an appreciation for the healing power of working the soil. When, against all odds, she recovered from her illness, she remembered the book and wanted to share her new love of the land with her clients in the jail.

Although she had no tools or money to start with, she convinced the sheriff to let her dig a garden plot out of the brambles on some idle acres adjoining the jail. In the meantime she went to a horticultural school in England and took a six-month course at the University of California. Even though they had to pull up blackberry brambles with no tools other than their hands, the prisoners who volunteered to help her create the garden responded enthusiastically, and before long there was a waiting list of prisoners who wanted to join the program. Sneed sought donations from local businesses for the purchase of tools, seeds, and watering equipment, and the Horticulture Project was soon producing organic radishes, lettuce, leeks, and strawberries for the jail kitchen, for local soup kitchens, and for pantries of organizations that assisted people with acquired immunodeficiency syndrome (AIDS). Today more than 100 prisoners at a time learn the rudiments of gardening, gaining a sense of pride and self-discipline and a chance to participate in healthy physical labor and the experience of nurturing and long-term planning. In addition, before discovering how good fresh produce tasted, many of the prisoners had lived on junk food, and Sneed has come to believe that a poor diet contributes to many drug users' cravings for drugs.

Along the way, Sneed realized that the Horticulture Project was not enough to keep prisoners from returning to crime or drug abuse once they were released from the jail. In 1990, sponsored by the San Francisco Sheriff's Department and with donations from local businesses, Sneed launched a postrelease bridge program called the Garden Project. It offers structure and support to former offenders through on-the-job training in gardening and tree care, counseling, and assistance in continuing education. Former inmates are given the opportunity to gain work experience and earn a living wage; they are able to live in two nearby drug-free homes and also to bring their children to work at the garden with them. They are required to work at least 16 hours a week and to earn a high school diploma if they do not already have one. Reliable work in the Garden Project often leads to jobs elsewhere through another of Sneed's projects—the Green Teams, which contract with businesses and the city to plant trees and gardens and to provide community cleanups. Besides providing a source of fresh, locally grown organic produce for area restaurants, the project has been successful at crime prevention and environmental awareness. The recidivism rate for those who have worked at the Garden Project is less than half that of the average postrelease population. The original Garden Project has now expanded to a second site, near the jail.

In 1993 Sneed came up with the idea of selling the organic Garden Project produce at a local farmer's market, held on Saturday mornings. Now, in addition to selling to local restaurants, the project sells to the public, giving employees an opportunity to interact directly with clients and to see a different side of city life than they see in their normal daily lives. Sneed received the U.S. M.F.K. Fisher Award in 1996, and in 1997, she received the Harry Chapin Self-Reliance Award; the work of the Garden Project has also received numerous awards and has been featured in national and international publications. The U.S. Department of Agriculture found the program so successful that it distributed a report describing it all across the country.

Sneed has now set up a project called Tree Corps that has planted over 2,000 trees along streets in San Francisco, many of which were treeless. She continues her work with the Garden Project, helping former prisoners restore their lives, their communities, and their families. She lives in San Francisco.

BIBLIOGRAPHY

Clavin, Tom, "As I Watch Them in the Garden, I Think, This Is Where It Begins, Pride and Respect," *Family Circle*, 1994; "The Garden Project," http://www.saturdaymarket.com/garden. htm; Sneed, Cathrine, "The Garden Project," *Whole Earth*, 1998.

Snyder, Gary

(May 8, 1930–)
Poet, Writer, Translator

The writings of poet Gary Snyder incorporate his knowledge and studies of Amerindian folklore, comparative mythology, Eastern religions, and natural history to articulate ecological ways of living on earth. As a young man he participated in the San Francisco poetry renaissance and later was a leader of the West Coast beat generation writers. A poet, scholar, citizen, and naturalist with a reverence for all things autochthonic, Snyder helped popularize notions of "stewardship," "reinhabitation," "bioregion," and "watershed" in literary and public policy circles and was an early protagonist in the back-to-the-land movement of the late 1960s and early 1970s. Since the first Earth Day in 1970, he has been active in the international environmental movement and in the environmental issues that have come up for the Yuba River watershed of Nevada County, northern California. Among his more than 20 books are *Mountains and Rivers without End* (1970), *Earth House Hold* (1969), the Pulitzer Prize–winning *Turtle Island* (1975), *The Practice of the Wild* (1990), and *The Gary Snyder Reader* (1999).

Gary Snyder was born on May 8, 1930, in San Francisco, California, to Harold Snyder and Lois Wilkey. His family moved from San Francisco when Gary was a few months old, and he spent the remainder of his childhood on a small farm north of Seattle. During World War II, the family moved to Portland, Oregon. His parents maintained a readerly household, and young Snyder was studious. As a teen, Snyder worked at a camp in Spirit Lake, Washington, and mingled with Northwest lumberjacks, Wobblies (members of the Industrial Workers of the World), and Indians, and he developed an interest in Amerindian folklore and myths. He attended Reed College on scholarship, graduating with a B.A. in anthropology in 1951. At Reed, he met two other young poets, Lew Welch and Philip Whalen, who, with Snyder, would become popular "beat" Buddhist poets of the mid-1950s San Francisco literary scene.

Gary Snyder (Courtesy of Anne Ronan Picture Library)

Snyder married Alison Gass as a senior in college in 1951, but by 1952 they had divorced, and he decided, in some sense, to marry himself to a contemplative life, seeking wisdom and serenity from nature and from his growing knowledge of Eastern religious practices. As a fire lookout and trail crew member in Washington State and Yosemite National Park, he read and meditated, and by the age of 24, Snyder had become the bohemian independent scholar and outdoorsman later mythologized by novelist Jack Kerouac as Japhy Ryder in *The Dharma Bums* (1958).

Snyder continued his formal studies in oriental languages at the University of California at Berkeley (UCB), during which time he participated in the circles of the San Francisco poetry renaissance and redefined his poetics under the influence of Kenneth Rexroth, Robert Duncan, and others of the Bay area cultural world. He was also part of a new generation of American poets that included Jack Kerouac, Allen Ginsberg, Michael McClure, and Lawrence Ferlinghetti, who inspired young people to question authority and traditional notions of progress in the United States. Snyder's aesthetic is rooted in what he called, in *A Controversy of Poets* (1965), "the most archaic values on earth. They go back to the late Paleolithic: the fertility of the soil, the magic of animals, the power-vision in solitude, the terrifying initiation and rebirth; the love and ecstasy of the dance, the common work of the tribe. I try to hold both history and wilderness in mind, that my poems may approach the true measure of things and stand against the unbalance and ignorance of our times."

Without completing the master's program at UCB, Snyder shipped out for Japan in 1956 on a scholarship from the First Zen Institute of America. He stayed for nearly 12 years of part-time monastic life, practicing zazen, translating classic Chinese and Japanese poetry, and writing much of the poetry that would appear in his books, *Riprap and Cold Mountain Poems* (1965) and *Mountains and Rivers without End* (1970). Masa Uehara gave birth to Snyder's and her son, Kai, in 1968, and they moved to northern California to take up residence in the Sierra Nevada foothills in 1970. He built a house in the ponderosa pine and black oak woods where he would ripen as a wilderness philosopher, poet, academic, and activist.

Snyder would redefine political boundaries by bioregion or watershed. he trained to be the poet/shaman who writes songs that heal and he has written prayers, hymns, and spells, making traditional native literary forms contemporary and American. He practices "singing a mountain range," literally using the peaks and valleys as a musical scale to create topographical notes. In discussing his poems about animals, he talks about getting inside the minds of animals through the shamanic practice of mimicry. His "Smokey the Bear Sutra" was widely published in the alternative press in the United States and Canada and was later included in *A Place in Space: Ethics, Aesthetics, and Watersheds* (1995).

Smokey the Bear
A handsome smokey-colored brown bear
  standing on his hind legs, showing that
  he is aroused and watchful.
Bearing in his right paw the Shovel that
  digs to the truth beneath appearances;
  cuts the roots of useless attachments,
  and flings damp sand on the fires of
  greed and war;

His left paw in the mudra of Comradly
Display—indicating that all creatures
have the full right to live to their limits
and that of deer, rabbits, chipmunks,
snakes, dandelions, and lizards all grow
in the realm of the Dharma . . .

Snyder is best known for the creativity of his expressions and the solutions he proposes. In 1972 Snyder joined STEWART BRAND, Joan McIntyre, and Project Jonah to lobby the United Nations Environmental Conference in Stockholm, Sweden, on behalf of whales. His report to the UN, "Mother Earth: Her Whales," appeared in the *New York Times* on July 13, 1972, as a poetic diatribe against policies of the power hungry in Brazil, Japan, China, and North America.

Snyder has always sought to apply his craft as a writer to a larger mission, "the real work," to be a worthy spokesman for the wild, as he attested in a 1973 interview: "My political position is to be a spokesman for wild nature. I take that as a primary constituency. And for the people who live in dependence on that, the people for whom the loss of that would mean the loss of their livelihood, which is Paiute Indians, Maidu Indians, Eskimos, Bushmen, the aborigines of New Guinea, the tribesmen of Tibet, to some extent the Kurds, people all over the world for whom that's their livelihood."

In his Pulitzer Prize–winning book *Turtle Island* (1975), Snyder compiles a range of work from snapshot-like images from nature in poems such as "The Night Herons," "Straight Creek–Great Burn," and "The Real Work," to his brief expository essays explaining his ecological views. The writing is embedded with his knowledge of folklore, archaeology, botany, and an ear tuned to the American idiom.

Apart from the Pulitzer Prize in 1975 for *Turtle Island*, Snyder is the recipient of a Bollingen Grant, a Guggenheim Fellowship, and the John Hay Award for Nature Writing. In 1998, he became the first literary figure in the United States to receive the prestigious Buddhism Transmission Award by the Japan-based Bukkyo Dendo Kyokai Foundation. Snyder is a member of the American Academy of Arts and Letters. Since 1985 he has taught at the University of California at Davis, where he was instrumental in founding the Nature and Culture program, an undergraduate academic major program for students of society and the environment. He gives public readings of his work when he can get away from his chores on the land he stewards with his wife since 1991, Carole Koda, in the Yuba River watershed of Nevada County, California.

BIBLIOGRAPHY

Faas, Ekbert, ed., *Towards a New American Poetics: Essays & Interviews*, 1978; Schuler, Robert, *Journeys Toward the Original Mind: The Long Poems of Gary Snyder*, 1994; Snyder, Gary, *The Gary Snyder Reader: Prose, Poetry and Translations 1952–1998*, 1999; Snyder, Gary, *Myths and Texts*, 1978; Snyder, Gary, *The Practice of the Wild*, 1990; Snyder, Gary, *Real Work: Interviews and Talks*, 1980; "UCD English Department," http://wwwenglish.ucdavis.edu/faculty/snyder/snyder.htm.

Soleri, Paolo

(June 21, 1919–)
Architect

A visionary architect who seeks to develop ecologically sound cities, Paolo Soleri is the author of the philosophy of "arcology," which unites architecture and ecology. Soleri is building a city in the Arizona desert, Arcosanti, which he hopes will embody the principles of arcology. Arcosanti is designed to make maximum use of sun, wind, and other renewable resources and bans cars and urban sprawl in favor of compact, vertical use of space.

Paolo Soleri was born on June 21, 1919, in Turin, Italy. He was awarded his Ph.D. in architecture from Torino Politecnico in 1946, with the highest honors. He emigrated to the United States in 1947. He studied with Frank Lloyd Wright in Taliesen, Arizona, from 1947 to 1949. During this time he captured international recognition for a bridge design, displayed at the Museum of Modern Art in New York and published in Elizabeth Kassler's *The Architecture of Bridges* (1949). In 1950 he returned to Italy to study ceramics and was commissioned to design the Solimene ceramics factory in Vietri-sul-Mare, Italy. He completed the commission in 1955 and returned to the United States, settling in Scottsdale, Arizona.

In 1956 Soleri and his wife, Corolyn, began the Cosanti Foundation to serve as a teaching compound in Scottsdale. Soleri had been influenced by the ideas of Frank Lloyd Wright, but where Wright's designs emphasize horizontal lines, Soleri's are vertical. The 1950s were a period of suburban sprawl; the booming economy promised a house with a two-car garage for every middle class family in the United States. As part of the greater Phoenix area, Scottsdale provided an example of sprawl, right on Soleri's doorstep. He began to see that urban design was tending to the needs of the automobile over human needs. In Phoenix, for example, more than 50 percent of the horizontal space was dedicated to moving and storing vehicles. Although Soleri believed that in principle cities were fundamental to human civilization and evolution, he saw that in the United States of the 1950s, they had become the source of spiritual alienation and ecological harm due to pollution, natural resource depletion, and a shrinking amount of land available for farming.

Soleri developed the concept of arcology in hopes that architecture and urban planning could work in harmony with, rather than against, the grain of ecology. Arcology is based in part on the idea that as organisms evolve, they tend toward increasing complexity, and Soleri argues that complexity should be tied to miniaturization. Large systems tend to dissipate energy, where small systems concentrate and conserve. As human society gets more complex, it should also, in a sense, get smaller. Urban sprawl works against this principle, spreading out rather than condensing, using resources in inefficient, wasteful ways. Soleri began to envision cities growing up rather than out, using vertical organization to contain the increasing complexity of human civilization.

In 1969 Soleri published *Arcology: The City in the Image of Man*, in which he laid out the principles of his urban designs. Soleri envisions cities without cars, where walking would be the principal mode of transportation. People would live close together, in integrated, vertical communities. Cities would have farms and gardens, in order to be as self-supporting as possible, and would rely on renewable sources of energy. Soleri's arcology was a response to the many problems of urban civilization, a recognition that these problems could not be solved without radical redesign of urban landscapes. In 1970, the Corcoran Gallery of Art in Washington, D.C., staged the widely attended exhibit "The Architectural Visions of Paolo Soleri," which eventually made a successful tour of the country. Soleri's designs were startling and spellbinding. They reenvisioned human civilization on a massive scale, and Soleri was described as both prophet and megalomaniac. The exhibit coincided with the initiation of work on Arcosanti, Soleri's model city at Cordes Junction in central Arizona, on 4,060 acres along the Agua Fria River. The city was designed to house 7,000 people, using only 25 acres of the 4,060-acre site for actual buildings. Once it is completed, Arcosanti will use only 2 percent of the average landmass covered by conventional towns of comparable size. The city is being constructed of modular forms, made of concrete, which Soleri calls the most flexible kind of masonry. Designs use passive solar principles and aim to conserve space and energy.

Building Arcosanti has been a slow process, in part because of limited funding. Arcosanti has been funded in large part by sale of ceramic and bronze windbells, designed and cast by Soleri, drawing on his experience with ceramics in Italy. Labor has been largely donated, often by architecture interns who pay a stipend to work on Arcosanti as part of their course of study. The city does have a restaurant, a bakery, a greenhouse, and several performance spaces and is a popular site for visiting tourists. The only residents of Arcosanti today are builders, who stay for several weeks at a time. Estimates for completion costs and dates vary wildly. Soleri continues to develop Arcosanti and divides his time between the city and his compound in Scottsdale.

BIBLIOGRAPHY

"Arcosanti," www.arcosanti.org; Blumenthal, Ralph, "Futuristic Visions in the Desert," *New York Times*, 1987; Sheppard, Harrison, "In Arizona, Architect's City of Dreams Remains a Mirage," *Boston Globe*, 1998; Wall, Donald, *Visionary Cities*, 1971.

Soulé, Michael

(May 28, 1936–)
Conservation Biologist, Founder of the Society for Conservation Biology, Cofounder of the Wildlands Project

Founder of the Society for Conservation Biology and the Wildlands Project and hailed as the "Father of Conservation Biology," Michael Soulé is one of the preeminent conservation biologists in North America. Through his research, work, and writing, Soulé strives to make conservation biology and the restoration of ecosystems accessible to a broader audience. His work with the Wildlands Project has been based in "rewilding" North America. Specific goals include the reintroduction of extirpated species; the design of connecting corridors (especially through areas with significant human obstacles); overcoming fragmentation and achieving habitat connectivity, wildlife population viability, and control of exotic species.

Michael Ellman Soulé was born Michael Herzoff, the son of Herman and Berenice (Ellman) Herzoff on May 28, 1936, in San Diego, California. His father died when he was two, and his mother married Alan Soulé, who adopted young Michael. Although neither of his parents was an environmentalist, naturalist, or outdoors recreation enthusiast, Soulé developed a passion for nature at a young age. Growing up in San Diego, close to the desert, the mountains, and the sea, Soulé developed a keen interest in wildlife, particularly butterflies and lizards. His parents supported his interest, to the extent that his mother, who was afraid of snakes, even allowed him to keep them in the house.

Soulé's transition from naturalist to conservationist occurred gradually as he witnessed the increase of urban development sprawl in San Diego. Graduating from high school in 1954, Soulé started studying at the University of California at Berkeley, then moved to the University of California at Santa Barbara, and then finally to California State University in San Diego, where he received his B.A. in zoology in 1959. He did his graduate work in population biology and evolution at Stanford University under PAUL EHRLICH, earning an M.A. in 1962 and his Ph.D. in 1964.

Soulé spent the next two years in Malawi as a lecturer in zoology at the University of Malawi in Limbe, before returning to the United States and joining the Department of Biology at the University of California at San Diego (UCSD) in 1967. He left UCSD in 1979 as a full professor to become the director of the Kuroda Institute for the Study of Buddhism and Human Value in Los Angeles. Soulé's temporary hiatus from academia was to pursue his interest in Buddhism. While at the Zen Center of Los Angeles, he finished two books, among other scientific activities. Soulé also directed the Zen Center's medical clinic for a year or so and even directed the Zen Center for a period of time. He returned to academia as a visiting and adjunct professor at the University of Michigan in the School of Natural Resources in 1984. Soulé founded the Society for Conservation Biology in 1985 and served as its president from 1985 to 1989, at which point he left the University of Michigan to become chair and professor of environmental studies at the University of California at Santa Cruz (UCSC). Soulé retired in 1996 and has since acted as a research professor for the UCSC environmental studies program.

In 1991, he helped to found the Wildlands Project, serving as its president from 1996 to 1998. He has been science director since then. The Wildlands Project, a nonprofit organization based in Tucson, Arizona, is a group of conservation biologists and citizen conservationists from across the continent devoted to developing a North American wilderness recovery strategy. The Wildlands Project sees its role as educating the public, the environmental movement, government agencies, the academic community, and others about the importance of biodiversity and wilderness and what we all need to do to protect it. It works in cooperation with independent grassroots organizations throughout the continent to develop wildland restoration proposals for each bioregion. Draft wildlands network design proposals are developed through discussion and conferences that bring together regional activists, conservation biologists and other scientists, and conservation groups across the spectrum of the movement. The Wildlands Project supports this process through funding, networking, and offering technical expertise.

Soulé has published more than 150 journal and symposium articles, editorials, commentaries, book reviews, forewords, and open letters. In addition, he has published several books on biology, conservation biology, and the social context for contemporary conservation. His main research interests have revolved around three related themes: the application of island biogeography theory to conservation, mesopredator release, and conservation genetics. The island biogeography model, designed by entomologist E. O. WILSON with mathematician Robert MacArthur, is used to explain species-area relationships and is used to predict the number and percentage of species that would become extinct if habitats were isolated. Originally used for islands, this model has since been adapted to explore national parks and nature reserves that are surrounded by damaged habitat. These reserves can be considered habitat islands, surrounded by unsuitable habitat. With respect to mesopredator release, Soulé contends that extinction of large carnivores results in increased problems from smaller carnivores. Soulé's work in conservation genetics has revolved around applications to maintaining genetics diversity, which in turn reduces the risk of species extinction.

In 1995, Soulé coedited with Gary Lease an attack on postmodern deconstruction and social relativism, which, he argued, could be as destructive to nature as bulldozers and chain saws. *Reinventing Nature?: Responses to Postmodern Deconstruction*, a collection of essays from a symposium funded by the University of California Humanities Research Institute, rejects the rhetoric of postmodern deconstructionism that questions the concepts of nature and wilderness, sometimes in order to justify further exploitation of nature and its resources for the sake of economic development.

Soulé is also a popular keynote speaker on conservation biology in academic and nonacademic settings and has served on advisory boards or the board of directors for several ecosystem recovery projects, journals, and conservation biology councils, including the Marine Conservation Biology Institute, the Rocky Mountain Biological Laboratory, and the Southern Rockies Ecosystem Project. Soulé currently resides in western Colorado, where he spends his time learning about the land, working with many conservation organizations, advising Ph.D. students at UCSC and the University of Colorado, doing research with a number of collaborators, and writing about biology, ethics, education, and conservation.

BIBLIOGRAPHY

Frankel, O. H., and Michael E. Soulé, *Conservation and Evolution*, 1981; Soulé, Michael E., "Conservation Biology in Context: An Interview with Michael Soulé," *Environmental Policy and Biodiversity*, R. E. Grumbine, ed., 1994; Soulé, Michael E., ed., *Conservation Biology: The Science of Scarcity and Diversity*, 1986.

Speth, James Gustave

(March 4, 1942–)
Cofounder of the Natural Resources Defense Council, Founder of the World Resources Institute, Dean of the Yale School of Forestry and Environmental Sciences

Cofounder of the Natural Resources Defense Council (NRDC) in 1970 and founder of the World Resources Institute (WRI) in 1982, James Gustave Speth led President Carter's Council on Environmental Quality (CEQ) and the United Nations Development Program and currently is dean of The Yale School of Forestry and Environmental Sciences. Speth has been a strong proponent of sustainable development; he believes that the best way to address the world's severe environmental problems is by integrating environmental, economic, and developmental concerns.

James Gustave ("Gus") Speth was born on March 4, 1942, in Orangeburg, South Carolina. As a youth he spent much time fishing and hunting. He attended Yale College, where he studied political science, graduating *summa cum laude* in 1964. A Rhodes Scholar, Speth studied economics at Oxford for two years, earning an M.Litt. in 1966. He returned to Yale Law School, graduating with a J.D. in 1969, and then clerked one year for U.S. Supreme Court Justice Hugo Black. Thinking about the roles of the American Civil Liberties Union (ACLU) and the National Association for the Advancement of Colored People (NAACP) legal fund, he realized that the environment needed its own legal defense fund, and he recruited classmates and faculty at Yale to cofound with him the Natural Resources Defense Council in 1970. He worked as senior attorney for NRDC for seven years, leading its programs in energy and water.

Pres. JIMMY CARTER recruited Speth in 1977 for the CEQ, which he chaired in 1979 and 1980. He helped produce the influential *Global 2000 Report*, whose forecast of spiraling population growth, environmental degradation, and worsening poverty has largely proven true. From

1980 to 1982, Speth taught environmental and constitutional law at Georgetown University. In 1982, Speth founded the World Resources Institute, with the goal of bringing to the attention of world leaders the problems of natural resource destruction, pollution, and degradation of the world's ecosystems. The center seeks to provide irrefutable scientific information and policy analysis to these leaders and help them include solutions to environmental problems in their political agendas. Ambitious in scope, WRI has achieved broad recognition and success; during its first two decades it expanded to work with partner institutions in more than 50 countries and influence the governments of many more countries. A biennial report, *World Resources*, provides information on the state of natural resources and the environment in every country in the world. Speth served as WRI's president until 1992, when he left to become senior adviser to Pres. Clinton's transition team, heading the team on natural resources, energy, and the environment.

In 1993, Speth was tapped to serve as administrator of the United Nations Development Program (UNDP), the UN agency founded in 1970 to provide technical assistance to developing countries. As with the approach of the World Resources Institute, Speth integrated environmental concerns into UNDP poverty-alleviation projects. One example of the sustainable development projects that Speth's administration promoted was the Songhai Center in Benin, where local people raised domesticated fowl for slaughter and sale, using waste products to grow fruits and vegetables and for aquaculture. Not only is such a project environmentally friendly and an efficient

use of resources, he wrote in *Foreign Affairs*, it is "an integrated, sustainable facility that trains people and produces income." Upon his departure from the UNDP, United Nations Secretary-General Kofi Annan thanked him for his work to "reform, reshape, and revitalize" an agency that had been plagued by inefficiency. Annan said that Speth had advanced "a vision of development that is both sustainable and centered on the real-life experience of human beings."

In 1999, Speth was recruited to become dean of the Yale School of Forestry and Environmental Studies, founded in 1900 with a grant from the family of GIFFORD PINCHOT as the nation's first institution to teach the principles of forest management and conservation. Speth's expertise in sustainable development was seen as an important qualification for the job, as Yale's leaders recognized that students need to understand the international, economic, and developmental context in order to design effective environmental solutions.

At a celebration for the Environmental Law Institute (ELI) lifetime achievement award he was given in 1999, Speth summed up the 30-year evolution of his thinking: "In 1969, we created a separate environmental sector; today we must make every economic sector an environmental sector. Every agency must be an environmental protection agency. The pollutant-by-pollutant, problem-by-problem approach must be replaced by a holistic approach that sees the challenge of sustaining the biosphere in all its complexity and richness and responds accordingly."

In addition to the ELI award, Speth has received numerous others, including the National Wildlife Federation's Resource Defense Award in 1975 and the Keystone Center's National Leadership Award in

1991. He is an honorary trustee of the NRDC and serves on the board of directors of numerous environmental organizations, including WRI, the U.S. Committee for UNDP, the Environmental Law Institute, and the Woods Hole Research Center. Speth resides in Guilford, Connecticut, with his wife, Cameron. They have three children.

BIBLIOGRAPHY

Fellman, Bruce, "The Forest and the Trees," *Yale Alumni Magazine*, 1999; Speth, James Gustave, "The Plight of the Poor," *Foreign Affairs*, 1999; "United Nations Development Program," http://www.undp.org/; "World Resources Institute," http://www.wri.org/; "Yale School of Forestry and Environmental Studies," http://www.yale.edu/forestry/.

Standing Bear, Chief Luther

(1868–February 20, 1939)
Chief of the Lakota Sioux, Writer

Luther Standing Bear was the chief of the Lakota Sioux from 1905 until his death in 1939. During his leadership of the Lakota people he worked ceaselessly to improve conditions on the reservation and restore a sense of pride to his people. He wrote four books, all about Indian life and the connection of that life to the natural world. He was an educator of his people and strove to enlighten the White man about the way of the Lakota.

Luther Standing Bear was born Ota K'te (Many Kill) on the Pine Ridge Reservation in South Dakota in 1868. His father was Chief Standing Bear the first of the One Horse band, and his mother was Pretty Face of the Swift Bear band. Standing Bear was reared in a traditional Sioux manner and trained from a young age to be a Sioux warrior, although by the time he reached fighting age the Lakota no longer engaged in physical warfare. In 1879 Standing Bear was sent to Pennsylvania and became a member of the first class to attend the Carlisle Indian School.

Much of what he wrote about in his autobiographical books was influenced by his experiences at Carlisle, where Indian children were forced to cut their hair, wear the White man's clothes, and give up their customs, religion, families, and language.

Standing Bear spent five years at Carlisle, then returned to Pine Ridge and resumed living in the Lakota way. He was shocked to see students returning from the East who had lost all knowledge of their native language and customs. For a short time Standing Bear became a teacher at the reservation school. Soon, however, all Lakota teachers were replaced by outsiders, White people who had no knowledge of or interest in anything Lakota.

In 1898 Standing Bear joined Buffalo Bill's Wild West show, traveling as far as England to perform before the king. The following year he was badly injured in a train wreck that killed members of the troupe. His experience with the show led him to California, where he joined a lecture circuit and became a member of the Indian Actor's Association in Hollywood. He also appeared in a few movies, although he opposed the way Native people were portrayed in these films.

Although he lived in California for some years, Standing Bear remained a leader of his people. He was continually appalled by the degradation on the reservation. He was angered and saddened by seeing his people, once strong and self-sufficient, belittled to the point of extreme poverty and helplessness. He became a champion of the Lakota in approaching the government for improved conditions and in striving to display to the White man that the Sioux were not ignorant savages as they were so often portrayed.

In 1928 Standing Bear published *My People the Sioux*, his first book about Sioux life. In this book he attempted to explain his people to a world tainted by an image told only from the White point of view. *My Indian Boyhood* followed in 1930 and painted a picture of what it was to grow up as a Lakota. Here Standing Bear describes childhood in a traditional Lakota setting. The boys learned through various games to become hunters and warriors. They also learned to care for and respect all members of the tribe and to appreciate and learn from all things in nature. Standing Bear often compared the Lakota ways to those of the White man, especially in the understanding of and love for nature: "The Indian tried to fit in with Nature and to understand, not to conquer and to rule. We were rewarded by learning much the white man will never know."

In 1933 Standing Bear wrote to Pres. Franklin D. Roosevelt to lobby for a bill that would make Indian history, religion, art,

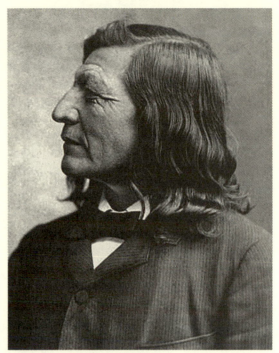

Luther Standing Bear (Corbis)

philosophy, and culture a part of all public school curricula. His letter remains unanswered by the U.S. government. The same year, Standing Bear's third book, *Land of the Spotted Eagle*, was published. This was another general description of Lakota life and values, but Standing Bear openly criticized government policy as it applied to the education and overall treatment of the Lakota people. Standing Bear strongly believed that Lakota children should have Lakota teachers and should be taught not only the White man's ways and history but their own as well.

Standing Bear claimed that "the white man has come to be the symbol of extinction for all things natural to this continent" and lamented that, by the White man, his people were "thrust . . . from [their] age-old mode of living into one that was foreign to [them] in every respect; religious, tribal, and social . . . placed in the impossible position of trying to remake or remould himself into a European." *Land of the Spotted Eagle* was the first book in which he so openly criticized the U.S. "Indian policy."

Standing Bear's fourth book, *Stories of the Sioux*, was published in 1934. This was a summary of many traditional Sioux legends and stories.

Standing Bear died on February 20, 1939.

BIBLIOGRAPHY

"Native American Authors," http://nativeauthors.com; Standing Bear, Luther, *Land of the Spotted Eagle*, 1933; Standing Bear, Luther, *My Indian Boyhood*, 1930; Standing Bear, Luther, *My People the Sioux*, 1928; Standing Bear, Luther, *Stories of the Sioux*, 1934.

Steel, William

(September 7, 1854–October 21, 1934)
Commissioner of Crater Lake National Park, First President of the Mazamas

Progressive conservationist William Steel, the first president of the Mazamas, a mountaineering and conservation organization based in Portland, Oregon, is popularly known as the "father of Crater Lake." Steel fought for 17 years to obtain national park designation for Crater Lake, Oregon's only national park. He spent the last 20 years of his life as superintendent and commissioner of the park, actively promoting the area and working to facilitate public access. He was also a participant in the campaign that led to the establishment of the Cascade Forest Reserve and was well known in the Northwest as a publicist of outdoor recreation and nature appreciation.

William Gladstone Steel was born on September 7, 1854, in Stafford, Ohio. His father, William, was a Scottish immigrant; his mother, Elizabeth, a native of Virginia. The Steel family moved to Kansas in 1868, and it was there that Steel first read about Crater Lake. He determined to see the spectacular lake in the wilds of far-off Oregon for himself. His family moved to Portland, Oregon, four years later, where they joined two of Steel's brothers, who had become successful financiers. Steel graduated from high school in Portland and began a three-year apprenticeship as a pattern maker for Smith Brothers Iron Works.

During this period, Steel acquired two interests that he would maintain for the rest of his life, writing and mountaineering. In 1879 he established the *Albany (Ore.) Herald*, but after a couple of years as a newspaper publisher, Steel returned to Portland, where he began work as a substitute letter carrier. He worked for the post office for 14 years, twice filling the post of superintendent of letter carriers. In 1883, he organized the first letter carrier's association in the United States.

In his spare time, Steel went mountaineering, climbing most of the peaks of the Cascades, including Mt. Hood, which he climbed more than 30 times in his life. He first visited Crater Lake in 1885. At that time very few people had heard of it—it was seven years after his arrival in Oregon before Steel met someone who had actually been there. The moment he saw the lake, Steel determined to have it protected as a national park so that all could enjoy its beauty. The next year he succeeded in obtaining an executive order from Pres. Grover Cleveland that withdrew land in the vicinity of the lake from public entry, but it would be 17 years before Steel's efforts to gain national park status for Crater Lake reached full fruition.

Two years after visiting Crater Lake, Steel organized the Oregon Alpine Club to attract attention to the Northwest's mountain scenery. In 1889, he published a collection of pamphlets in a promotional book, *The Mountains of Oregon*, with descriptions of the club's activities. Financial difficulties limited the organization's success, however, and in July 1894 Steel helped formed the Mazamas, a mountaineering club that is still active today. The club was named after Mount Mazama, the dormant volcano in whose crater Crater Lake has formed. It was Steel's idea to make the ascent of a glacier-clad peak a membership requirement that he felt would make the Mazamas a more successful organization than the moribund Oregon Alpine Club. He was elected the organization's first president.

With the support of the Mazamas, Steel continued in his efforts to make Crater Lake a national park. During the 1880s and 1890s, Steel engaged in and supported scientific study of the lake. He also successfully fended off repeated attempts to establish a state park there, be-

lieving the state could not provide the necessary maintenance and protection. He was particularly concerned about the impact of sheep and other livestock. Whether or not to allow livestock grazing was a key issue in another campaign Steel participated in, the establishment of the Cascade Forest Reserve. Created in 1893, this 300-mile-long reserve, which included Crater Lake, stretched from the Columbia River to the California border.

In 1895, Steel traveled to Washington, D.C., where he helped foil an attempt to rescind the executive order that established the Cascade Reserve. He also lobbied senators and representatives in an attempt to convince them that Crater Lake deserved national park status. "I was about as popular with them as the plague," he related in a column 18 years later. "When they asked me what my motive was and I said 'My love for Oregon and my desire to be of benefit to future generations is my only motive,' they would look at me, shake their heads and instruct their doorkeepers not to admit me again as I was a crank." His persistence finally paid off, however, when he convinced Secretary of the Interior Ethan A. Hitchcock to lend his support to the plan. U.S. Forestry Division chief GIFFORD PINCHOT also enthusiastically endorsed national park status for the lake, which he considered one of the great natural wonders of the continent. Pres. THEODORE ROOSEVELT signed the bill making Crater Lake a national park on May 22, 1902.

Steel's efforts to protect and promote Crater Lake did not end in 1902, however. He spent the rest of his life lobbying for funds to develop the park. He believed that everyone should have a chance to enjoy the area, not just the privileged few. He became the park's first concessionaire in 1907, providing transportation for tourists, a tent camp at Annie Springs, and boat tours of the lake, in which he had planted the first trout. A couple of years later he provided funds for the construction of Crater Lake Lodge. In 1913, Steel replaced W. F. Arant as superintendent of the park. Three years later, after the establishment of the National Park Service and the first federal appropriations to the park, Steel was promoted to commissioner of the park. Soon after his promotion, he began a campaign to raise $700,000 to establish a highway to Crater Lake. Although it would be several years before the highway was built, the first road encircling the rim of the lake was completed in 1918. The road-building efforts at Crater Lake were linked to the See America First campaign, spearheaded by STEPHEN MATHER, first supervisor of the National Park Service.

In 1921, Steel rejoined the Mazamas after a 15-year estrangement sparked by a dispute about his role in the organization and the nomination of Pres. Roosevelt as an honorary member. Steel also continued his publicity efforts in the 1920s, publishing an occasional pamphlet series and working for a time as editor of the Grants Pass *Courier*. In 1929, he moved to Medford, where he worked to expand and develop Oregon's state park system. He made his last visit to Crater Lake in the summer of 1932. Lydia, his wife of more than 30 years, died shortly thereafter, and Steel, too, fell ill.

Steel passed away on October 21, 1934, after two years of failing health. His pall bearers were rangers from Crater Lake National Park, and he was interred in Medford's Siskiyou Memorial Park wearing his Park Service uniform. He was survived by his daughter, Jean.

BIBLIOGRAPHY

Lockley, Fred, "In Earlier Days," *Oregon Journal*, 1913; Mark, Stephen R., "Seventeen Years to Success: John Muir, William Gladstone Steel, and the Creation of Yosemite and Crater Lake National Parks," *Mazama*, 1990; "Mazama's Main Menu," http://www.mazamas.org/mainmenu.htm; Steel, W. G., *The Mountains of Oregon*, 1889; "William G. Steel, Father of Crater, Called by Death," *Medford Mail Tribune*, 1934.

Stegner, Wallace

(February 18, 1909–April 12, 1993)
Writer, Historian

A prolific and successful author, Wallace Stegner won nearly every major award given a writer, except the Nobel Prize, and penned works spanning everything from fiction to biography and history. He is remembered as an eloquent voice for the early environmental movement as it gained public recognition and momentum through such contentious issues as the attempted damming of Dinosaur National Monument. He was a consistent and vehement supporter of federal wilderness preservation and management.

Wallace Stegner was born February 18, 1909, on a farm near Lake Mills, Iowa. He was the second son of George and Hilda Paulson Stegner. He spent his early days with his mother and brother, Cecil, following along on his father's epic traipse through the West in search of what Stegner would later term "The Big Rock Candy Mountain, where something could be had for nothing, where life is effortless." For Stegner this journey included Iowa, Washington, and the Canadian province of Saskatchewan, effectively ending in 1921 in Salt Lake City, Utah, where he completed his high school education in 1925 and enrolled in the University of Utah, obtaining a B.A. degree in English in 1930. For George, this journey would not end until the summer of 1940, when he killed himself in a hotel room in Salt Lake City.

Stegner continued his education at the University of Iowa, completing his master's thesis in 1932. It comprised a series of three short stories, two of which were later published. He then went to the University of California, Berkeley, to begin his doctorate work but found the program unsatisfactory. He moved back to Iowa, finishing his graduate work by the summer of 1935. During this time he met Mary Stuart Page, whom he married in 1934. Together they had a son, Page, born in 1937.

He wrote his first novel in 1936 after reading an advertisement for a novel-writing contest sponsored by the Little, Brown Publishing Company in Boston. The prize was for $2,500. At this time Stegner was an instructor at the University of Utah, making only $1,700 a year. He wrote *Remembering Laughter*, submitted it, and won. Thus began a writing career that would lead to 34 books and many essays and articles.

Wallace Stegner (AP Photo/HO)

Stegner was an influential author/historian/biographer, who was often decades ahead of the general thinking. *One Nation*, published in 1945, is a collection of essays and photographs originally produced for *Look* magazine. It confronts the race problems faced by the United States, which was in the process of fighting World War II with a racially segregated army and was busy interning Japanese Americans by the thousands. Stegner describes the "wall" splitting the United States. On one side are the white Protestant gentiles, and on the other are those who, because of race, religion, or color, are not considered full U.S. citizens.

Stegner was unique as a writer from, and grounded in, the American West. This region had a profound effect on the way he thought and wrote about the natural environment. The vast, wide-open spaces left an indelible impression on him, defining him as a person and as an artist. He later spoke of the immense distances, of the roads and telephone lines that disappear on the horizon, of a massive circle of observation with him always at the focal point. He saw wilderness preservation as a necessity because to him wilderness was nothing less than "the challenge against which our character as a people was formed."

Stegner is mainly remembered as a voice for a sensible western land ethic. He was a staunch supporter of federal management of public lands, stating that "without the federal bureaus, the west would be a wasteland." And though he thought of himself primarily as an author, he also wore the hat of sometime activist. In 1954 DAVID BROWER of the Sierra Club enlisted his help in the campaign to prevent the damming of the confluence of the Green and Yampa Rivers at Echo Park in Dinosaur National Monument on the Colorado-Utah border. This proposed Bureau of Reclamation project would have flooded a spectacular area of the monument with 500 feet of water. Stegner edited and contributed to a book entitled *This Is Dinosaur: Echo Park Country and Its Magic Rivers*, which was presented to members of Congress in an attempt to keep the project from going forward. The campaign was successful, and Dinosaur was not submerged. The victory was only partial, however, for the compromise that was reached with the environmentalist lobby allowed for a dam to be built farther downstream on the Colorado River, flooding Glen Canyon, a part of the Colorado that was nearly inaccessible and familiar to only a few people. Stegner was one of these few, having traveled this stretch of the river in preparation for his biography on JOHN WESLEY POWELL: *Beyond the Hundredth Meridian: John Wesley Powell and the Second Opening of the American West.*

Stegner's views on wilderness preservation are best expressed in a letter he wrote in 1960 to David E. Pesonen, consultant to the Outdoor Recreation Resources Review Commission, which was in the process of reviewing for Congress the need for wilderness legislation. In this letter Stegner stresses the necessity for preserving wilderness, stating that its worth cannot be measured only by its usefulness. He concludes by saying: "We simply need that wild country available to us, even if we never do more than drive to the edge and look in. For it can be a means of reassuring ourselves of our sanity as creatures, as a part of the geography of hope."

Wallace Stegner died April 12, 1993, of injuries sustained in a car accident in Santa Fe, New Mexico.

BIBLIOGRAPHY

Benson, Jackson, *Wallace Stegner, His Life and Work*, 1996; Colberg, Nancy, *Wallace Stegner, A Descriptive Bibliography*, 1990; Rankin, Charles E., ed., *Wallace Stegner: Man and Writer*, 1996; Simpson, Richard H., *Dictionary of Literary Biography: American Novelists 1910–1945*, Vol. 9, 1981.

Steingraber, Sandra

(August 27, 1959–)
Ecologist, Poet

Sandra Steingraber is an ecologist and poet who holds degrees in both biology and creative writing. She has conducted field studies in Costa Rica, Africa, and the United States and has published a volume of poetry. Recently she tackled the issue of environmental contamination in her book, *Living Downstream: An Ecologist Looks at Cancer and the Environment* (1997), which traces alarming worldwide patterns of cancer incidence to the use of agricultural chemicals and other toxins. A cancer survivor herself, she brings a personal perspective to her crusade to expose agricultural and industrial recklessness and its effect on human health.

Sandra Kathryn Steingraber was born on August 27, 1959, in Champaign, Illinois, and was adopted by Wilbur and Kathryn Steingraber, both teachers. She grew up with a younger sister in the town of Pekin, Illinois, surrounded by farmland. When she was 15 years old, her mother was diagnosed with metastatic breast cancer. Although her mother survived it, the experience had a profound impact on Steingraber and provoked her interest in biology. In 1979, when she was a biology student at Illinois Wesleyan University, she was diagnosed with bladder cancer herself. Having grown up in a county with 15 hazardous waste sites and several carcinogen-emitting industries, she became interested in whether the toxins she was exposed to as a child could have contributed to her contracting cancer.

Steingraber finished her B.A. in biology and graduated from Illinois Wesleyan University in 1981 and then received her M.A. in creative writing from Illinois State University in 1982. In the same year she married, and a few years later she accepted a fellowship to study ghost crabs in Costa Rica. In 1989, by the age of 29, she had earned her Ph.D. in biology from the University of Michigan. Upon finishing her graduate studies she was still concerned with the connections between human health and the environment, and she decided that someday she would become a scientist within the activist community. She started teaching at Columbia College in Chicago in 1990 and stayed for three years. Then in 1993 a Bunting Fellowship allowed her to start writing fulltime, and she completed a book of poetry called *Post-Diagnosis* in 1995. Her next book brought her instant acclaim and provided crucial information that was, until then, not easily available. Called *Living Downstream: An Ecologist Looks at Cancer and the Environment* (1997), it brought together information gathered under new right-to-know laws on toxic releases and newly released cancer registry data. From this and other lines of evidence, she concluded that environmental influences are more crucial than heredity in causing cancer. Her own experience provides an example of this. She and several other members of her family, including her mother, an aunt, and many uncles, all contracted cancer— leading one to suppose that it simply runs in the family. Yet she was adopted and shares no genetic history with any family members. She also uses the example of immigrants coming to this country with little or no history of cancer, who gradually assimilate until their cancer rates equal those of their new host country. Clearly, environment plays a large role.

In *Living Downstream*, Steingraber frequently refers to RACHEL CARSON, who died of breast cancer herself shortly after her influential book *Silent Spring* was published. In *Silent Spring*, Carson called for a systematic evaluation of the contribution of toxic chemicals to increased human cancers. Over 30 years later, Steingraber repeats this call and emphasizes that Carson's warning has not been heeded. She cites the fact that between 45,000 and 100,000 chemicals are now in common use, and only 1.5 to 3 percent of them have been tested for carcinogenicity. While there are regulations in effect that require the evaluation of old, untested pesticides, the tests themselves will not be completed until 2010. Until then the pesticides can still be sold and used. In a country where more than 750 million tons of toxic chemical wastes have been discarded since the late 1950s and where detectable pesticide residues are found in 35 percent of the food consumed, Steingraber insists that uncertainty should not be an excuse to do nothing. She advocates a human rights approach, one that shows respect for life on the planet by not allowing untested chemicals free access to the environment.

She is a passionate and popular speaker and has given talks on environmental health issues at Harvard University, Yale University, Cornell University, Woods Hole Oceanographic Institute, the First World Conference on Breast Cancer, and many other venues. At the end of one of her keynote addresses at a conference on pesticides, she reached behind the podium and showed her audience a jar of human breast milk, explaining that it contained more dioxin, more polychlorinated biphenyls (PCBs), and more dichlordiphenyltrichlor (DDT) than any food that was allowed on the market. Her message is that the current system of regulating the manufacture and release of known and suspected carcinogens is intolerable.

Her efforts have earned her rewards and recognition. *Ms. Magazine* named her a woman of the year for 1997, and in 1998 she won the Altman Award for the inspiring and poetic use of science to elucidate the causes of cancer. She recently remarried (after a divorce in 1994) and had a daughter and is working on a new book: *The Ecology of Pregnancy and Childbirth*. She lives in Somerville, Massachusetts.

BIBLIOGRAPHY

Gross, Liza, "Rachel's Daughter," *Sierra Magazine*, 1999; Lindsey, Karen, "Sandra Steingraber," *Ms. Magazine*, 1998; Steingraber, Sandra, *Living Downstream: An Ecologist Looks at Cancer and the Environment*, 1997.

Stone, Christopher

(October 2, 1937–)
Legal Scholar

Author of the 1972 classic *Should Trees Have Standing?*, Christopher Stone has helped shape the terms of how the law addresses environmental issues. In that work Stone argued that natural objects—rivers, oceans, forests, and the environment itself—should be granted legal standing as "persons" with rights, specifically the right to bring suit if threatened with harm. Though this argument has never been fully accepted in the courts, Stone's work has helped expand the grounds on which environmental issues can be fought in the courts, by weakening the necessity for plaintiffs to suffer immediate personal harm in order to have standing to sue. In this and other writing, Stone has influenced the development of environmental ethics.

Christopher Stone was born on October 2, 1937, in New York City. Stone is the son of the independent journalist I. F. Stone and inherited from his father an interest in philosophy and progressive politics. He graduated from Harvard University in 1959 with an A.B. in philosophy and from Yale University with an LL.B. in 1962. After serving as a fellow in law and economics at the University of Chicago from 1962 to 1963, Stone was hired by the School of Law at the University of Southern California in Los Angeles. He was granted tenure in 1966. In 1969 the U.S. Forest Service granted Walt Disney Enterprises the right to build a resort complex in Mineral King Valley, located in the Sierra Nevada. The Sierra Club went to court to block the development and won a temporary restraining order. That initial success was overturned, however, when the court ruled that the club did not have legal standing to bring the suit. The Sierra Club would not be directly injured by the resort's destruction of Mineral King Valley, the court ruled, and therefore could not pursue its claim. Stone intervened to challenge this narrow view of standing. What if, Stone argued, the valley itself was seen to have standing, and the Sierra Club to serve as its guardian? Stone rushed his ideas into print through the *University of Southern California Law Review*, in hopes of influencing the Sierra Club case, then pending before the U.S. Supreme Court. Though the Sierra Club lost the case, "Should Trees Have Standing?" was cited favorably in the dissenting opinion of Supreme Court Justice WILLIAM O. DOUGLAS; through his advocacy, it was picked up by other lawyers and the media. Stone's argument began to influence the debate over environmental rights, particularly the question of whether the issues could be more effectively argued from a human or nonanthropocentric viewpoint.

Stone made the case for an environment-centered approach in three main parts, the first demonstrating that conceiving of the environment as having legal rights could be no more "unthinkable" than rights for women or children had been at other points in Western history. In the second part, Stone addressed the "legal-operational aspects" of his argument and demonstrated that while natural objects currently had no legal rights per se, there were legal precedents and structures in place that could be usefully and reasonably reinterpreted so as to confer such rights on the environment. Finally, Stone addressed the "psychic and socio-psychic aspects" of his argument, specifically the need to reimagine the relationship between humankind and nature less as "man over nature" than with a view of humanity as part of the interconnected, organic whole of nature. In the years following its publication, Stone's argument began to be used in actual cases. Complaints were filed in the name of several nonhumans, including a polluted river in southern Connecticut, Death Valley Monument in southern California, and the endangered palilla bird in Hawaii. These cases, however, involved joint plaintiffs, wherein the nonhuman complainant was linked with a human individual or group, so the principle of full legal rights for natural objects remains unestablished. Stone's ideas are in circulation, but not fully in practice.

Since the publication of *Should Trees Have Standing?* in book form in 1972, Stone has contributed two further volumes to the discussion of environmental ethics. 1987's *Earth and Other Ethics* follows up on Stone's earlier work in an extended meditation on the limitations of what he terms "moral monism." Stone argues that conventional ethics tends to put forth a single principle—for example, the greatest good for the greatest number—as the key to all ethical dilemmas. Stone argues that this framework inherently privileges human over nonhuman actors and therefore should be replaced with a pluralist approach, recognizing

differences among situations, actors, and acts. Stone suggests that this approach can give greater weight to the standing and rights of natural objects. *The Gnat Is Older than the Man: Global Environment and the Human Agenda* (1993) is less philosophically and more practically oriented. In particular Stone addresses finding solutions to pressing issues of global environmental degradation, with a particular emphasis on the need for international treaties. Treaties, he argues, offer greater hope for constructive environmental action than the courts, which are slow to respond to immediate danger. Stone argues for the importance of what he calls the Global Commons, the skies and the oceans, the places where human and natural interests transcend national sovereignties.

The Gnat Is Older than the Man was influenced by Stone's work as a researcher and consultant to governmental and nongovernmental agencies, including the American Bar Association, the National Science Foundation, and the Department of Energy. He is the Roy P. Crocker Professor of Law at the University of Southern California and lives in Los Angeles.

BIBLIOGRAPHY

Staff Editors, "Developing an Environmental Ethos," *Tennessee Law Review*, 1988; Wenham, Brian, "Towards a Treaty of Life for the Global Commons," *Financial Times of London*, 1993; Westing, Arthur, "Should Trees Have Standing?" *Environment*, 1997.

Subra, Wilma

(August 14, 1943–)
Analytical Chemist, Citizen Advocate

Chemist and MacArthur Foundation award winner Wilma Subra has helped innumerable communities in her native Louisiana and 20 other states to study the health risks posed by local environmental contamination. She also helps them work toward cleaning up the damage.

Wilma Subra was born in Morgan City, Louisiana, on August 14, 1943, the eldest of six sisters. She was raised with a strong Samaritan ethic: her family helped whoever needed a hand—relatives, friends, neighbors. She became interested in chemistry as a child working for her father, who owned a lab that ground oyster shell into fine powder that could be used in paint, makeup, pharmaceuticals, even as a filler for tires.

Subra attended the University of Southwestern Louisiana in Lafayette, earning a B.S. in 1965 and an M.S. 1966, both in microbiology and chemistry. She taught mathematics to junior high school students for one year and then in 1967 began working for the Gulf South Research Institute, a research laboratory that contracted with government agencies and private companies to do cancer and viral studies, pharmaceutical testing, environmental evaluations, environmental impact statements, and other similar work. One memorable contract was from the Environmental Protection Agency (EPA) to visit Love Canal in 1980 for a quick response project, a rapid evaluation of the community's health problems. During this and other similar evaluations, Subra found herself in the uncomfortable position of knowing more about the resi-

dents' health problems through blood analyses and air sampling results than the residents knew themselves, yet her employer prohibited her from informing them of what she knew.

Subra found it more satisfying to be able to communicate openly with people subject to environmental contamination and offer her assistance, and began doing this in the mid-1970s. She and her colleagues at Gulf South were often approached by private citizens living in the area who were concerned about environmental contamination near their homes. Subra did chemical analyses of soil, air,

Wilma Subra (Courtesy of Anne Ronan Picture Library)

and water at the laboratory, and evenings would hold meetings with these citizens to explain the results. If they were interested, Subra would instruct them on how to learn more, accompanying them to state regulatory agencies to find out which local industries had permits to pollute, which were discharging waste into the air or waterways, and which ones were exceeding the limits of their permits.

Subra has worked in this capacity with 400 grassroots citizens groups in many areas of Louisiana. She has helped communities around Lake Charles succeed in having commercial hazardous waste facilities cleaned up, preventing new ones from opening, tightening permit regulations for industry, and relocating people living on top of contaminated groundwater plumes. She worked with the citizens of Calcasieu Parish to demand an end to the contamination of the Calcasieu estuary. Her tests of contaminated soil served as evidence to convince the EPA to establish three new Superfund sites in the heavily contaminated Vermilion Parish. The Houma Indians asked her to help monitor pollution in their community at Grand Bois, where land farming of commercial oilfield wastes (packing these toxic wastes into the soil) was poisoning land and air. She worked with residents of her native Morgan City to prevent the reopening of the Marine Shale Processors plant, the largest hazardous waste incinerator in the United States, after it had been shut down by the government for exceeding its emissions limits.

Subra has also served on transition teams for environmental issues for two Louisiana governors, Buddy Roemer, elected in 1987, and Edwin Edwards, elected in 1991. The transition teams toured the state to meet with local citizens to learn about their environmental problems, wrote a report, then toured the state again to present the report. She serves on various Louisiana state government committees that work with regulatory agencies and on several EPA committees as well, advocating for grassroots citizens groups. Subra estimates that one-third of her time is spent volunteering for the EPA, and that does not include the time she spends meeting with citizen groups to learn what they want from the EPA.

Subra's work is not limited to Louisiana. She has worked in more than 20 states since 1987 with the National Citizens Network on Oilfield Waste, reviewing deficiencies in state regulations for oilfield waste treatment and offering technical assistance to citizens groups living near oilfield waste–processing plants.

Subra has owned a chemical laboratory and environmental consulting firm, Subra Company, Inc., since 1981 and does receive some technical assistance grants for evaluating site investigation data at Superfund sites. But Subra spends most of her work weeks—20 hours a day, seven days a week—offering volunteer assistance to victims of environmental contamination.

Subra received a 1999 MacArthur Foundation "genius" grant of $370,000, which she uses to fund even more work for environmental causes. Other awards she has received include the 1989 Woman of Achievement award from Connections, a Lafayette, Louisiana, organization; and the Louisiana Wildlife Federation Governor's Conservation Achievement Award, the top environmental award in the state, also in 1989. Both were given to her in recognition for her work with citizens' groups on environmental issues, as well as work on Governor Roemer's transition team and her participation in various state committees. In 1999 she was recognized by the Coalition to Restore Coastal Louisiana for her work with coastal communities on environmental issues.

Subra has been married since 1965 to Clint Subra, a medical technician whose work includes bloodwork analysis when necessary for people who live near toxic contamination sites. Together they have three grown children, who grew up accompanying Subra to meetings and on trips throughout the United States as Subra collected data for new projects.

BIBLIOGRAPHY

Dunne, Mike, "La. Activists, Scientists Making a Difference," *Baton Rouge Advocate*, 1999; Schultz, Bruce, "La. Environmentalist Wilma Subra Receives MacArthur Grant," *Baton Rouge Advocate*, 1999; Schwab, Jim, *Deeper Shades of Green*, 1994.

Suckling, Kierán

(October 11, 1964–)
Cofounder of Center for Biological Diversity

Kierán Suckling founded and heads the Tucson-based Center for Biological Diversity, a nonprofit environmental organization that successfully uses the court system to fight for habitat and wildlife protection. Between 1991, when it was founded, and 1999, the Center for Biological Diversity filed over 132 lawsuits and won 86 percent of them. It has also succeeded in adding dozens of species to those protected by the Endangered Species Act.

Kierán Francis Suckling was born on October 11, 1964, in Winchester, Massachusetts, to an Irish mother and English father. He moved frequently as a child, because his father worked for an engineering company that built large industrial plants throughout the world. A self described "moderate achiever and fairly constant discipline problem" in high school, Suckling was nonetheless hungry for knowledge—especially in the fields of linguistics and philosophy—and devoured a collection of philosophy books that he inherited from an uncle's Jesuit seminary. Suckling became involved in nuclear disarmament and Central American justice issues during his high school years. He studied at Salve Regina College in Rhode Island and Worcester Polytechnical Institute in Massachusetts, where he set up a chapter of Student Pugwash, an organization founded by Albert Einstein that explores the social implications of technology and works against the arms race. He transferred to the philosophy program at the College of the Holy Cross in Worcester, Massachusetts, graduating *magna cum laude* in 1987, with a B.A. in philosophy, and winning the school's McCarthy Award in philosophical research.

Upon graduation, Suckling set off to explore the West. After hiking the canyons and mountain country of western North America, he enrolled in the State University of New York (SUNY) at Stony Brook's graduate program in philosophy, focusing on Greek philosophy, phenomenology, deconstruction, and language extinction. Because he was convinced that there was a relationship between extinction of languages and of species, Suckling surveyed endangered spotted owls and northern goshawks for the U.S. Forest Service (USFS) in Arizona and New Mexico. He became so dismayed by the evidence that these birds were in danger of extinction that he vowed to save them. Suckling and his survey partner, biologist Peter Galvin, joined with emergency room doctor Robin Silver in 1990 to brainstorm the Southwest Center for Biological Diversity, an organization dedicated to saving endangered species and habitats. Suckling completed his coursework at SUNY, passed his history of philosophy exam with honors in 1991, and returned in 1991 to direct the Southwest Center (now called the Center for Biological Diversity; CBD).

The CBD quickly became one of the region's most active and best known environmental groups. It seeks to protect habitat and wildlife of the wild Southwest, fighting ranchers, loggers, the government agencies that enable them, and any other agency that destroys or develops wildlands. Eschewing the current trend of working with all parties concerned with the land, in order to arrive at consensus or a compromise, the CBD is adamant about environmental preservation. Cofounder Galvin, who currently works as CBD conservation biologist and

litigation coordinator, explained the CBD's position to *Outside Magazine:* "The developers and the extractors have eaten nine pieces of a ten-piece pie . . . and they want to negotiate about the tenth piece. I'm happy to stick my fork in their hand."

During the CBD's early years, Suckling and Galvin were tutored in environmental litigation by Biodiversity Legal Foundation director JASPER CARLTON. Funded by memberships and such foundations as the Pew Charitable Trust and the Turner Foundation, the CBD works by filing lawsuits to prevent environmentally destructive development projects. They monitor the activities of the U.S. Forest Service and the U.S. Fish and Wildlife Service (USFWS), filing Freedom of Information requests to obtain the information they need. In August 1995, at a time when the Republican-dominated U.S. Congress had placed a year-long moratorium on new endangered species listings and a salvage logging rider had lifted all timber-cutting regulations in national forests, the CBD convinced a judge to halt all logging in the Southwest until its effects on the threatened Mexican spotted owl could be investigated. The injunction on cutting in national forests was lifted when the USFS agreed 16 months later to abide by the USFWS Mexican spotted owl recovery plan.

The CBD has also tried to reform the USFS. Suckling told reporter Michael Kiefer of the *Phoenix New Times*, "We've got an economic system designed to make money by destroying nature, whether it's timber sales or grazing allotments. A lot of groups like ours that traditionally did not deal with economics, are starting to look at ways to both expose and to take advantage of the economic system. Because the way it's set up now, they're just giving away trees and giving away grass to the ranchers and the loggers." The CBD collaborated with environmental groups in the Northwest on pushing for "unlogging" permits: leases on national forest lands through which the purchaser would pay for the right to *not* log the forests, but in February of 1997 USFS employees rejected the idea without consulting their supervisors at the U.S. Department of Agriculture. The rejection was rescinded, and the issue is still under review.

The CBD counts among its many successes the salvation of the San Pedro River, a 130-mile-long tributary of the Gila River on whose banks half of all the 800 bird species known in North America have been sighted. *Birding* magazine has called it the world's best birding area, and the Nature Conservancy has declared it one of the world's eight "Last Great Places." In addition to birds, the San Pedro harbors the second-highest concentration of mammalian species in the world, behind Costa Rica's montane rain forests. It is so biologically rich that it has been declared one of only two National Riparian Conservation Areas in the country. The San Pedro was in danger of hydrological collapse because of its proximity to the United States Army's Fort Huachuca and the nearby town where Fort Huachuca employees live, Sierra Vista, both of which pump water from the aquifer below the San Pedro. Filing suits on behalf of many endangered species that inhabit the San Pedro, the CBD persevered through five years of litigation until Fort Huachuca agreed in 1999 to reduce its water use and support reductions throughout the river basin.

For their feisty and successful approach to conservation and their persistence, Suckling and Galvin won recognition as "Deep Ecologists of the Year" from the Oregon-based Fund for Wild Nature in 1996, and in 1997 the *Arizona Business Journal* highlighted Suckling as one of the eight people most likely to shape Arizona's future. CBD's unique brand of activism has been featured in the *New Yorker*, *Wall Street Journal*, *Washington Post*, *San Francisco Chronicle*, *High Country News*, *Backpacker*, *Outside*, and many other magazines and newspapers. Suckling resides in Tucson, Arizona.

BIBLIOGRAPHY

Aleshire, Peter, "A Bare-knuckled Trio Goes after the Forest Service," *High Country News*, 1998; "Center for Biological Diversity," http://www.sw-center.org/; Kiefer, Michael, "Cow Punchers: Ranching Takes a Blow in the Courts, an Uppercut to the Bottom Line and a Jab in the Marketplace," *Phoenix New Times*, 1997; Kiefer, Michael, "Owl See You in Court," *Phoenix New Times*, 1996; Skow, John, "Scorching the Earth to Save It," *Outside Magazine*, 1999.

Swearingen, Terri

(November 24, 1956-)
Grassroots Toxics Activist

During the 1990s, registered nurse and dental technician Terri Swearingen became a powerful opponent of a toxic waste incinerator that was built in her hometown of East Liverpool, Ohio. Even after she exposed the dangers of toxic emissions being released only a few blocks away from an elementary school, revealed the corruption and disregard for the law that led to the permits being granted by the Environmental Protection Agency (EPA), and led thousands of local residents to protest the siting, the incinerator was built and is in operation today. Although Swearingen and the others fighting the Waste Technologies Industries (WTI) incinerator did not succeed in stopping its construction, they are credited for forcing the EPA in 1993 to set national standards for siting hazardous waste management facilities.

Teresa Joyce Swearingen was born in East Liverpool, Ohio, on November 24, 1956, one of five siblings whose family has a long history in the Ohio River Valley. Her father worked in a steelmill, and her mother worked in an elementary school cafeteria. Swearingen studied nursing at the Ohio Valley Hospital School of Nursing, graduating first in her class in 1978. Also a trained dental technician, she worked in the dental office of her husband, Lee Swearingen. She was actively involved in various committees at her church, where she also taught Sunday School. In her free time, she enjoyed making stained glass windows and other crafts.

All of that changed one day in 1982, when a patient of her husband told her about plans to construct one of the world's largest hazardous waste incinerators in a poor minority neighborhood in East Liverpool. The burner, permitted to release 4.7 tons of lead and 1.28 tons of mercury annually, was to be built 400 yards from a 400-student elementary school and 320 feet from the nearest home. From her medical training, Swearingen knew that lead exposure was dangerous, especially to children during critical periods of growth and development. Mercury exposure can cause irreversible nerve damage and many other serious problems, particularly for children. Swearingen was especially shaken by this news because she was pregnant with her first and only child, Jaime.

Swearingen began talking to friends and neighbors about the proposed incinerator, writing to elected officials and to local newspapers. She continued to have faith that the EPA would never permit such a dangerous project. But in 1989, Waste Technologies Industries began preparing for construction. It was then that Swearingen realized that the project was indeed moving forward and that if she did not get more involved, she could not expect anyone else to either. So she jumped to action.

Swearingen began a process of self-education, starting with the book, *Rush to Burn: Solving America's Garbage Crisis*, a 1988 publication compiling articles by reporters from *Newsday*. She tracked down everyone cited in the book: citizen activists, politicians, and experts. They referred her in turn to more articles and studies, and she called everyone who authored or was cited in those as well. Soon Swearingen was well connected with many other grassroots toxics activists and very well versed about the hazards of incineration. She and other local activists hosted town meetings and conducted public information sessions to spread the word about the incinerator. In 1990 she cofounded the Tri-State Environmental Council (TSEC) to coordinate the activities of those in Ohio, Pennsylvania, and West Virginia who opposed the incinerator.

When, in January 1991, the EPA held a public hearing about a modification to their permit for the incinerator, Swearingen invited nationally known experts to testify about its dangers. They included dioxin expert PAUL CONNETT, EPA whistleblower HUGH KAUFMAN, and several others. Since each member of the public was allowed five minutes to speak, local attendees would stand up every five minutes to take the microphone and cede their time to the expert currently testifying. Upset that the meeting had effectively been seized by the incinerator's opposition, EPA officials changed the rules for testimony mid-meeting—after most of the anti-incinerator experts had testified but before the WTI expert even arrived. Swearingen and her colleagues realized that in spite of their success at getting accurate information communicated during the hearing, it was basically a sham. The EPA was ready to rubber-stamp the modified WTI plant design. Later hearings and meetings with the EPA showed the activists that EPA officials were unable to dispute damning evidence presented by activists and experts yet still refused to recall the permit. TSEC began to pursue other tactics to express their opinion.

During the early 1990s, Swearingen and others organized dozens of rallies and protest marches, attended by thousands of local residents, most of whom had never before been involved in protests. Their tactics were creative. An early action, dubbed "Hands Across the River," was a march by 1,000 people from Pennsylvania, Ohio, and West Virginia through all three states, culminating on the Jennings Randolph bridge over the Ohio River, to express tri-state opposition to the incinerator. On one of their frequent trips to Ohio governor George Voinovich's mansion, protesters erected "For Sale" signs on his lawn because they believed he had sold out to toxic polluters.

The group was very disappointed in the Clinton-Gore team, which had made WTI a campaign issue and had made WTI the very first environmental issue they addressed after their election. Once in office, however, they failed to keep their promise not to allow WTI to operate. This was especially ironic because AL GORE's book *Earth in the Balance* specifically criticized incineration. In late 1993 the group "recalled" Gore's environmental treatise and deposited hundreds of copies with a guard at the White House gate. During other protests, Swearingen and others were arrested for illegal trespassing onto the incinerator site and in Washington—including once inside the Clinton White House—to protest the lack of action taken by federal authorities.

Although TSEC has not succeeded in stopping the WTI incinerator in East Liverpool yet, their efforts have halted other commercial incinerators from being built around the country. Ohio enacted a moratorium on the construction of new hazardous waste incinerators. The protesters motivated Congress to conduct its first-ever hearing to look at the ways the EPA bent the rules to help the industries they are supposed to regulate. They forced an overhaul of federal combustion regulations, including the development of more stringent emission limits for toxic heavy metals and the first-time emission limit for dioxin under the Resource Conservation and Recovery Act (RCRA; the Federal hazardous waste law). The day after a mass arrest of 54 protesters, including Swearingen along with members of other citizens' groups and Greenpeace, who had chained themselves to a mock incinerator complete with a belching smoke stack outside the White House gates, the EPA announced a new policy governing hazardous waste incineration and declared an 18-month freeze on new incinerator construction. The May 18, 1993, announcement gave satisfaction to the group, but Swearingen, barred from attending the press conference, could not help noticing the irony of EPA administrator CAROL BROWNER's top priority for the new combustion and waste management strategy: "increased public participation."

Although Swearingen and the others fighting WTI have not enjoyed victory in their own community, they have compelled the government to acknowledge the serious risk that pollution poses to the food chain, and they have been credited as the driving force behind the EPA's action to implement first-time national siting standards for hazardous waste management facilities.

Swearingen is now world famous for her leadership of the struggle against toxic incinerators and their siting in poor and minority neighborhoods where people are less likely to protest. She is the re-cipient of many awards and recognitions, including the 1999 William E. Gibson Achievement Award presented during the 211th General Assembly of the Presbyterian Church (USA) by Presbyterians for Restoring Creation and a 1999 Women at Their Best Award Winner from Cover Girl and *Glamour* magazine. She was the recipient of the 1997 Goldman Environmental Prize for North America and the 1997 National Peace Award from Clergy and Laity Concerned. She was named one of six "Ecowarriors" by *marie claire* magazine in 1997, was a TOYA (Ten Outstanding Young Americans) Award nominee in 1995, and was named the Ohioan of the Year in 1994. Also in 1994, *Time* magazine named Swearingen, along with John F. Kennedy, Jr., Oprah Winfrey, and Bill Gates, as one of the 50 most promising leaders in America. In 1993 she was inducted into the Citizens Clearinghouse for Hazardous Waste Grassroots Honor Roll and Hall of Fame. In the same year, she received the Joe A. Calloway Award for Civic Courage administered by the Shafeek Nader Trust for the Community Interest.

Currently, Swearingen and another local citizen face libel charges brought on by a SLAPP suit (Strategic Lawsuit against Public Participation) filed by WTI. She now lives across the Ohio River from East Liverpool in Chester, West Virginia, with her husband and daughter. There she continues her work in the environmental and social justice movement.

BIBLIOGRAPHY

Brown, T. C., "A Slow Burn," *Cleveland Plain Dealer Sunday Magazine*, 1998; Gorisek, Sue, and Ellen Stein Burbach, "Ohioan of the Year: The Housewife That Roared," *Ohio Week Magazine*, 1994; Rembert, Tracey C., "Terri Swearingen: The Long War with WTI," *E Magazine*, 1997; Schneider, Keith, "For Crusader against Waste Incinerator, a Bittersweet Victory," *New York Times*, 1993; Schwab, Jim, *Deeper Shades of Green*, 1994.

Tall, JoAnn

(1953–)
Environmental Activist and Organizer, Cofounder of Native Resource Coalition,
Cofounder of Indigenous Environmental Network

JoAnn Tall, Oglala Lakota, is an environmental activist whose work has focused on environmental justice issues at her native Pine Ridge Reservation in South Dakota. Since the 1980s, she has led grassroots struggles to protect Pine Ridge and other reservations from such environmental threats as nuclear testing and toxic waste dumping. Tall is an active member of the American Indian Movement (AIM), she helped to create the Native Resource Coalition and the Indigenous Environmental Network, and she serves on the board of directors of the Seventh Generation Fund.

Joann Tall is an Oglala Lakota (Sioux), born in 1953 on the Pine Ridge Reservation in South Dakota. Her introduction to activism came in 1973 when she participated in the American Indian Movement's occupation of the village of Wounded Knee on the reservation. Wounded Knee was the site of an 1890 massacre in which more than 200 unarmed Native American men, women, and children were slaughtered by the U.S. Seventh Cavalry (the same regiment Gen. George Custer had led into Little Big Horn in 1876). The cavalry also succeeded in killing 25 of their own soldiers in crossfire. Despite this heavy casualty toll in the face of an unarmed and fleeing enemy, the regiment received 20 Congressional Medals of Honor. In his book, *In the Spirit of Crazy Horse*, PETER MATTHIESSEN writes, "After Wounded Knee, the soldiers were replaced by bureaucrats, including 'educators' whose official task was to break down the cultural independence of the people." The Lakota were forbidden to participate in traditional spiritual ceremonies, were not allowed to wear Indian dress, and were discouraged from speaking their own language. Their children were taken and relocated to government boarding schools. This type of oppression characterized the U.S. government's relationship to the Lakota over the course of the next century.

AIM was founded in the late 1960s with the idea that Native Americans needed to find ways to solve their own problems, rather than continuing to rely on the destructive supervision of the U.S. government. On February 28, 1973, a group of several hundred AIM members and supporters took over the historically significant community of Wounded Knee to protest the U.S. government's treatment of Native Americans on the Pine Ridge Reservation. AIM issued a public statement demanding an investigation of the corrupt Bureau of Indian Affairs (BIA) and a hearing on the Fort Laramie Treaty of 1868 (this treaty granted the Lakota "absolute and undisturbed use of the Great Sioux Reservation" and has been ignored and broken by the United States ever since). Tall participated in the 71-day sit-in at Wounded Knee, during which the community was surrounded by U.S. marshals, Federal Bureau of Investigation agents, and tribal police, who exchanged gunshots with the occupiers, killing two of them. After the protesters had held out for more than two months, the U.S. government agreed to AIM's demands on the condition that Wounded Knee be vacated. Once AIM had left Wounded Knee, hearings on the Fort Laramie Treaty never took place, nor was an investigation into the BIA ever undertaken.

Tall's environmental activism spans more than 20 years and is grounded in her people's reverence for, and respect of, the natural world. She has a spiritual connection with "Grandmother Earth" that is outside of the experience of many Americans. Her motives for protecting the earth are not so much a desire to "conserve" or "preserve," as they are a moral imperative. Her belief system simply requires her to respect and appreciate the life-sustaining land. Tall has been guided by dreams since she was a young child and is quoted on the "Greening of the Future" web site as saying, "One [dream] was of going into this forest, when a doe came towards me in tears. I followed her to a clearing. One of her fawns was dying, the other was trying to get up. I looked around their habitat. It was contaminated. I told her 'I will not be able to help you, but I am going to find out who did this and deal with them the best I can.' That dream has followed me over the years in health, land and environmental issues."

In 1987, a defense contractor, Honeywell, announced plans to conduct weapons testing in a Black Hills canyon on the Pine Ridge reservation, sacred to the Lakota. Tall utilized the locally owned and operated Pine Ridge radio station, known as KILI, to inform reservation residents of the threat to their ancient burial ground and ceremonial site. Then, at a public meeting in which Honeywell planned to convince local resi-

dents that the testing would be harmless, Tall "tore into the PR guy." A *Ms. Magazine* article reports her demanding, "How dare you try to desecrate our church? That's what the Black Hills mean to us: they're our church. . . . We're going to win this war." Tall then organized a "resistance camp" of more than 150 people at the proposed testing site in the Black Hills. They camped out for three months, until Honeywell abandoned the project.

Tall followed this success by helping to found the Native Resource Coalition (NRC), a group whose first action was to protest the creation of a 5,000-acre landfill and incinerator site on the Pine Ridge Reservation. The NRC found that over a two-year period more than 60 Indian reservations nationwide had been targeted by the waste industry as sites for disposal. This was due to the fact that, having sovereign status, many reservations had much less stringent regulations than the U.S. government about the treatment of toxic waste. Especially attractive to the waste industry were Native American tribes for whom a language barrier prevented their full understanding of what they were agreeing to. These people had no words in their traditional languages for poisons such as polychlorinated biphenyls (PCBs) or dioxins.

In 1991, NRC joined forces with other grassroots groups to organize a conference in the Black Hills, the purpose of which was to educate and inform the many tribal groups that were being subjected to the environmental racism and economic blackmail of the waste disposal industry. Largely thanks to this conference and to Tall's leadership, the native councils rejected proposals for toxic waste disposal sites on the Pine Ridge, Rosebud, and other nearby reservations. It was at this conference that Tall helped to organize the Indigenous Environmental Network, which now consists of more than 50 member organizations, working to educate indigenous people about environmental threats in the United States and in Canada.

Tall is the mother of eight children. She looks to the future with optimism. As she says on the "Greening of the Future" web site, "When I look at all of us with our work, our sacrifices, there's hope. You always have to have hope, because we talk about the children, the future generations. My dream tells me there is hope for the future." She continues to live on the Pine Ridge Reservation where, increasingly, she is taking on the role of elder, adviser, and educator. She serves on the board of directors of the Seventh Generation Fund, a foundation and advocacy organization that supports Native Americans through grants, technical training, and issue advocacy. In 1993, she received an environmental hero award from the Goldman Environmental Foundation.

BIBLIOGRAPHY

Breton, Mary Joy, *Women Pioneers for the Environment*, 1998; "The Greening of the Future," http://www.motherjones.com/motherjones/MA95/heroes.sidebar.html; Matthiessen, Peter, *In the Spirit of Crazy Horse*, 1980; Taliman, Valerie, "Saving Native Lands," *Ms. Magazine*, 1994.

Tchozewski, D. Chet

(January 25, 1954–)
Anti-Nuclear Activist, Executive Director of Global Greengrants Fund

Chet Tchozewski was an important protagonist in the antinuclear movement of the late 1970s and 1980s, organizing mass protests for ten years at the Rocky Flats Nuclear Weapons plant in Colorado until it was closed in 1989. Currently Tchozewski is the founder and executive director of Global Greengrants Fund of the Tides Foundation, which channels donations from U.S. funders to grassroots environmental organizations abroad.

Darrell Michael Tchozewski was born on January 25, 1954, in Shelby, Michigan, and grew up with six siblings in nearby Montague, on the banks of Lake Michigan. His father worked at the Hooker Chemical Company chlorine plant, and Tchozewski remembers that at least once a week the whole town reeked of the chemical. No one worried about it then, but after toxic releases decimated the area's fish and local wells were found to be contaminated, both the Hooker plant and a neighboring DuPont Chemical freon factory were shut down and declared Superfund sites. Tchozewski attributes his environmentalism to this early influence.

Tchozewski was the first in his family to attend college. He started at a community college in Muskegon, continued at Michigan State University, and then in 1974 moved west to Colorado and attended classes at the University of Colorado. When the protests at the Rocky Flats Nuclear Weapons Plant just northwest of Denver began in 1978, Tchozewski joined them. The first protest he attended in April of that year was planned as a 24-hour vigil in which 250 protesters were to symbolically block the railroad tracks used for shipping out the nuclear triggers manufactured at the plant. After one night on the tracks, Tchozewski and a group of 35 others decided to stay there. That marked the birth of the Rocky Flats Truth Force (RFTF), a vanguard grassroots nonviolent direct action group dedicated to closing the Rocky Flats Plant through civil disobedience.

Tchozewski helped set up an office for the RFTF. He worked there full-time and spent the summer of 1978 learning about the nationwide antinuclear movement by traveling to protests at nuclear power plants all over the United States. By the spring of 1979, when the Three Mile Island nuclear reactor in Harrisburg, Pennsylvania, melted down and public opinion turned solidly against nuclear power, the focus for protests shifted to nuclear weapons. Tchozewski worked through the RFTF and later the Rocky Flats Project of the American Friends Service Committee to organize major protests attended by thousands of people, as well as smaller on-going vigils on the tracks. One of the most effective events was a 20,000-person encirclement of the plant in 1983; others involved marches from Boulder, 12 miles away, to the plant; others were combined with huge concerts on the plant's grounds by such antinuclear musicians as Jackson Brown and John Denver.

In 1983, Tchozewski cofounded the Rocky Mountain Peace Center in Boulder to work on closing Rocky Flats and on other peace and environmental issues. In the spring of 1986, Tchozewski learned that despite Soviet president Gorbachev's promise not to test Soviet nuclear weapons unless the United States tested its weapons, the Department of Energy (DOE) was testing nuclear weapons underground at its Nevada Test Site near Las Vegas. Greenpeace activists had attempted to prevent an earlier test by sneaking onto the test site and alerting the DOE of their presence; Tchozewski wanted to do the same. So one afternoon, after learning that another test was scheduled for sometime that week, he recruited three friends for the expedition. By 7:30 that evening they were on a plane to Las Vegas. A contact picked them up and drove them to the test site, and at midnight they were hiking toward Ground Zero. The contact alerted the local press and DOE officials, who searched the area by helicopter. Once the DOE believed the protesters were not at Ground Zero, the bomb was detonated. As it turned out, Tchozewski and his companions were still miles away from Ground Zero when the bomb went off, but when they hiked off the site 36 hours later they were greeted by a large group of sympathizers. During the next year or so the Rocky Mountain Peace Center coordinated activists who took turns making well-publicized forays onto the test site to prevent or delay nuclear testing.

Meanwhile, back at Rocky Flats, after 11 years of almost continuous protests, the Environmental Protection Agency (EPA) in 1989 discovered that the contractors who operated Rocky Flats had been engaged in many illegal activities. Through overhead infrared nighttime photos of the plant, the EPA obtained proof that illegal incineration was occurring, and in June 1989 the Justice Department authorized the Federal Bureau of Investigation to raid the plant and shut it down. Since then, the EPA and the Colorado Department of Health and Environment have overseen a cleanup of all of the radioactive waste contaminating the plant site.

Tchozewski left Colorado in June 1989 for the San Francisco office of Greenpeace, which hired him as its Southwest regional director. He worked for three years supervising 200 employees working in the southwest part of the United States and in Latin America and the Pacific Rim. At this point, during the resurgence of the environmental movement marked by the *Exxon Valdez* oil spill, the growing awareness of global warming and the hole in the ozone layer, and the 20th anniversary of Earth Day, Greenpeace was the largest environmental organization in the world, with a peak membership of three million.

In 1992 Tchozewski and his wife, Susan Carabello, left for a year-long sabbatical trip through south and southeast Asia. During that year, Tchozewski realized that the environmental issues facing poor countries were even more serious and complicated than those of the United States. There were grassroots groups to address the problems, but their lack of funding was crippling. By the time the couple returned to Colorado in 1993, Tchozewski had brainstormed a solution. He would start a foundation that could serve as a conduit for U.S. donations to grassroots groups in developing nations around the world. Global Greengrants was established at the Tides Foundation in 1993, funded by donors

whom Tchozewski had met over the years, including some antinuclear protesters from the 1970s who had since made lots of money.

The Global Greengrants Fund works with an advisory board consisting of U.S.-based nonprofit groups that work on international environmental issues (Friends of the Earth, Earth Island Institute, Rainforest Action Network, Pesticide Action Network, the Pacific Environment and Resources Center, and the International Rivers Network). Because these organizations are in close contact with grassroots environmental groups abroad, they can help determine which groups should receive small donations that together amount to between $100,000 and $150,000 per year. Tchozewski continues to be amazed at how much a small group abroad can do with as little as $750; he believes that grassroots groups are probably the most economical and efficient environmental organizations in existence.

Tchozewski lives in Boulder, Colorado, with his wife and his daughter, Tian.

BIBLIOGRAPHY

"Global Greengrants Fund," http://www.greengrants.org; Hickman, Jason, "Rocky Flats Protesters Meet Again Twenty Years Later," *Boulder Daily Camera*, 1998; Human, Katy, "Flats Activists Look Back," *Boulder Daily Camera*, 1998; Levitt Ryckman, Lisa, "Crossroads at Rocky Flats, Yearlong 'Occupation' 20 Years Ago This Month Challenged Public Acceptance of Nuclear Facility," *Rocky Mountain News*, 1998; "University of Colorado Library—Western History Archives: Collection on Environmentalism—Chet Tchozewski Collection," http://www-libraries.colorado.edu/ps/arv/col/lists/ecology.htm.

Tewa, Debby

(June 16, 1961–)
Solar Electrician

Debby Tewa is a solar electrician who directs the Hopi Foundation's NativeSUN project. The project brings solar electricity to people living off the grid on the Hopi and neighboring Diné Reservations.

Debby Tewa of the Coyote Clan was born on June 16, 1961, in Phoenix, Arizona. She grew up in her grandmother's traditional stucco and sandstone home on the ceremonial plaza of Hotevilla, a village on the Hopi reservation. She describes her childhood as adventurous: solitary hikes around the silent plateau lands of the reservation stimulated her curiosity, imagination, and sense of exploration. She recalls, even as a child, inquisitively examining the gears on her bicycle, demonstrating her love of the mechanical. In 1979, Tewa graduated as valedictorian from Sherman Indian High School, a boarding school run by the Bureau of Indian Affairs in Riverside, California. She attended Northern Arizona University for two years as a liberal arts major. She earned her certificate as an electrician at the Gila River Career Center. Tewa worked at commercial outfits and for government housing authorities in the Phoenix area for several years until 1989, when the Hopi Foundation offered to send her to a photovoltaics workshop put on by Solar Energy International and held in Carbondale, Colorado. She tried working as a solar technician on the Hopi Reservation soon after the training, but sales were too slow. She returned to Phoenix and worked there until 1991, when funding became available for her to work on the Reservation.

One-third of the villages on the Hopi Reservation refuse to connect to the local electric company's grid. They dislike the visual pollution of power lines. They do not want to cede a right-of-way to the power company for installation and repair. They worry that if people cannot pay their bills, the power company may retaliate by claiming Hopi land as collateral. Some residents use propane or generators, but they must haul fuel, maintain the generators, and deal with noise and safety. For this reason, individually owned solar energy systems offer an ideal way to provide electricity without compromising traditional values.

NativeSUN offers low-interest loans to people who want to purchase a solar electricity system for their home. Once the loan (usually between $5,000 and $6,000) is repaid, the owner can enjoy as much free electricity as his or her panels can generate. Tewa has installed about 300 systems, including some upgrades of previously existing systems.

Tewa recognizes that solar energy cannot satisfy all of the requirements of modern life, since the four- to eight-panel systems do not run washers, dryers, or freezers. Some people combine their solar energy with propane or generators. Tewa knows that, for convenience, some of her neighbors would prefer to connect to the power grid. She believes that even if her village signs up with the power company, people will conserve electricity. Dependence on solar energy "teaches you to be conservative, because you're getting your power from batteries," she told WINONA LADUKE, author of *All Our Relations*.

Tewa has lectured throughout the United States about the Hopi Foundation and NativeSUN at conferences and solar demonstrations, colleges and universities, high schools and summer camps. She has appeared on numerous radio shows. She was featured in the Public Broadcasting Service documentary "Honey We Bought the Company." In 1998 and 1999 she traveled to the SUN21 Conference in Switzerland to share the successes of NativeSUN and learn about similar projects in other countries. She has traveled to Ecuador twice, in 1992 and 1994, with the Center for International Indigenous Rights and Development to teach indigenous people in small villages how to build solar ovens as an alternative to gathering the little available firewood. In 1996, NativeSUN received a Renew America award and was a runner up for the Christopher Reeves Environmental Award. Over the years, they have received grants from many organizations, including the Hitachi Foundation, the Joyce Mertz Gilmore Foundation, the Charles Stewart Mott Foundation, and the Lannan Foundation.

Tewa's interests outside solar energy are archaeology and art.

BIBLIOGRAPHY

Cole, Nancy, and P. J. Skerrett, "Debby Tewa, Profiles in Activism," *Solar Today*, 1995; LaDuke, Winona, *All Our Relations: Native Struggles for Land and Life*, 1999; LaDuke, Winona, "Debbie Tewa—Building a Future with her Community," *Indigenous Woman*, 1994; Leiva, Miriam, and Richard G. Brown, *Geometry: Explorations and Applications*, 1998; Stone, Laurie, "Amazon Power: Women and PV," *Home Power*, 1997–1998.

Thompson, Chief Tommy Kuni

(Mid-1850s to early 1860s–April 12, 1959)
Chief of the Wyams, Fishing Rights Advocate

Chief Tommy Kuni Thompson, salmon chief of the great fishery at Celilo Falls and leader of Celilo Village, a small Native fishing community 11 miles east of The Dalles, Oregon, was a tireless defender of both Columbia River salmon and Indian fishing rights during the first half of the twentieth century. His life represents an important chapter in the history of Native peoples' efforts to protect diverse and productive ecosystems across North America. While ultimately unsuccessful in preventing the damming of the Columbia and the destruction of what was perhaps the most productive inland fishery in the world, his integrity as a leader is recognized to this day by Indians and non-Indians alike.

Kuni, which means "full of knowledge" in the Sahaptin language, was born by the banks of Nch'i-Wána, later known as the Columbia River, sometime between the mid-1850s and the early 1860s. He was told that his ancestors had always lived and fished at Wyam, which means "the echo of falling water." His great-uncle was the renowned Chief Stocket-ly, who signed the 1855 Middle Oregon treaty for the Wyams and was killed nine years later while acting as a scout for the United States Army. Kuni's father died when he was still an infant, and his mother died a few years later while on a berry-picking expedition. Before she died, she urged Kuni to listen carefully at the councils so that he might grow up to be a great chief like his uncle Stocket-ly.

His mother's advice came to fruition when, despite his repeated protestations, both the Wyams and the Skinpa, who lived across the river, elected him their chief. Kuni, whose White playmates many years earlier had named Tommy Thompson, was only about 20 years old, the youngest chief in memory.

As a result of the massive migration of Whites into the region during Chief Thompson's youth, Celilo Village had changed dramatically by the time he assumed the chieftainship. What had been a community of 600 to 700 people, and host to thousands more who came from all over the region to trade and visit, had shrunk to fewer than 200. While most of the Indians moved to one of the region's reservations, many of the River People remained by the banks of the Columbia. The Wyams and other river Indians adapted to the new economy the Whites developed, working in the orchards and railroads and selling fish to the canneries that cropped up along the mid-Columbia in the last two decades of the nineteenth century.

Celilo was the center of the Indian contribution to the commercial fishing industry. The miles of rapids, eddies, and narrow channels leading up to the great falls concentrated the migrating salmon, making them easy targets for experienced dipnetters. The fishery at Celilo be-

came especially important during the twentieth century owing to the gradual destruction of traditional fishing sites and salmon runs in such tributary streams as the Yakima, Clearwater, and Umatilla. Chief Thompson struggled to maintain traditional management of the fishery at Celilo, where, by the 1940s, dipnetters scooped more than 2.5 million pounds of salmon out of the river every year. In addition to determining which families had rights to specific fishing places, as salmon chief, Chief Thompson opened and closed the fishery on both an annual and a daily basis. However, as Columbia River salmon runs declined and increasing numbers of both Indians and Whites came to Celilo to fish, traditional management practices became harder to enforce.

In 1934, several dozen Indian leaders from the Warm Springs, Yakama, and Umatilla reservations met to discuss the conflicts over the Columbia River salmon fishery between Whites and Indians and among Indians themselves. Chief Thompson urged the attendees to fish in harmony and to heed fishing traditions as handed down by their ancestors. He lamented the fact that fishermen at Celilo were exhibiting selfish attitudes, claiming priority of rights to productive fishing stations. He declared that he was often troubled to maintain peace and a cooperative spirit among the growing number of Indian fishermen.

The gathered leaders decided to form an intertribal fishery management agency, the first of its kind. They dubbed it the Celilo Fish Commission and elected Chief Thompson as the first chair. Though the members became embroiled in a bitter debate over whether the Nez Perce had rights at Celilo, the commission's primary objective was to promote and protect Indian fishing rights. The tribes' right to fish at off-reservation fishing places like Celilo, a right they reserved in the treaties of 1855, was threatened on two sides. The fish and game agencies of Oregon and Washington saw the Indians as competitors in a zero-sum game and used discriminatory regulations in an attempt to minimize Indian harvest. More alarming, however, were the federal government's plans to dam the Columbia River, which threatened to destroy entirely the River People's most sacred resource, the salmon.

Upon learning of the Army Corps of Engineers' plans to build Bonneville Dam 54 miles downstream from Celilo, Chief Thompson urged U.S. attorney Carl C. Donaugh to protest the dam as a violation of Indian fishing rights. Despite firm tribal opposition, the Corps commenced construction of Bonneville in 1934, using Public Works Administration funds meant to help ameliorate the nation's dire economic situation. In addition to interfering with salmon migration and destroying important main-stem spawning habitat, the dam flooded numerous Indian homes, as well as the Great Cascades, an important Indian fishery.

A few years after the Corps finished Bonneville, the state fishery agencies attempted to close the river above the dam to commercial fishing. Though they claimed it was for conservation purposes, it was clear they were targeting the Indians, who made up the majority of commercial fishermen above Bonneville. Chief Thompson voiced his opposition to Congress and started a petition campaign against Washington governor Mon C. Wallgren's scheme to buy out Indian fishing rights. The chief argued forcefully in a statement to Congress, translated from his native Sahaptin, that tribal fishing rights included the right to sell part of their harvest, since their ancestors had always traded fish. The tribes successfully stopped the states from closing their commercial fishery, but only temporarily. Only the *U.S. v. Oregon* (1969) and *U.S. v. Washington* (1974) decisions would come close to addressing the problem of discriminatory regulation of the salmon fishery, which still occasionally occurs.

The greatest challenge Chief Thompson faced in preserving salmon and Native fishing rights came in the form of the Dalles Dam, a power and navigation project that destroyed the Celilo dipnet fishery in 1957. The tribes of the Nez Perce, Umatilla, Warm Springs, and Yakama reservations, as well as unaffiliated River People such as the Wyams, fought a long and ultimately unsuccessful battle against this dam. Chief Thompson testified numerous times in opposition to the project. He conducted a prayer and song ceremony in a final attempt to save Celilo Falls, reminding Congress that "the Almighty took a long time to make this place." When other Indian leaders finally relented, accepting a $27 million settlement, Chief Thompson adamantly refused to "signature his salmon away." His life was the river, the great falls, and the salmon. He neither knew nor wanted any other.

On March 10, 1957, the Corps of Engineers filled the reservoir of the Dalles Dam, flooding Celilo Falls and old Celilo Village, resulting in the dispersal of more than half the residents. Indians gathered from around the region to mourn the loss of their most important fishery. The elderly chief was confined to a nursing home, however, and did not witness the destruction. He died on April 12, 1959. Flora, his wife of 20 years, was convinced the loss of Celilo killed him. More than a thousand people came from around the region to pay their respects to the great chief. He was survived by his wife and many children, grandchildren, and great-grandchildren. Despite their loss, many of Chief Thompson's people continue to fish for salmon, and Celilo Village, though much reduced, still sits by the side of Nch'i-Wána.

BIBLIOGRAPHY

Allen, Cain, "They Called It Progress: Salmon, Indian Fishing Rights, and the Industrialization of the Columbia River," master's thesis, Portland State University, 2000; Barber, Katrine, *After Celilo Falls: The Dalles Dam, Indian Fishing Rights, and Federal Energy Policy on the Mid-Columbia River*, Ph.D. dissertation, Washington State University, 1999; McKeown, Martha, "Celilo Indians: Fishing Their Way of Life," *The Oregonian*, 1946; Ulrich, Roberta, *Empty Nets: Indians, Dams, and the Columbia River*, 2000; Wagenblast, Joan Arrivee, and Jeanne Hillis, *Flora's Song: A Remembrance of Chief Tommy Kuni Thompson of the WyAms*, 1993; "Where Have All the Fishes Gone?" *Wana Chinook Tymoo*, 1992; Woody, Elizabeth, *Seven Hands, Seven Hearts*, 1994.

Thoreau, Henry David

(July 12, 1817–May 6, 1862)
Writer

Henry David Thoreau wrote *A Week on the Concord and Merrimack Rivers* (1849) and *Walden* (1854). Although these were the only two books he published in his lifetime, they, along with his poetry, essays, and extensive journals, provided a philosophical and aesthetic base for the later movement to preserve wilderness. "In Wildness is the Preservation of the World," he wrote in "Walking," and his lifetime of observations helps support this truth. Yet his best writing reaches beyond "facts" and "common sense": "The verses of Kabir have four different senses; illusion, spirit, intellect, and the esoteric doctrine of the Vedas; but in this part of the world it is considered ground for complaint if a man's writings admit of more than one interpretation," he complains in the Conclusion of *Walden*. "While England endeavors to cure the potato-rot, will not any endeavor to cure the brain-rot, which prevails so much more widely and fatally?" This element in Thoreau—that his writing always has more senses than are immediately apparent—is part of why readers were at first slow to accept him and why his work is still pertinent today and alive to further interpretation.

Henry David Thoreau was born in Concord, Massachusetts, on July 12, 1817. His father, John Thoreau, was an easygoing store keeper who, with his son's help, had some success in the pencil business. His mother, Cynthia Dunbar

Thoreau, often more than supplemented the family's income by taking in boarders. The Thoreaus were a close family with four children: Helen, John, Henry, and Sophia. Henry was sent to Harvard, where he earned a bachelor of arts degree in 1837. Back in Concord, he began to meet with RALPH WALDO EMERSON's circle of transcendentalists. Emerson, almost 15 years older than Thoreau, had been a Unitarian minister and was by that time a well-known writer. Emerson encouraged Thoreau to write and later to publish. The two men became friends for life.

Transcendentalism was then used as a loose term for free-thinking alternatives to local notions of spirituality. The term derived from the philosophy of Immanuel Kant and philosophical and aesthetic systems opposed to the "rationalism" of John Locke and other founders of the "scientific method" (with their belief that knowledge comes only through the senses, and so on). Whereas Western orthodox sciences and religions tend to subjugate nature to man, God, or both—claiming "dominion"—transcendentalists generally argue for a unity of nature and spirit and for enlightenment through intelligent communion with the natural world. In this sense the Concord transcendentalists had sympathies in common with Native American spiritualism, but they also freely imported Eastern spiritual traditions, European romanticism, natural philosophy, and anything else that interested them.

In 1839, having tried a few trades (school teaching, pencil making, farm labor), Thoreau took a "fluvial excursion" with his brother, John, that would provide material for *A Week on the Concord and Merrimack Rivers*. When John died of an infection three years later, Thoreau

Henry David Thoreau (Corbis)

began expanding even more on his journal accounts, turning them into a book that was (among other things) a tribute to his late brother. Publishers eventually refused to finance it, even with Emerson's strong recommendations. Although they were impressed with its "nature writing," most found the heresies of Thoreau's "pantheism" unacceptable. Thoreau paid to have it published in 1849, losing more money than he could afford in exchange for hundreds of unsold copies in the family attic. That and *Walden* would be the only two books Thoreau published during his lifetime, and together they would bring paltry financial returns. Nor was he paid for lyrics and essays published in the transcendentalists' magazine, *The Dial*, though he derived some small fame and profit from lecturing.

By 1841 Thoreau had moved into the Emerson household, serving as a handyman in exchange for room and board. In 1843 he went to New York to make literary contacts and tutor Emerson's brother's children, but he was unhappy with city life and returned to Concord early in 1844.

In July 1845 he moved into a hut he had built himself on some of Emerson's remote land near Walden Pond. He would spend the next two years there, working on *A Week on the Concord and Merrimack Rivers*, writing in his journals, and taking close note of what happened in the pond and in its immediate surroundings.

Walden, after eight drafts with many revisions and additions, was published in 1854. In it, Thoreau imparts what he sees as the unity of all life, all spheres of existence, involved in a vast and intricate economy. At the center of *Walden* is the pond: "earth's eye; looking into which the beholder measures the depth of his own nature." Thoreau makes the pond into an eye, a mirror, a window, a lens—mingling visions and insights through the complex optical device of his inner and outer eye(s). Opening a window through the pond's ice, Thoreau reports that "Heaven is under our feet as well as over our heads." Describing a "concord" within a "thrilling discord" of owls on a winter night, he adds: "I also heard the whooping of the ice in the pond, my great bedfellow in that part of Concord . . ." letting his reader draw the connection from the owl to Minerva, wisdom, tradition—and back to the immediate "eye" and "I" of Walden Pond in a vision that unites wildness, wisdom, owl, goddess, ice, pond, author, and reader . . . in an amazed concordance. Through and throughout *Walden*, Thoreau sees the unity of all life, all spheres of existence, involved in a vast intricate economy.

"Economy," the first and longest chapter in *Walden*, begins to develop the intricacy and extent of the term as Thoreau uses it, expanding the root (Gr. *oikos:* house, home) and meaning toward the later term: *ecology*. The argument is fundamentally Confucian: Home economics is the foundation of our treatment of nature (the ecosystem). If men's hearts, homes, villages (in that order) are in order, then the state, and the world, will be in order. True order is not imposed from above; it comes from the hearts of individuals, as plants grow from seeds. This was also a premise of transcendentalism, and originally of democracy, though hierarchies of church, business, and government continually reverse the natural sequence. In Thoreau's unities, arrogant enforcement of power, economic and ecological imbalance—or any exploitation of a natural economy—damages individuals and society as a whole.

By "wildness" Thoreau meant more than "wilderness." His rambles through the woods and studies of nature were a large part of his passion, but humankind too was part of his nature. "Great persons are not soon learned, not even their outlines, but they change like the mountains and the horizon as we ride along," he wrote in his journal in 1842. His friends were required to have open minds. He admired such "wild" contemporaries as WALT WHITMAN, Joe Polis (an Indian guide), and abolitionist John Brown. He verbally assaulted churches, states, and fixed doctrines of any sort. His stance was poetic and transcendental: "If you stand right fronting and face to face with a fact, you will see the sun glimmer on both its surfaces, as if it were a cimeter, and feel its sweet edge dividing you through the heart and marrow," he wrote in *Walden*.

Although Thoreau was not widely known in his lifetime, his writings have become classics, and their influence has been vast. His essay, "On the Duty of Civil Disobedience" (1849), written after one night he spent in jail for not paying the church tax, and other essays on civil resistance became important to Mahatma Gandhi, Martin Luther King, and other subsequent reformers. JOHN MUIR and others of the nineteenth-century preservation movement drew inspiration from what Muir called "the pure soul of Thoreau." Thoreau's works were republished in 1906, leading to appreciation of his ideas for a new generation. They have been and continue to be essential to those who love the environment and aspire to preserve it.

On May 6, 1862, at the age of 44, Henry David Thoreau died of tuberculosis.

BIBLIOGRAPHY

Atkinson, Brooks, ed., *The Writings of Ralph Waldo Emerson*, 1940; Bode, Carl, ed., *The Portable Thoreau*, 1967; Harding, Walter, *The Days of Henry Thoreau*, 1962; Krutch, Joseph Wood, *Henry David Thoreau*, 1948; Richardson, Robert D., *Henry David Thoreau: A Life of the Mind*, 1986; Thoreau, Henry David, *Walden*, 1854.

Thorne, Oakleigh, II

(October 12, 1928–)
Ecologist, Educator, Founder of Thorne Ecological Institute

In 1954, Oakleigh Thorne, II founded the nonprofit Thorne Ecological Institute in Boulder, Colorado, a pioneer environmental education organization that offered ecology classes for children and ran corporate ecology seminars. The same year, he incorporated Thorne Films, Inc., which produced hundreds of instructional films on ecology, biology, history, elementary science, art, and architecture. Previous to this, while a graduate student at Yale University, he coordinated a successful effort to preserve the Sunken Forest on Fire Island, New York.

Oakleigh Thorne, II was born October 12, 1928, in New York City, the youngest of six children. His family lived at Brookwood, an 80-acre estate on Long Island, with streams, woods, and a lake. His father introduced him to fishing and duck hunting, his uncle started him collecting birds' eggs, and bird artist John Henry Dick encouraged him by giving him ducks and pheasants to raise. When hawks arrived to eat his pigeons, Thorne became interested in raptors. At the age of ten, he purchased a stuffed passenger pigeon, now an extinct species, that he still treasures today.

At Millbrook School in upstate New York, renowned for its ornithology and natural history programs, Thorne studied with biology teacher Frank Trevor, who taught him to band birds and make films. After high school he summered at the

Valley Ranch in Cody, Wyoming, where he became familiar with western wildlife species and wilderness survival skills that he learned from Roy Glasgow, an old mountain man who was part American Indian. Thorne returned to the ranch every summer through his college years, developing a strong, spiritual connection to nature. One summer (1949), however, he stayed in Connecticut and worked at the Audubon Center in Greenwich, where he met ROGER TORY PETERSON, RICHARD POUGH, and Charles Mohr, all of whom deepened his understanding of ecology, conservation, and ornithology.

Thorne earned a B.S. in biology at Yale in 1951 and continued there to study with Paul B. Sears and Albert E. Burke in the Yale Conservation Program, from which he received an M.S. in 1953. He stayed at Yale for an extra year to become musical director of the Yale "Whiffenpoofs," the men's a capella singing group founded in 1909. Concurrently with his master's studies, Thorne worked on his first conservation project: to save the Sunken Forest on Fire Island from development. He obtained $15,000 from the Old Dominion Foundation, with the help of Pough, RACHEL CARSON, and John Oakes (a well-known *New York Times* editorial writer). Thorne eventually raised $45,000, which was passed through the recently formed Nature Conservancy. This was the conservancy's first effort to preserve natural areas after receiving its tax exempt status in 1952.

Thorne moved to Boulder, Colorado, in 1954, where that same year he opened Thorne Ecological Research Station and incorporated Thorne Films, Inc. His first film was about the Sunken Forest, a unique ecosystem composed mainly of American holly trees behind sand dunes on a barrier island off Long Island. Another early film, "Arctic Wildlife Range," was made together with Bob Krear, who had photographed northeast Alaska on the famous Murie expedition to the Brooks Range. He traded a copy of this film to DAVID BROWER, who used it as he successfully lobbied for the preservation of this area through the Sierra Club.

One of the most effective products of Thorne Films were film loops, which were popular through the 1960s. These were three- to five-minute silent movies in a cartridge that was as simple to operate as an audiocassette: children of grade-school age could project the film loops against the wall at the back of the room without disturbing the rest of the class. Teachers could use them during their lectures, and there was no language barrier because they were silent. A film loop about a starfish was as interesting to college students as it was to kindergarten children. Thorne saw the "silent moving image" as "a powerful teaching tool . . . a lesson I learned from Charlie Chaplin and Stan Brakhage." At its peak, Thorne Films had 40 employees and was selling thousands of film loops each year. It was bringing in $500,000 annually because the U.S. Department of Health, Education, and Welfare (HEW) was paying half the cost of the projectors and films for schools, but after President Nixon vetoed the HEW bill in 1971, the company quickly went out of business.

The nonprofit Thorne Ecological Institute survived and grew during the next decades. One of its most successful programs has been the Thorne Natural Science School, which since 1957 has helped to connect children to nature through hands-on field trips. More than

20,000 students have been served by this program alone.

Thorne Ecological Institute was one of the first environmental groups in the world to assist industries and public agencies in applying the principles of ecology to environmental problems. Thorne, with Bettie Willard, who had begun working for the institute during the early 1960s, helped start the Colorado Open Space Council, an environmental group that met regularly with the Colorado Association of Commerce and Industry. By 1967 Thorne and Willard developed Seminars in Environmental Arts and Science (SEAS), whose objective was to teach ecological principles to the heads of key corporations. These yearly programs were held in Aspen, Colorado, from 1967 to 1984. This was the first attempt to connect ecology with economics. Thorne Ecological Institute also helped Elizabeth Paepcke form the Aspen Center for Environmental Studies (ACES). Charles Luce of ConEdison came to a SEAS meeting and went back to New York City and had "Conserve Energy" painted on all of the company's trucks. General Cassidy of the Army Corps of Engineers attended a seminar, returned to Omaha, hired an ecologist, and set up the National Environmental Advisory Board for the corps. Stan Dempsey of American Metals Climax (AMAX) came to SEAS and later requested help from the institute before developing the Henderson molybdenum mine in central Colorado. A group of ecologists and engineers met for two years in the late 1960s to figure out how to reduce the environmental impact of the mine. Their "experiment in ecology," as it was called, resulted in the first environmental impact statement nationally, produced before such a statement was required by the passage of the National Environmental Policy Act (NEPA) in 1969.

In 1969 Thorne bought the Valley Ranch in Cody, Wyoming. More than 3,000 people visited the ranch and learned from him both ecological principles and wilderness philosophy. Over the years, however, it was never able to attract enough paying guests, and the property was lost through foreclosure in 1987.

Thorne Ecological Institute currently focuses on environmental education for children. After two years of negotiations, the institute worked out a unique, cooperative plan with the City of Boulder Open Space Department and the Boulder Valley School District to preserve the Sombrero Marsh near Boulder and create an ecological instruction program there. Thorne himself continues to teach bird banding to teenagers through the institute and ecological principles to college students in the University of Colorado's Capital Alumni Student Mentoring Program. In addition, he teaches nature photography at New Vista High School and is mentor to several undergraduate singing groups at the university. He strongly believes that providing environmental education opportunities to young people will help save the earth and better the odds for survival of the human species.

In addition to its work in environmental education, Thorne Ecological Institute helped to cofound local chapters of the Audubon Society, The Nature Conservancy, and the Sierra Club. In fact, Thorne became the first official Colorado representative for The Nature Conservancy when he moved west in 1954. He also collaborated with Boulder environmental activists to form People's League for Action Now (PLAN)–Boulder County, a citizen's planning group that helped pass the innovative open space sales tax in 1967. The tax, which has been emulated by many other communities nationwide, is set up to preserve unique open space sites and prevent urban sprawl.

Thorne continues to live in Boulder, Colorado, and his four children, Jonathan, Sarah, David, and Schuyler, live in Colorado and Wyoming. His oldest daughter, Susan, died in 1993 from mental illness. He spends half of each workday at Thorne Ecological Institute and has absolutely no plans to retire.

BIBLIOGRAPHY

Human, Katy, "Thorne Ecological Institute to Celebrate 45th Anniversary," *Daily Camera*, 1999; Roberts, Chris, "Instituting Knowledge of Nature," *Daily Camera*, 1996; Shindelman, Rachael, "Swallow Tale: Bird in the Hand Proves Students' Worth," *Daily Camera*, 1998; Shindelman, Rachael, "Thorne's Birds Educate," *Boulder Planet*, 1999; Smith, Sue-Marie, "Oakleigh Thorne, People Making a Difference," *Daily Camera*, 1998; Thorne, Oakleigh, "Sombrero Marsh—A Shallow Wetland," *Boulder County Kids*, 2000; "Thorne Ecological Institute," http:/www.thorne-eco.org.

Thorpe, Grace

(December 10, 1921–)
Antinuclear Activist, Cofounder of National Environmental Council of Native Americans

Grace Thorpe, Oklahoma Sauk and Fox, convinced her tribe in 1993 to reject a proposal from the Department of Energy (DOE) to store radioactive waste on tribal lands. She and others went on to found the National Environmental Council of Native Americans (NECONA), which helps other tribes keep radioactive waste off their land. With support from NECONA, more than 75 Indian nations in Alaska, Canada, and the lower 48 states have declared their lands "nuclear-free."

Grace Frances Thorpe, Oklahoma Sauk and Fox, was born in Yale, Oklahoma, on December 10, 1921. Her father, Jim Thorpe, was a 1912 Olympic decathlon and pentathlon champion, star football and baseball player, and first president of the organization that became the National Football League. She attended Indian boarding schools in Kansas and Oklahoma and joined the Women's Army Corps during World War II. She was stationed in New Guinea when the first atomic bomb was dropped on Hiroshima in August 1945. Thorpe and her fellow soldiers were delighted to hear that the bomb had ended the war, but once she moved to Japan afterwards she was shocked by the horror of atomic war and felt guilty for her country's actions. Thorpe remained in Japan during the postwar occupation, working for Gen. Douglas MacArthur as a personnel interviewer. In 1946 she married Fred Seely, a paratrooper she had met in New Guinea. They had two children before divorcing, and she returned to the United States in 1950 with the children.

Thorpe became involved with the Indian rights movement of the 1960s, acting as public relations liaison for the occupation of Alcatraz in 1969 and the takeover of the Fort Lawton, Washington, surplus military base in 1970. She worked with the Pit River Indians of California to defend their land against Pacific Gas and Electric and helped to found the Deganawidah-Quetzalcoatl University in Davis, California, in 1970. During the 1960s and 1970s, Thorpe organized economic development conferences for the National Congress on American Indians and served as legislative aide to the U.S. Senate Subcommittee on Indian Affairs; member of American Indian Policy Review Board of the U.S. House of Representatives; and director of the Return Surplus Lands to the Indians Project. She was named an urban fellow at the Massachusetts Institute of Technology in 1972. In 1974 she obtained a paralegal certificate from Antioch School of Law and in 1976 earned a B.A. in history from the University of Tennessee at Knoxville.

Thorpe retired in 1976 to Oklahoma, where she took up the artistic pursuits of painting and ceramics and served the Sauk and Fox people as a tribal court judge and a member of the Health Commission. In 1992, her retirement abruptly ended when she learned that the Sauk and Fox tribal council had accepted a $100,000 Department of Energy grant to study the feasibility of establishing a storage site for nuclear waste on tribal land. The DOE sought temporary storage for high-level radioactive waste from nuclear power plants in a "monitored retrievable storage" (MRS) system and offered research funds to any community willing to consider setting up such a site. DOE negotiators knew that the money would be enticing to impoverished tribal councils and also targeted Indian reservations because their sovereign status allowed state environmental laws to be bypassed.

After researching the problem and becoming convinced that transportation and storage of spent nuclear fuel rods was potentially disastrous, Thorpe sprang into action. She was outraged that her tribe had accepted the money, and she gathered enough signatures to request a special meeting of the tribe. Seventy-five tribal members attended the special meeting in February 1993, and 70 of them—all but the five tribal council members present—voted to return the uncashed check to the DOE.

After this victory, the Indigenous Environmental Network invited Thorpe to be the keynote speaker at its Third Annual Gathering in 1993. Soon she realized that many other Native antinuclear activists were in the same predicament: with the DOE on one side trying to convince tribes to consider housing the waste and the tribal councils, on the other side, tempted to accept the huge sums of money being offered. In 1993, Thorpe cofounded the National Environmental Council of Native Americans, whose goals are to educate Indians about the health dangers of radioactivity and the transportation of nuclear waste by rail and road, develop networks of Indian and non-Indian environmental groups to counteract the efforts of the nuclear industry, and encourage the creation of nuclear-free zones on Indian land across the country. Thorpe has served as president of NECONA since its founding.

Thorpe's speeches explain what environmental racism is and appeal to Native Americans to resist it. Thorpe, whose Indian name, No-ten-oh-quah, translates as

"Woman with the Power of the Wind that Blows Up before a Storm," told the National Congress of American Indians at its December 1993 conference:

> It is wrong to say that it is natural that we, as Native Americans, should accept radioactive waste on our lands, as the U.S. Department of Energy has said. It is a perversion of our beliefs and an insult to our intelligence to say that we are "natural stewards" of these wastes. The real intent of the U.S. government and the nuclear industry is to place this extremely hazardous garbage on Indian lands so they can go and generate more of it. They are poisoning the earth for short term financial profit. They try to flatter us about our ability as "earth stewards." They tell us, when our non-Indian neighbors object to living near substances poisonous for thousands of years, that this is an issue of "sovereignty." It is not! It is an issue of the earth's preservation and our survival.

So far, thanks to grassroots activists and NECONA coordination, 75 tribes have declared themselves "nuclear free," and all but one of the tribal councils that had accepted DOE grants to study MRS feasibility have rejected it. The Skull Valley Band of Goshutes of Tooele, Utah, 70 miles southwest of Salt Lake City, was at this writing the only tribe still considering nuclear waste storage and was negotiating with a group of 11 nuclear plants. MRS study grants were discontinued in 1993, after Thorpe appealed personally to Sen. Jeff Bingaman from New Mexico. However, the federal government is still pursuing a plan to move nuclear waste to a permanent underground storage site at Yucca Mountain, located on Western Shoshone lands in Nevada. (The Western Shoshone oppose the plan and argue they have never sold their land to the U.S. government.) Thorpe and other NECONA activists believe that such a storage system is not only dangerous, but also encourages the nuclear industry to continue generating the deadly waste. In the speech cited earlier, Thorpe later asked "What kind of society permits the manufacture of products that cannot be safely disposed of? Shouldn't we have a basic law of the land that prohibits the production of anything we cannot safely dispose?" Her three-part goal for radioactive waste is: "Leave it where it is. Secure it. Stop producing it." She believes that it is important to promote alternative energy sources such as hydroelectricity and solar and wind power.

In addition to her work on nuclear waste issues, Thorpe is a tireless promoter of her father's memory. She successfully pushed to have her father's 1912 Olympic gold medals returned to the family—they had been taken back by the Olympic Committee when it was learned that Thorpe had played minor league baseball in 1909 and 1910—and convinced the U.S. Post Office to issue a Jim Thorpe postage stamp. She has worked since 1996 to proclaim her father athlete of the century, a goal that was realized when the American Broadcasting Companies (ABC) voted to bestow this title on him at the 2000 Super Bowl in Atlanta. She currently works with Miss Indian USA, Anna McKibben, Quapaw Oklahoma, to place Thorpe's picture on the Wheaties box.

Thorpe works with several other organizations in addition to NECONA. She serves on the board of directors of the Nuclear Information and Resource Service and on the Greenpeace American Indian Advisory Committee; she is a member of the Alliance for Nuclear Accountability, based in Seattle. She was a delegate to President Clinton's 1995 White House Conference on Aging, was a member of his Aging advisory group, and lobbies on issues of importance to the elderly, veterans, and American Indians. She was awarded the Nuclear-Free Future Resistance Award in Los Alamos, New Mexico, in 1999. Thorpe resides in Prague, Oklahoma, with her daughter Dagmar Thorpe Seely and granddaughter Tena Malotte.

BIBLIOGRAPHY

"NECONA," http://www.alphacdc.com/necona/; Rogers, Keith, "Thorpe Battles Nuclear Waste," *Las Vegas Review-Journal*, 1996; Thorpe, Grace, "The Jim Thorpe Family History," *Chronicles of Oklahoma*, 1981.

Tokar, Brian

(April 9, 1955–)
Writer, Educator, Cofounder of Vermont Greens

Author of *Earth for Sale: Reclaiming Ecology in the Age of Corporate Greenwash* (1997), Brian Tokar has been writing and agitating for environmental change since the 1970s. Active in the antinuclear movement, the Green Party, and campaigns against genetically altered food, Tokar brings to his writing the experience of a grassroots activist. His first book, *The Green Alternative* (1987; revised 1992), is a comprehensive examination of Green politics in the United States, while *Earth for Sale* looks at the corporate sellout of the environmental movement by mainstream, national groups. As teacher, journalist, organic gardener, and activist, Tokar works to develop green alternatives to market-dominated, consumer culture.

Brian Tokar was born on April 9, 1955, in Brooklyn, New York. He grew up in New York City and attended Stuyvesant High School in Manhattan during the most turbulent years of the 1960s. As a student at the Massachusetts Institute of Technology (MIT) during the early 1970s, he became more deeply involved in the anti–Vietnam War movement, community organizing, the early food co-op movement, and antinuclear activism. He graduated from MIT in 1976 with a B.S. in biology and physics and earned an M.A. in biophysics from Harvard in 1981, specializing in the neurophysiology of the visual system. Tokar then moved to Vermont and became active in local progressive politics, including a founding role in the Vermont Greens and neighborhood organizing in Burlington's old North End. He began teaching at the Institute for Social Ecology in Plainfield, Vermont, in 1983. The institute had been cofounded in 1974 by MURRAY BOOKCHIN and Daniel Chodorkoff; it brings together a community of scholars and activists from various fields, including anthropology, ecology, feminism, agriculture, education, architecture, social theory, and biology. The influence of this community is clearly visible in *The Green Alternative*, in which Tokar stresses the diverse roots of Green politics. With the European Green movement as a jumping-off point, Tokar uses examples from a wide variety of groups and actions across the United States to illustrate the underlying principles of the Greens. He emphasizes the central role played by feminism in the Greens and suggests that Green politics involve not only ecological justice but also a commitment to social and economic justice, as well as peace and nonviolence. The Greens are not united by a single issue, or even a single name, but rather by a vision of a more just and ecologically balanced society. In Tokar's characterization, "The Green movement is working to evolve a broad vision for a transformed society that can thrive in harmony with the rest of nature and that fosters harmony, equality and freedom among its citizens." *The Green Alternative* was one of the first books in the United States to look specifically at the broad agenda of both the European and U.S. Green movements and helped shape the direction of grassroots politics in the late 1980s and early 1990s.

Earth for Sale looks at evolutionary patterns in environmental politics in the 1990s and comes to the conclusion that the major national environmental groups are no longer able to effectively advocate for ecological justice. Tokar identifies three major trends that have led to what he and others have called the "greenwashing" of American culture, that is, the way in which the political and corporate United States pay lip service to environmental ideals without making any real progress toward changing ecologically destructive practices. Tokar argues first that mainstream environmental groups such as the Sierra Club and Environmen-

tal Defense have traded environmental convictions for a seat at the table of Washington political power. These groups have been co-opted by party politics and corporate-funded foundations, so that often their own funding depends on the financial success of antienvironment corporations such as oil companies and timber interests. Second, Tokar outlines the emergence of "corporate environmentalism," wherein corporations conduct public relations campaigns, touting sponsorship of Earth Day events or other superficial environmental commitments, as cover for their underlying assaults on the environment. Corporate environmentalism was on display at the Earth Summit in Rio de Janeiro in 1992, when groups like the Business Council for Sustainable Development used its power behind the scenes to quiet discussion of the role played by transnational corporations in worldwide environmental degradation. Third, Tokar critiques "green consumerism," the tendency to reduce environmental problems to the level of consumer choice. Green consumerism is based on the myth that individuals can solve environmental problems by choosing "natural" and recycled products while still maintaining consumerist lifestyles. Tokar argues that the emphasis on green products masks the underlying economic structures that have driven the world to the brink of ecological collapse.

As an antidote to the "selling of the earth" in the name of market forces, Tokar offers a hopeful assessment of grassroots environmental activism. While mainstream environmental groups have grown closer to corporate interests, grassroots activists have given their energy to a broad variety of local initiatives, including campaigns for environmental justice and radical, effective forest-protection projects. This kind of activism, Tokar argues, has the potential to offer a thorough critique of corporate culture and to develop a strong, competing system of ethics in which human needs and the integrity of natural ecosystems take precedence over market profit. Grassroots activists can reclaim ecology from corporate greenwash, and Tokar continues to be committed to this work. Recent activism includes speaking out at antiglobalization protests and organizing Biodevastation 2000, a March 2000 gathering of activists opposed to genetic engineering, which took place in Boston. Tokar writes regularly for *Z Magazine, The Ecologist, Food & Water Journal,* and other national and international publications. His 1998 exposé of Monsanto's aggressive history of biotechnology and corporate irresponsibility won a 1999 Project Censored award. He serves on the boards of the Native Forest Network and the Edmonds Institute, an international organization focused on biosafety and biotechnology issues.

Tokar teaches at the Institute for Social Ecology and Goddard College and lectures internationally on environmental issues and movements. He lives in Plainfield, Vermont.

BIBLIOGRAPHY

"Biodevastation 2000," http://www.biodev.org; "The Institute for Social Ecology," http://ise.tao.ca/; "Native Forest Network," http://www.nativeforest.org; Tokar, Brian, "Monsanto: A Checkered History," *The Ecologist,* 1998; "Z Articles," http://www.zmag.org/zmag/zarticles.htm#T.

Toor, Will

(October 21, 1961–)
Activist, Mayor

Elected mayor of Boulder, Colorado, in 1998, Will Toor is a long-time peace and environmental activist. During the 1980s Toor helped organize a long series of actions against the Rocky Flats nuclear weapons plant and worked at Eco-Cycle, one of the most successful recycling centers in the United States. Toor has also been an advocate for affordable and cooperative housing projects. As a city council member and mayor, Will Toor has lobbied for alternative transportation, open space and wildlife protection, and managed growth policies.

Will Toor was born in Pittsburgh, Pennsylvania, on October 21, 1961, the son of two college professors. Early in life he exhibited phenomenal ability in math and science and started attending college seminars at age 11. He graduated from Carnegie Mellon University in Pittsburgh in 1979 with a B.S. in physics. In 1980 Toor was hitchhiking across the country when the car in which he was riding broke down in Boulder. He liked the college town and decided to stay. During his early years in Boulder, Toor was active in the campaign against nuclear weapons production at nearby Rocky Flats, a campaign that involved many of the region's progressive activists. Rocky Flats was finally shut down as a weapons facility in 1989. He also ran an outreach campaign for a successful municipal initiative against U.S. policies in El Salvador and Nicaragua. From 1982 to 1985 Toor worked at Eco-Cycle, a pioneering community nonprofit recycling center. As operations supervisor, Toor helped develop curbside recycling services that were run with a shoestring budget, volunteer labor, and aging school buses. Eco-cycle has grown into a model agency that now runs the city's multimillion-dollar recycling center.

In 1985, Toor enrolled at the University of Chicago to resume his work in physics. He completed an M.S. in 1987 and his Ph.D. in 1992, conducting his dissertation research on electro-rheological fluids, which have important applications in robotics. While in Chicago, Toor was involved in a variety of environmental groups, including the Student Environmental Action Coalition, the 1990 CATALYST conference, a national convention of student environmental activists, and the University of Chicago Environmental Concerns Organization, of which he was a founding member. From 1989 to 1991 he directed the University of Chicago recycling program.

In 1992 he returned to Boulder as director of the Environmental Center at the University of Colorado. The program promotes recycling, alternative transportation, and energy conservation and develops student environmental leadership. In 1997 Toor ran for Boulder's city council, on a platform of a healthy environment, economic opportunity, fiscal responsibility, prevention of urban sprawl, and re-duction of traffic pollution and congestion. He won election with the highest vote total among the candidates. In 1998 the city council elected Toor to the post of mayor.

Boulder, along with much of the rest of the Front Range region of Colorado, is facing unprecedented urban growth. Boulder has been in the forefront in the preservation of open space in the area, and Toor has been a strong advocate for keeping land free from development. He has played a major role in the purchase of strategic open space parcels to prevent large-scale development of open lands in adjacent counties. Growth has put pressure on housing and transportation, two of the most volatile issues in Boulder. The city of Boulder is facing an explosion in housing costs. Many poor and middle-income citizens have been forced out of the city, unable to find affordable housing. Toor has voted consistently to fund low-cost housing programs and has been an outspoken advocate for laws to support cohousing and other alternative housing arrangements, in part because he has often lived in cooperative living situations. Until 1996 the city had a law that barred more than three unrelated persons from sharing living quarters, and Toor was instrumental in overturning that rule. Toor has also been a strong advocate for alternative transportation, calling for free bus passes for all the city's residents. Toor is an outspoken critic of plans to develop more highways and expand lanes on the region's already-existing major arteries, arguing that the money could be more wisely spent on railways, high-speed bus lanes, and other alternative transportation projects. In the spring of 2000 Toor went to Brazil to study the transportation system in Curitiba, which is known as Brazil's "green" city in part because of its model bus system.

Toor continues to work as the director of the University of Colorado Environmental Center. He lives in Boulder with his wife, Mariella Colvin, and their son, Nikolaus.

BIBLIOGRAPHY

Dizon, Kristin, "From Activist to Mayor," *Boulder Daily Camera*, 1998; "Our Leaders," http://www.cit.boulder.co.us/cmo/citycouncil/leaders.html; Toor, Will, "A Boulder Vision," *Boulder Daily Camera*, 2000; "Welcome to the CU Environmental Center!" http://www.colorado.edu/StudentAffairs/cuenvironmentalcenter/.

Turner, Frederick Jackson

(November 14, 1861–March 14, 1932)
Historian

Considered by many to be the father of the history of the American West, Frederick Jackson Turner articulated ideas that have dominated the way twentieth-century historians have interpreted American history. One of his many important contributions was to emphasize the significance of the frontier to the identity of the United States. In his view, the presence of an ever-receding frontier, where land and resources were free and abundant, was a major influence on this country's national character. This thesis offered a theoretical base to the environmental consciousness emerging during his lifetime.

Frederick Jackson Turner was born on November 14, 1861, in Portage, Wisconsin, to Andrew Jackson Turner, a newspaper editor and lover of the outdoors, and Mary Hanford, a former schoolteacher. He completed undergraduate work in the ancient classical course at the University of Wisconsin at Madison in 1884 and master's work in history, also at the University of Wisconsin, in 1888. He was awarded a Ph.D. in history from Johns Hopkins University in 1891. Turner's dissertation, on the Wisconsin Indian trade, was an elaboration of the master's thesis he completed at Wisconsin. By 1889, the year he married Caroline Mae Sherwood, Turner's influential 35-year career as an academic was underway. He and his wife had three children, only one of whom survived past childhood.

Turner's ideas about American history, and specifically his frontier thesis published in *The Significance of the Frontier in American History* (1893), became the prominent interpretation of the history of the United States for more than 50 years. In his thesis Turner argued that the social and political uniqueness of the United States derives from the settlement and development of the American frontier. According to Turner, "The existence of an area of free land, its continuous recession, and the advance of American settlement westward, explain American development." Thus, Turner claimed, American development did not merely advance along a single line; it evolved as a return to primitive conditions on a continually ad-

Frederick Jackson Turner (Bettmann/Corbis)

pages into *The Significance of the Frontier in American History*, would such gifts of free land offer themselves. Although the frontier offered new opportunities to citizens, they came at a price. The problem of the United States, once the frontier was gone, would be how to save and wisely use the remaining resources.

In *The Significance of the Frontier in American History* Turner also introduced sectionalism, a concept specifying that residents of geographic regions (for instance, New England, the South, the West) develop different identities, desires, and political interests. In essence, Turner's sectionalism, which became an emergent field in the study of history, foreshadowed the importance today's environmentalists place on the stability fostered by citizens caring for their own local communities. Turner believed that such a geographic identity would provide "a highly organized provincial life to serve as a check upon mob psychology on a national scale, and to furnish that variety which is essential to vital growth and originality."

As an academic, Turner was always torn between his own scholarly work and his teaching duties. He taught at the University of Wisconsin, Madison, where he was named head of the history program, from 1899 until 1910. In 1910 he accepted an appointment as a distinguished chair of history at Harvard University, a decision that Turner half regretted since it led to an increased teaching load and since he felt that his ideas about western history and sectionalism were better received by students at Wisconsin than by students at Harvard. At both Wisconsin and Harvard, Turner was a popular and well-loved teacher; he was especially well regarded by graduate students, many of

vancing frontier line and a new development of that area. In other words, social development in the United States, unlike the development of European countries, began over and over again on the frontier as settlers moved westward, always attracted to the free land and abundant resources just out of their reach. Turner asserted that this process shaped the American character by promoting self-reliance, individualism, mobility, materialism, resourcefulness, originality, and optimism in its citizens. These same qualities, while essential for forming a strong economy and young democracy, could also lead, Turner noted, to the exploitation and depletion of natural resources. But never again, Turner predicted a dozen or so whom became prominent historians themselves. Thanks in part to Turner's efforts and talents, Wisconsin and Harvard housed two of the great history departments of the twentieth century.

Turner was always plagued by not having written a major book, an accomplishment generally expected of someone of his stature. His essays, which appeared in some of the nation's most influential and scholarly journals and which were revisions of lectures Turner had already delivered, were collected by Turner and others throughout his career and after his death. *The Frontier in American History*, first published in 1920, is perhaps his most important collection. *The Significance of Sections in American History* (1932) was posthumously awarded the Pulitzer Prize in 1933.

Although Turner's frontier thesis remains an important explanation of American development even to this day, criticism of Turner's ideas began to appear in the 1930s. Recent critics have been especially hard on Turner, noting, for example, his easy rejection of Europe's influence on the early development of the United States and his dismissal of the emergence of urbanization in this country. Such attention, albeit negative, only highlights Turner's position as one of the most influential American thinkers of the twentieth century.

After retiring from academics in 1924 due to poor health, Turner worked as a senior research associate at the Huntington Library in San Marino, California, until his death on March 14, 1932.

BIBLIOGRAPHY

Billington, Ray A., *Frederick Jackson Turner: Historian, Scholar, Teacher*, 1973; Bogue, Allan, *Frederick Jackson Turner: Strange Roads Going Down*, 1998; Farragher, John Mack, ed., *Rereading Frederick Jackson Turner: The Significance of the Frontier in American History and Other Essays*, 1998; Jacobs, Wilbur R., *On Turner's Trail: 100 Years of Western Writing*, 1994; Limerick, Patricia Nelson, Clyde A. Milner II, and Charles E. Rankin, eds., *Trails toward a New Western History*, 1991.

Turner, Ted

(November 19, 1938–)
Media Magnate, Philanthropist

Ted Turner created the Turner Broadcasting System cable channel (TBS) in 1976 and Cable News Network (CNN) in 1980; he has long been interested in utilizing mass media to bring about positive social change and environmental awareness. His cable channels have led the way in environmental programming, airing such popular shows as *Earthwatch* and environmentally focused documentaries. Turner is the largest private landowner in the United States, and takes personal responsibility for protecting and restoring more than one million acres of land in the United States and South America. He also founded the Turner Family Foundation, which gives $25 million every year to nonprofit environmental organizations.

Robert Edward (Ted) Turner III, the first child of Ed and Florence (Rooney) Turner, was born on November 19, 1938, in Cincinnati, Ohio. Florence Turner came from a family of Cincinnati grocers, and Ed Turner was a salesman for a billboard advertising company. By 1939, he had created his own billboard advertising company, the Turner Advertising Company. Ted Turner spent the first six years of his life in Cincinnati with his family. He was a handful as a young boy, amusing himself with activities such as spreading mud all over the neighbor's freshly washed sheets as they hung on the line drying. In 1944, Turner's father volunteered for the United States Navy and was assigned to a station on the Gulf Coast. Turner's parents decided to leave him at a boarding school in Cincinnati, but they took his younger sister along with them. This early abandonment had a significant negative impact on young Ted. Turner attended two separate military academies, having generally negative experiences until his last couple of years at the McCallie School in Chattanooga, Tennessee. There he came into his own, helping to organize a sailing club (he would eventually become a world-class sailor, winning the Americas Cup in 1977), joining the Debate Club, and winning a state debating contest.

After graduating from the McCallie School in 1956, Turner wanted to attend the U.S. Naval Academy; however, his father steered him to Brown University in Providence, Rhode Island. Turner was suspended for a semester because of the part he played in a ruckus at the neighboring women's college, Wheaton. He spent six months in the Coast Guard,

then returned to Brown, but he was expelled for good halfway through his junior year after being caught with a woman in his dorm room, a violation of Brown's rules of the time.

Turner moved to Florida, and after fulfilling his obligation to the Coast Guard—he still owed them a tour of duty—he began working as a junior salesman for his father's advertising company in 1960. Three years later, in 1963, Turner's father committed suicide. He left his son in charge of a large but ailing billboard advertising company, which Ted was able to revitalize over the next decade. In 1970, Turner bought two television stations, one in Atlanta, Georgia, the other in Charlotte, North Carolina. Once these stations began turning a profit, he purchased two Atlanta professional sports teams, the Braves (baseball) and the Hawks (basketball), ensuring that he would be able to air their games on his stations. In 1972, he turned his attention to cable television. He invested in the equipment necessary for broadening cable broadcast ranges with satellites, and in 1976, he began broadcasting his original Atlanta television station around the United States, thus launching the cable channel TBS. This channel would air many environmentally focused programs, including the popular *Earthwatch* series, which began in the early 1990s. In 1980, Turner created CNN, which became a vastly successful 24-hour news network. To these he added Turner Network Television (TNT) in 1988, and in 1994, Turner Classic Movies (TCM). In 1996, Turner merged his media holdings with Time-Warner, a move that made Time-Warner the largest media company in the world. His current position involves overseeing the cable operations of Time-Warner.

1985 was an important year for Turner. Prior to that year, many people believed that he was on a path similar to the one that had led his father to suicide. He talked about suicide himself, and according to a 1992 *Time* article, he "was doing his best to imitate his father." In 1985, Turner sought help from an Atlanta psychologist who helped him to move out from beneath his father's shadow. The doctor also began prescribing the drug lithium, which according to those nearest to Turner, brought about miraculous changes in his behavior. Strengthened by his newfound sense of permanence and purpose, Turner apparently began looking outside of himself for ways to make a difference in the world.

In 1985, he founded the Better World Society, whose mission was one of "Harnessing the Power of Television to Make a Better World" and whose logo was a heart-shaped earth. It utilized information from such sources as the Worldwatch Institute to develop television programming that dealt with environmental and social global crisis. The society produced over 30 documentaries before folding in 1991. Turner also created *Captain Planet*, a cartoon show featuring an ecologically minded hero and five multiracial Planeteers who battle against overpopulation, global warming, the depletion of endangered species, and other such environmental evils.

In a 1988 speech to the Hollywood Radio and Television Society, Turner presented a list of ten "voluntary initiatives" designed to replace the "obsolete" Ten Commandments. Included on the list are "I promise to have love and respect for the planet earth and the living things thereon, especially my fellow species—humankind," "I promise to have no more than two children, or no more than my country suggests," and "I pledge to use as little non-renewable resources as possible." In 1991, Turner created the Turner Family Foundation, which donates $25 million each year to nonprofit groups. "Then," Turner told *Audubon* in 1999, "we stay out of their way." The foundation gives to about 500 organizations annually, primarily organizations focused on the environment.

Turner is the largest private American landowner. He owns 1.35 million acres of land in Montana, Nebraska, New Mexico, Colorado, Georgia, South Carolina, and Argentina. None of this land, Turner has pledged, will ever be developed. Turner is committed to restoring his property to its natural state. He has rid it of domestic livestock, such as cattle and sheep, replacing them with native species, such as bighorn sheep and bison in his western holdings. He has also helped reintroduce native predators. His 107,000-acre ranch outside of Bozeman, Montana, is representative of his efforts in this area. Turner bought the ranch in 1989, and in the years since, he has removed 250 miles of fence, implemented a burning program that ignites 5,000 to 10,000 acres a year, and spent almost $140,000 a year in an attempt to eradicate nonnative, invasive weeds. "All we are doing," he says, also in the 1999 *Audubon* article, "is allowing the ecosystem to be as natural as possible. We are trying to replace as many missing pieces to the environment as we can . . . we're trying to save the natural world."

Turner is also a humanist. He staged the first Goodwill Games in Moscow in 1986 in an effort to ease Cold War tensions. The games have taken place at an interval of every four years since then. The most recent were held in New York City in 1998. In 1997, Turner pledged one billion dollars to the United Nations, an organization he believes must play a critical role in the rapidly developing global society. Turner has received many awards for his efforts, including being named Man of the Year by *Time* in 1992. Turner is married to actress Jane Fonda and has five children from two previous marriages: Laura Lee, Robert Edward, IV, Beauregard, Rhett, and Jennie.

BIBLIOGRAPHY

Bibb, Porter, *It Ain't as Easy as It Looks*, 1993; Golberg, Robert, and Gerald Jay Golberg, *Citizen Turner*, 1995; Lanham, Julie, "The Greening of Ted Turner," *The Humanist*, 1989; Painton, Priscilla, "The Taming of Ted Turner," *Time*, 1992; Webster, Donovan, "Welcome to Turner Country," *Audubon*, 1999.

U

Udall, Morris

(June 15, 1922–December 12, 1998)
U.S. Representative from Arizona

During his 30 years in the House of Representatives, Morris Udall, the younger brother of Pres. John F. Kennedy's secretary of the interior, STEWART UDALL, became known as an environmentalist and a reformer. Morris Udall was the catalyst of several bills for the protection or restoration of the nation's natural resources, including bills concerned with land use planning and strip-mining. His concern for Native Americans and love of the environment resulted in numerous pieces of legislation moving through Congress. His greatest environmental achievement might be considered his efforts to pass the Alaskan Wilderness bills. A contender for the Democratic nomination for president in 1976, Udall preached environmental responsibility and awareness that natural resources had a finite capacity and must, therefore, be used more sparingly and thoughtfully.

Morris King Udall was born on June 15, 1922, in St. Johns, Arizona, to Levi Stewart, a chief justice of the Arizona Supreme Court, and Louise (Lee) Udall. Udall attended public schools in St. Johns. When he was seven, he lost his right eye in an accident, but this loss did not prevent him from becoming cocaptain of his high school basketball team, quarterback for the football team, student body president, and valedictorian; nor from playing professional basketball with the Denver Nuggets and serving in the U.S. Army Air Force during World War II. In 1949, after graduating with an LL.B. with distinction from the University of Arizona, he was admitted to the bar of Arizona and entered into a partnership with his brother in Tucson, which lasted until Stewart Udall became secretary of the interior in 1961.

"Mo" Udall was elected to the U.S. House of Representatives in 1961, in a special election to replace his brother's vacant position. Like the majority of first-time representatives to Congress, he struggled to adapt to the complex protocols of the House during his first term. During his next term in office, he organized a workshop for incoming congressmen. Out of this experience came Udall's book, *The Job of the Congressman*, which is still generally regarded as an essential primer for freshman congressmen.

Udall was one of the most productive members of Congress in the latter part of the twentieth century and quickly established a reputation as an environmentalist and a reformer. His work for political reform resulted in the Campaign Finance Law of 1974 that established a campaign spending limit as well as the public funding of presidential elections. But his environmental influence had an even more lasting legacy. Among his chief accomplishments was the Alaska Lands Act of 1980, which effectively doubled the size of the national park system and tripled the size of the national wilderness system. Other significant legislation includes the Central Arizona Water Project (though he would later publicly regret his support for it), the Strip Mining Reclamation Act, the Nuclear Waste Management Policy Act, the Arizona Wilderness Act,

Morris Udall (Corbis)

Abraham Lincoln, because of his height (he was six feet five inches tall), and Will Rogers, for his story-telling abilities. By the end of 1975, Udall was considered one of only a few candidates who had a legitimate chance at winning the nomination. With the support of many moderate and liberal Democrats, he fared well in the 1976 primaries.

Udall's environmental advocacy was also a significant part of his presidential campaign. His platform concentrated on what he called "the three E's": energy, environment, and economy. On environmental issues Udall insisted upon careful and planned use of natural resources. During his campaign he warned that "we have come to the end of an era: the era of cheap, abundant land, oil, timber, minerals, food, and water. The age of unlimited growth is over." With respect to the other two "E's," Udall's energy platform advocated the breaking up of big oil companies into smaller, more competitive companies and supported programs to develop other energy sources. To strengthen the nation's economy, Udall proposed a government-guaranteed full employment program, a cut in the defense budget, and the federalization of the welfare system. He also supported the Equal Rights Amendment and a national health insurance plan. Such policies earned Udall the respect of many politicians and media people who considered him the only candidate to pledge to be an activist president.

In spite of his strong results in the early primaries, Udall never actually won a primary. He finished second so many times that he jokingly called himself "ol' second-place Mo." After losing the Ohio primary in 1976, Udall withdrew from the race. While he was widely recognized as

the Indian Gaming Act, the Arizona Desert Wilderness Act, and the Tongass Timber Reform Act.

In 1974, Udall was approached by several colleagues who persuaded him to run for the Democratic presidential nomination in 1976. Although initially he was not taken seriously, being a relatively unknown figure from a small state, Udall became a serious contender as his campaign progressed. Throughout 1974 and 1975, Udall appeared at several Democratic fundraising events around the country, lending support to fellow Democrats in their campaign efforts as well as making himself better known among the party leadership. During his campaign he was often compared to both the second-place candidate, Udall knew he would not win the Democratic nomination. He also contended that a presidential campaign "tests your marriage, your sanity, your digestion, your sense of humor, and just about everything else." Further amusing anecdotes can be found in his autobiography, *Too Funny to be President*. Udall returned to the House and was named chairman of the House Committee on Interior and Insular Affairs (now Committee on Resources) in 1977; he held that position until 1991.

Udall's interest in the environment was not restricted to land use policy. He also enjoyed outdoor recreation; throughout his life, Udall was an avid hiker and mountain climber. In 1979, Udall was diagnosed with Parkinson's disease, a neu-rological disorder impairing movement and speech. Udall was hospitalized for the first four months of 1991, before stepping down in May 1991. He was reluctant to resign his seat in the House until he had served a full 30 years. He died on December 12, 1998, leaving six children, one stepson, and his third wife, Norma Gilbert Udall.

BIBLIOGRAPHY

Berry, James, "This Fella from Arizona," *Audubon Magazine*, 1981; Cook, James, "The Enigma of Mo Udall," *Arizona Republic*, 1971; Udall, Morris K., "A National Park for the Sonoran Desert," *Audubon Magazine*, 1966; Udall, Morris K., "Standing Room Only on Spaceship Earth," *Reader's Digest*, 1969.

Udall, Stewart

(January 31, 1920–)
Secretary of the Interior, U.S. Representative from Arizona

Nominated by Pres. John F. Kennedy as U.S. secretary of the interior in 1961, Stewart Udall served as secretary until 1969. At his post he was a staunch supporter of both conservation and the development of the nation's natural resources and authored many influential books and articles about conservation issues.

Stewart Lee Udall was born January 31, 1920, in St. Johns, Arizona, the son of Levi Stewart, a lawyer and state justice, and Louise (Lee) Udall. Udall's grandfather was David King Udall, a Mormon missionary who founded the town of St. Johns in 1880. Levi Stewart Udall was chief justice of the Supreme Court of Arizona, and Udall followed in his father's footsteps by pursuing a career in law. Udall attended Eastern Arizona Junior College and graduated LL.B. in 1948 at the University of Arizona.

Udall's education was interrupted by service with the U.S. Army Air Forces in Italy during World War II, where he served as a gunner in a B-24. After completing his college education when he returned home from the war, Udall was admitted to the Arizona bar in 1948. He practiced law in Tucson with his brother MORRIS UDALL for six years before running and being elected as a Democrat to the

The Reverend Ralph Abernathy, leader of the Poor People's Crusade, and Interior Secretary Stewart Udall come face-to-face on the problem of poverty at the Interior Department, 1 May 1968. (Bettmann/Corbis)

U.S. House of Representatives from the 2d Congressional District of Arizona in 1954. Udall took his seat in the 84th Congress in January 1955 and was reelected to the next three Congresses.

Early in his political career, Udall demonstrated a concern for conservation issues when, in 1955, he fought for the repurchase and return to the Coconino and Sitgreaves National Forests of 100,000 acres that had been detached from them

by a court order. Recognizing the dangers and severity of water shortage in the arid Southwest, Udall voted for a plan to develop the Colorado River Basin. In 1960, Udall fought for a bill to control water pollution and introduced another bill to provide federal aid to states for water desalinization plants.

After his appointment by President Kennedy as secretary of the Department of the Interior, Udall took charge of administering a staff of roughly 50,000 employees and an annual budget in excess of $800 million. Udall sought to advance a traditional conservation agenda—efficient resource management, public recreation, and expansion of the national parks—when the conservation movement was in a period of transformation. The Kennedy administration (and the subsequent Johnson administration) was subject to pressure from the emerging environmental movement that lobbied forcefully for wilderness preservation and environmental protection. Udall was not always sensitive to these rising concerns, but the Department of the Interior under Udall was more responsive than the previous Eisenhower administration had been.

As secretary of the interior, Udall was responsible for the custody of 750 million acres of federally owned lands; conservation and development of natural resources, mineral, animal, and vegetable; management of certain hydroelectric power systems; geological and topographical mapping of the nation; reclamation of arid lands through irrigation; and the administration of the nation's national parks. It was in these last two capacities that Udall was perhaps at his most prominent and controversial. In certain respects, cast in the same mould as GIFFORD PINCHOT, Udall was a utilitarian; he felt that nature should be used to human advantage. Nevertheless, he saw the national parks as a prime example of how nature was valuable if left alone. Udall argued that the national parks generated more income than their lumber ever could.

With respect to aridity and irrigation, however, Udall provoked considerable controversy and protest from conserva-

tion groups, most notably DAVID BROWER and the Sierra Club. In 1963, Udall made public the Bureau of Reclamation's billion-dollar Pacific Southwest Water Plan. In an effort to resolve the long-standing water shortage problems of the American Southwest, reclamationists proposed diverting water from the Pacific Northwest through a series of tunnels, ducts, and canals to the arid lower Colorado River basin. The scope of this ambitious plan was unprecedented. Dams would be built at Bridge and Marble Canyons in the Grand Canyon to finance the project and to generate hydroelectric power to pump water into central Arizona. The project would, however, affect 40 miles of the Grand Canyon National Monument and 13 miles of Grand Canyon National Park. That state and federal governments could destroy such a valuable cultural and ecological icon inflamed preservationists. Brower led an effective campaign against the proposed Pacific Southwest Water Plan. On February 1, 1967, Udall announced that the Johnson administration had changed its mind about the Grand Canyon dams. Later that year, Udall and his family visited the Grand Canyon; after a raft trip through the canyon, Udall conceded that he had erred in making an "armchair" decision about the dams.

Leaving his position as secretary of the interior in 1969, Udall spent a year as an adjunct professor of environmental humanism at Yale University and returned to writing, which he had started with his important and topical *The Quiet Crisis*, published in 1963. Several books on environmental issues would follow, including *The National Parks* in 1966, *America's Natural Treasures: National Nature Monuments and Treasures* in 1971, and

The Energy Balloon in 1974. While each book is poignant and well written, none matched the reception or popularity of *The Quiet Crisis*, in part an environmental history lesson, which effectively argued for contemporary expansion of the concept of conservation in the United States. *The Quiet Crisis* is equally pertinent and accessible today as it was almost 40 years ago.

Udall married Ermalee Webb on August 1, 1947. Together they had six children: Thomas, Scott, Lynn, Lori, Denis, and James. Not one to appreciate or work for the protection of the outdoors only from behind a desk, Udall has enjoyed hiking, hunting, fishing, swimming, horseback riding, and camping much of his life.

BIBLIOGRAPHY

Nash, Roderick, *Wilderness and the American Mind*, 3d ed., 1982; Smith, Thomas G., "John Kennedy, Stewart Udall, and New Frontier Conservation," *Pacific Historical Review*, 1995; Udall, Stewart Lee, "Human Values and Hometown Snapshots: Early Days in St. Johns," *American West*, 1982; Udall, Stewart Lee, and Jack Loeffler, "Stewart Udall: Sonoran Desert National Park," *Journal of the Southwest*, 1997.

Vogt, William

(May 15, 1902–July 11, 1968)
Writer, Ornithologist, Ecologist

Author of the 1948 classic *Road to Survival*, William Vogt was a writer, administrator, and ornithologist. As a fellow member of the New York birding community, he encouraged ROGER TORY PETERSON to create *A Field Guide to the Birds* and convinced Houghton Mifflin to publish it. He studied the guano birds of Peru and served as chief of the conservation section for the Pan American Union, forerunner of the Organization of the American States. *Road to Survival* was an influential treatise on the damaging effects of overpopulation, and following its publication Vogt served as national director of the Planned Parenthood Federation of America. As researcher, population expert, and naturalist, Vogt is seen today as a pioneer ecologist.

William Vogt was born on May 15, 1902, in Mineola, Long Island, New York. He contracted polio as a child and was left with a noticeable, somewhat limiting limp. He attended St. Stephens (now Bard) College in Annandale-on-Hudson, New York, where he edited the college literary magazine and graduated with a B.A. in journalism in 1925. While working as a journalist in New York City, Vogt got involved in the Linnaean Society of New York and joined the Bronx County Bird Club (BCBC), where he met Roger Tory Peterson. Peterson later traced the birth of his famous field guide to a suggestion made by Vogt during one of the BCBC's annual Christmas bird counts. In 1932 Vogt went to the office of Robert Moses, New York City's director of parks, to lodge a complaint, and Moses was so impressed with Vogt that he appointed him curator of the Tobay wildlife sanctuary at Jones Beach. Vogt served in this position from 1932 to 1935, during which time his small home at the sanctuary became a stopping point for many of the century's famous ornithologists.

In 1935 Vogt moved back into New York City, where he edited *Bird Lore* (later *Audubon*) and began to serve as a lecturer and field naturalist for the National Association of Audubon Societies. While editor of *Bird Lore*, Vogt instituted breeding bird surveys that continue to be one of the most important sources of avian population counts. In 1939 Vogt was ousted from his positions with the Audubon Society, after a dispute with Audubon president John H. Baker. Through fellow naturalist Robert Cushman Murphy, Vogt landed a post as consulting ornithologist to the Peruvian Guano Administration Company, and from 1939 to 1942 he lived and studied on Peru's guano islands. What Vogt found was a bird population unique in the world. Millions of birds lived along a narrow band of cold water along the Peruvian coast, producing enough guano to serve as one of Peru's major industries. The Guano Administration Company shoveled the guano into bags and barged it to the mainland, where it served as rich fertilizer for Peruvian and international agriculture. Vogt was called in because every seven years millions of birds died as a result of the El Niño warming in the Pacific Ocean. As the waterways around the islands warmed, the fish migrated away, and bird populations plummeted. Vogt's research documented the complex relations between species and their environment, with attention to concepts such as niche and competitive exclusion that mark the study as sensitive to ecological ideas far ahead of its time.

Vogt's work in Peru also contributed to *Road to Survival* in important ways. Watching the devastation among the island birds, Vogt began meditating on human overpopulation. If birds could be so dependent on their environment, and so rapidly depopulated, what was to prevent a similar disaster among human beings? In 1948 Vogt published his thoughts, to widespread acclaim and alarm. The book looks at the relationship between the carrying capacity of land area; the "biotic potential," the ability of the land to produce food, shelter, and clothing; and "environmental resistance," the limitations that any environment places on biotic potential. Vogt hoped to awaken the world's populace to growing food shortages as land reached its carrying capacity, in hopes of avoiding worldwide disaster. In the book's conclusion, Vogt implicates all human beings in the coming ecological crisis, but the bulk of the work concerns overpopulation in the Third World. Modern-day critics see this as a limitation in the work, in that it scapegoats the weakest members of the world's populace while largely letting wealthier countries and corporations off the hook. Critics have found the book both racist in some of its assumptions and pioneering in its look at the global ecological whole.

After finishing his work in Peru, Vogt worked for the Pan American Union from 1943 to 1950, directing conservation and research efforts in a number of Latin American countries. From 1951 to 1961 he directed Planned Parenthood. In 1960 he published *People!*, a follow-up to *Road to Survival;* the new book updated issues of population explosion and resource degradation. From 1964 to 1967 he served as secretary for the Conservation Foundation. He planned to use his retirement to return to Latin America but was prevented from doing this by a stroke. On July 11, 1968, suffering from depression and limited mobility, William Vogt took his own life.

BIBLIOGRAPHY

Duffy, David Cameron, "William Vogt: A Pilgrim on the Road to Survival," *American Birds*, 1989; Hammond, Richard, "William Vogt," *Biographical Dictionary of American and Canadian Naturalists and Environmentalists*, 1997; Peterson, Roger Tory, "William Vogt: A Man Ahead of His Time," *American Birds*, 1989; Stewart, Doug, Lisa Drew, and Mark Wexler, "How Conservation Grew from a Whisper to a Roar," *National Wildlife*, 1999.

Walter, Martin

(January 26, 1945–)
Mathematics Professor

Martin Walter is a University of Colorado at Boulder mathematics professor who developed an environmental math class that teaches the mathematical fluency needed to analyze environmental problems and the critical thinking skills necessary for identifying the faulty assumptions that have led to public misperceptions about environmental conditions and dangers. He believes that there are immense numbers of environmental problems that can benefit from a mathematical perspective.

Martin Edward Walter was born on January 26, 1945, in Lone Pine, California, to Clare and Karl Walter. He spent his early childhood on the eastern edge of Los Angeles, until his father learned that a freeway would be built in the family's backyard, at which time they moved to San Jacinto, at the edge of the Soboba Indian Reservation. Walter spent his free time wandering the remnants of the native chaparral ecosystem, and the family spent vacations visiting national parks. He obtained his B.S. in mathematics at the University of Redlands in Redlands, California, in 1966 and then went on to the University of California at Irvine, where he earned his M.S. in 1968 and his Ph.D. in 1971, both in mathematics. He worked as a research associate at Queen's University in Kingston, Ontario, from 1971 to 1973 and then moved to the University of Colorado, as an assistant professor from 1973 to 1977, an associate professor from 1977 to 1984, and a full professor since 1984, chairing the department from 1996 to 2000. During the early 1980s, he became involved with an ad hoc group of conservationists

fighting to preserve wilderness from logging in the lush Bowen Gulch area of Colorado, which borders Rocky Mountain National Park to the west and is home to the state's oldest and largest stand of Engelmann spruce trees. After almost a decade of massive letter-writing and public education campaigns, civil disobedience on site to prevent Louisiana Pacific from cutting trees, and a boycott of that company, Walter and his fellow activists convinced Congressman David Skaggs to obtain federal protection from logging for Bowen Gulch. The so-called Bowen Gulch Rescue Effort changed its name in 1990 to Ancient Forest Rescue and has engaged since then in a number of campaigns to prevent timber sales throughout the state. Walter has devoted much of his free time to this work.

In 1992, after spending many years teaching the basic algebra/trigonometry class to his students and realizing that the material did not inspire great interest or engagement, Walter designed a new class, "Mathematics for the Environment." This class, which he has taught every semester since the fall of 1992, teaches the language of math—mathese, as Walter calls it—through the study of social and environmental problems. Walter originally used a textbook entitled *Consider a Spherical Cow: A Course in Environmental Problem Solving*, written by University of California, Berkeley, professor John Harte (1988). However because most of the students in Walter's classes were not math majors, but rather environmental studies majors focusing on policy instead of science, they needed a text that presented complex mathematical concepts more gradually. So by 1997, Walter had written his own text for the course.

Walter's *Reader* provides a basic structure for the class. The first chapter provides a review of math and science, defining math as the search for and study of patterns and providing 100 pages of examples of how mathematical concepts can facilitate our understanding of the world. Its exercises are based both on pure math and on real-life topics, such as the kinship system of the Warlpiri aboriginal people of Australia, depletion of ocean fisheries, and the annual growth rate of brain cancer. The second chapter, "Media Literacy, Communication: Information, Honesty and Truth," is based on the ideas of media critics Ben Bagdikian and Noam Chomsky. Recounting stories that have been largely censored by the mainstream media (but have been published in such alternative media sources as Pacifica Radio News, *The Nation, Covert Action Quarterly*, and *Extra)* this chapter focuses on the importance of the views and facts that could widen Americans' thinking space models, correct common fallacies, and expose hidden assumptions. The third chapter deals with counting, statistics, and probability in environmental contexts, and the fourth, "Connections and Changes," works with population modeling, economics, food and health, and how human activities have altered the earth's ecosystems.

The class is designed to give a social and environmental context to show how math can be used as a tool to reveal a bigger picture. It emphasizes the importance of assumptions and how the most complete information possible can help one develop more accurate assumptions. Students learn that conclusions are almost always determined by the assumptions that one makes at the outset, and they are trained to spot when math is being misused or where arguments are fraudulent. Walter's class quickly became so popular that two sections, of at least 30 students each, are now taught each semester.

Walter has also supervised projects by math majors that are based on environmental problems: One M.S. thesis modeled the goshawk population on the Kaibab plateau near the Grand Canyon; another student's independent study used math to explore the patterns of what Walter calls "weatherquakes"— storms and other weather phenomena causing extreme destruction, happening increasingly on a global scale. In addition to Walter's focus on environmental modeling, his mathematical specialties are noncommutative harmonic analysis and self-organizing systems.

Walter has been a visiting professor in Trondheim, Norway, and Leuven, Belgium, and was an Alfred B. Sloan Fellow at the University of California, Berkeley, and the University of Pennsylvania in 1977 and 1978, respectively. He referees for a number of mathematical journals. In addition to attending conferences and giving talks on his mathematical specialties, Walter is well known locally for his environmental activism on campus and beyond. He received the Conservationist of the Year award from the Rocky Mountain Chapter of the Sierra Club in 1990 and was chosen as a CU Favorite Professor by alumni in 1996. Walter lives with his wife, Joy, in Boulder, Colorado.

BIBLIOGRAPHY

Boyd, David, "Teaching Math through Environment: Professor Reaches Out with Relevant Subject Matter," *Boulder Daily Camera*, 1995; "Marty Walter," http://www.Colorado.EDU/math/children/faculty/walter/.

Warburton, Barbara

(September 10, 1915–September 30, 1996)
Educator

Barbara Warburton taught three generations of secondary and college-level biology students in Brownsville, Texas; founded a biological station in the northeastern Mexican state of Tamaulipas; and fought for the conservation of a cloud forest there that has since become the nucleus of the United Nations Man in the Biosphere El Cielo reserve.

Barbara Ellen Taylor was born on September 10, 1915, in Tomah, Wisconsin. Her father, who worked for the John Deere farm equipment company, died in 1922 when she was seven years old. Her mother, left a widow with no one to help support the family, took a job as a teacher at schools on Indian reservations. Barbara and her sister and brother accompanied their mother to Carson City, Nevada, for one such job, but when she was hired to work on the Navajo reservation in New Mexico, the children were not permitted to accompany her. So they were sent to live with an uncle in southern Texas and visited their mother during the summers. Despite the economic hardship the family faced, Mrs. Taylor insisted that her daughters study. She felt that an education was the only thing that could not be lost or taken away in hard times. Barbara was sent to a high school run by Catholic nuns and then received a scholarship to Baylor Women's College. Her brother worked on ships off Galveston in order to support his sister; Barbara later helped her brother, too, by working as a teacher so he could attend college.

Warburton graduated from Baylor in 1941 with a B.A. in biology. She married Joe Warburton shortly after graduating and moved to La Feria, Texas, with him. In 1943, Warburton commenced what would be a career of more than 50 years teaching biology at Brownsville High School and Junior College, which later became Texas Southmost College and now shares a campus with the University of Texas at Brownsville. During the mid-1950s, Warburton earned a M.S. in biology at the University of Texas at Austin.

In the early 1960s, Warburton read an article in the National Biology Teachers Association magazine about how junior college biology students should be required to do fieldwork just as senior college students were. Knowing that some of her students were embarrassed to be attending a junior college, but fully aware that their education could be as solid as or even more so than that offered by a senior college, Warburton decided to fortify her curriculum with a field laboratory. She asked the college president for funds to build cabins in the nearby Sierra Madre Oriental of the northeastern Mexican state of Tamaulipas, 280 miles south of Brownsville. With the $2,500 from a private bequest left by a retired entomologist, Warburton and a group of enthusiastic students built three cabins themselves.

Once the facilities were set up, Warburton offered all of her students—not just the biology majors—the opportunity to take field trips to Rancho El Cielo. She would lead four-day trips during the spring and fall semesters and longer trips during the summers. Many undergraduates became so enamored with the beauty and diversity of the seven forest zones criss-crossed by rivers and surrounded by arid ecosystems that they changed their major to biology while there. The Mexican neighbors of the reserve were also impressed by its ecological value and worked to have it declared a biosphere reserve, earning it a higher degree of protection by Tamaulipas state authorities even than Mexican national parks. The Gorgas Science Foundation, which had been founded in 1947 by students at Southmost College, collaborates with Tamaulipas authorities and other educational institutions of the region to support the field station at Rancho del Cielo.

Warburton retired from teaching in 1978 but remained keenly interested in the Rancho del Cielo program. She volunteered her time in the Gorgas Science Foundation until her death. Her legacy manifests in hundreds of her students. In a videotaped interview produced by the University of Texas at Brownsville/Texas Southmost College, she and her former students recall how she instilled them with pride in their bilingualism and biculturalism and in the incredibly rich ecology that characterized their binational bioregion. She also encouraged a diligent work ethic, telling her students that the taxpayers were subsidizing their education and that they had to prove that their education was worthwhile. Many former students have pursued the sciences and are now professional scientists at work in the southeast region of Texas that she taught them to appreciate and value.

Warburton died in Brownsville on September 30, 1996, of complications following a heart attack. Her husband died two weeks later. They are survived by one son, Geoff.

BIBLIOGRAPHY

Medrano, Manuel, *Barbara Warburton*, 1998 (video-recording available through University of Texas–Brownsville/Texas Southmost College media services); Moreno, Jenalia, "Longtime Professor Warburton Dies at 81," *Brownsville Herald*, 1996; "University of Texas at Brownsville and Texas Southmost College Biological Sciences," http://unix.utb.edu/biology/.

Waring, George

(July 4, 1833–October 29, 1898)
Sanitation Engineer

George Waring used science and technology to clean up filthy, disease-infested cities of the 1800s. His economic city sewerage systems were the best early models, and as commissioner of New York City's Department of Sanitation from 1895 to 1898, he reduced that city's death rate by 6,000 a year.

George Edwin Waring was born on July 4, 1833, in Poundridge, New York, to George Edwin and Sarah Burger Waring. His father manufactured iron tools and stoves in Stamford, Connecticut, where Waring spent most of his childhood. After graduating from a private high school in Poughkeepsie, New York, in 1849, Waring tried selling hardware for a year and then managed a grist mill for two years. In 1853, he began studying methods of scientific agriculture with the well-known agricultural scientist, James Mapes. Within a year, Waring had learned enough that Mapes encouraged him to tour New England, giving lectures to farmers' clubs on scientific agriculture. Waring continued to do this during the winters of 1853, 1854, and 1855, and at the age of 21 he published his first book, *The Elements of Agriculture.*

Waring became manager of the Chappaqua, New York, farm of abolitionist publisher Horace Greeley in 1855 and then managed the Staten Island farm of Frederick Law Olmsted Sr. in 1857. That same year, Olmsted and Calvert Vaux, designers of New York City's Central Park, enlisted Waring to design the park's drainage system. Waring was the first to break ground when the landscaping began in 1859. He worked full-time on the project until the Civil War broke out. In 1861 he was commissioned major of the Garibaldi Guards, and he served until 1865.

Upon his return from the war, Waring became involved in unsuccessful oil and coal businesses and then in 1867 became manager of the Ogden Farm near Newport, Rhode Island, keeping that position for ten years. During his time there, he continued to pursue his interest in drainage and sanitation. He wrote *Earth Closets* in 1868, which described how human waste could be biodegraded and inoffensively integrated into family garden plots. In 1871 he was contracted to design the sewerage of Ogdensburg, New York, and in 1874 that of Saratoga Springs, New York. In 1877, as his fame in the sanitation field was growing, War-

ing abandoned farming as a profession to devote himself full-time to sanitation engineering.

As a member of the National Board of Health, Waring was invited to visit Memphis, Tennessee, one of the least sanitary cities in the United States, if not the world. A yellow fever epidemic that hit during two successive years in the late 1870s killed more than 5,000 of the city's 40,000 residents and led more than 20,000 others to flee. Waring found that the city's cisterns and wells for drinking water were almost all contaminated by the city's 6,000 cesspools and outhouses. Because the city was seriously in debt, the solution needed to be as economical as possible. Waring was hired to design a sewerage system. It had several features that were unique at the time: the pipes were narrower than most; they handled household sewage only—no roof or yard runoff was allowed to drain into the system; there were no manholes; and the pipes were well ventilated and flushed out every 24 hours by automatic flush tanks. Notes on the Memphis design were published in French, Dutch, Spanish, and German, and many more city sewerage systems were built following the Memphis model. Waring built two more sewerage systems in his life: one for Buffalo, New York, and another for Santiago, Cuba. He worked during the remainder of his life on perfection of sewage processing, strongly maintaining his faith that naturally occurring bacteria were the best agents for its decomposition. He told the graduating class of 1896 at Yale Medical School that chemical disinfection was "a clog in the wheel of nature's beneficent processes."

Waring is perhaps best remembered for his three years as commissioner of the Department of Sanitation of the City of New York. He took that position in 1895, appointed by reformist mayor William Strong, who had ousted the corrupt Boss Tweed and his Tammany Hall. Up until that time, New York's garbage collectors were hired by party bosses as a favor for their political support and spent most of their time on the job drinking in bars. New York was a filthy, plague-ridden city. Waring's early moves included cutting the pay of the garbage collectors to rid the force of those in there purely for political reasons and requiring the men to wear white uniforms that they purchased themselves. The uniforms gave the corps a military spirit and sense of dignity and might have discouraged the men from frequenting bars while on the clock.

Waring's tight organization and scientific methods worked. Immigrant children were recruited to join the effort, winning prizes and parties for their contributions. During Waring's time in office, 2,500 children joined 44 leagues. The garbage collectors, dubbed "White Angels" by New Yorkers, proudly performed their duties, and New York under Waring transformed into a much cleaner, health-

ier place. Six thousand fewer people died every year in New York each year Waring was in office. In addition to effective, efficient garbage collection, Waring's force also recycled and sold reusable elements of the solid waste stream. This reduced the amount of refuse to be dumped and raised money for the Department of Sanitation's work. Waring did not approve of dumping trash into the ocean and instead initiated incineration in New York.

Waring traveled widely and wrote prolifically. His books were on matters of sanitation as well as about his travels in Europe. In 1898 he traveled to Havana, Cuba, on a U.S. government mission. He was to report how yellow fever could be eradicated from the city of Havana. Waring contracted yellow fever on this visit and died in New York City on October 29, 1898. He left his third wife, Louise Yates, whom he had married earlier that year.

BIBLIOGRAPHY

FitzSimmons, Neal, "Pollution Fighter: George Waring," *Civil Engineering*, 1971; "George Waring: Giving Sanitation Status," *Civil Engineering*, 1976; Gottlieb, Robert, *Forcing the Spring*, 1993.

Warshall, Peter

(December 6, 1943–)
Editor of *Whole Earth*, Environmental Consultant

Peter Warshall is the editor of *Whole Earth*, a publication dedicated to fostering social change and environmental restoration through the introduction of diverse new ideas and trends.

In addition to his research and writings for the magazine, he works with a variety of communities on conservation, including indigenous people and ranchers in Arizona and Mexico and corporations seeking to reduce their impact on the environment. For the latter, he works through his environmental consulting firm, Peter Warshall and Associates. In addition, Warshall serves as chairman of Scientists for the Preservation of Mt. Graham. Although he harbors a concern for the environment as a whole, Warshall has focused much of his energy on watersheds, wastewater, and wildlife and is dedicated to encouraging and implementing practices that favor biodiversity and sustainability.

Peter Jack Warshall was born December 6, 1943, to parents Hymen and Beulah Warshall in El Paso, Texas. He spent his earliest childhood years in El Paso, moving with his family to Brooklyn, New York, as a young boy. He graduated from Stuyvesant High School in New York in 1960. Warshall then went on to Harvard University, from which he graduated with a B.A. in biology in 1965. From 1965 to 1967, he studied cultural anthropology, with an emphasis on Native American history and mythology, at École Pratique des Hautes Études in Paris, France. In 1968 he returned to Harvard to study biological anthropology, receiving his doctorate in 1970.

From 1972 to 1980 Warshall worked for the Bolinas (California) Public Utilities District, where he developed projects to improve the area's watershed management and water flow conservation. In this capacity, he designed a zero-discharge sewage treatment plant and constructed wetlands. He also initiated the protection of the Bolinas Lagoon, habitat to many shorebirds. During this time period, Warshall served on the governor's Emergency Task Force during the California drought in 1976, creating the first citizen's pamphlet on gray water systems and their reuse.

In 1975, Warshall began writing for *Co-Evolution Quarterly*, a magazine first published in 1974 that encouraged environmental consciousness by introducing concepts such as whole system thinking and voluntary simplicity to its readers. The magazine changed its name to the *Whole Earth Review* in 1985, then *Whole Earth* in 1997 when Warshall became editor. Warshall first wrote about land use and over the years has written on questions of watersheds and water resources, natural history, anthropology, and sustainability.

Warshall worked with the Office of Arid Lands in Arizona from 1982 to 1990, developing the first water hyacinth treatment plant in the state and initiating work on home-monitored use of water. In 1986 he became a research scientist for a proposed astronomical observatory site in the Sky Island region of Arizona and New Mexico. The University of Arizona, along with the Vatican, wanted to build 17 astronomical telescopes on Mount Graham, one of the "island" mountains surrounded by desert. Warshall was hired to provide the environmental impact study. During his time on Mount Graham, however, he encountered many rare species endemic to that small area. Perhaps the most notable of these was the Mount Graham red squirrel, formerly thought to be extinct. Warshall knew that developing the proposed telescopes would devastate the entire delicate ecosystem on Mount Graham. His *Whole Earth* articles during this time period mirrored his renewed dedication to wildlife and biodiversity conservation. In the Spring 1986 issue Warshall wrote about preserving wild lands such as Mount Graham through "wildlands philanthropy." He followed up with an article in the Summer 1986 issue on the Gulf of Mexico's diverse ecology and the connection of that ecology with the different cultures living in that bioregion.

Warshall united with the San Carlos Apaches and other environmental groups to fight for the preservation of Mount Graham. This quest led him to both the U.S. Congress and the Vatican and resulted in a decrease in the proposed number of telescopes to be built on Mount Graham. Warshall has also helped the Tohono O'odham people defend their land and water rights. He has worked in Ethiopia hunger camps for the United Nations High Commission for Refugees. Warshall has joined forces with the U.S. Agency for International Development to achieve biodiversity and better use of natural resources in ten other African countries.

Warshall has acted as a consultant to numerous large corporations, including General Mills, Volvo, Trans Hygga, Scandinavian Airlines System (SAS), and Clorox, helping them to reform their environmental practices. He has performed his consulting work through his own small firm as well as through the Global Business Network, an international collage of innovators, strategists, scientists, and organizations working toward improved global awareness of environment, business, and government. He has written about the effects of large corporations on the environment, most recently in the Spring 2000 issue of *Whole Earth* in an article about the World Trade Organization's effect on industry and pollution, as related to cultural and environmental change. Warshall often lectures at biodiversity and sustainability conferences, and in addition to his work for *Whole Earth*, his articles can be found in *San Francisco Chronicle*, *Orion*, *Animal Kingdom*, and *River Voices*. Warshall resides in Tucson, Arizona, and Sausalito, California.

BIBLIOGRAPHY

"Bioneers Home Page," http://www.bioneers.org/warshall.html; "Ecotech," http://www.ecotech.org/ecotech3/warshall.htm; "Global Business Network," www.gbn.org; "Whole Earth Home Page," www.wholeearthmag.com/about.html.

Waxman, Henry

(September 12, 1939–)
U.S. Representative from California

Henry Waxman has long been recognized as a champion of liberal causes and as an effective and passionate advocate for stronger government regulations in health and environmental issues. He was one of the primary authors of the landmark 1990 Clean Air Act and sponsored 1986 and 1996 amendments to the Safe Drinking Water Act. Waxman has fought on behalf of the elderly and poor and championed improvements in Medicare and Medicaid. He has fought hard for research on acquired immunodeficiency syndrome

(AIDS) and battled the big tobacco companies. Waxman was first elected to Congress in 1974 for California's 29th District, which includes the western portion of Los Angeles, including the coastal city of Santa Monica and affluent Beverly Hills.

Henry Arnold Waxman was born on September 12, 1939, in Los Angeles to Lou and Ester Waxman, who raised both Henry and his sister, Miriam, above the family grocery store in Watts. Both parents were staunch Democrats and very interested in the political scene. Waxman remembers not being poor at that time but certainly not wealthy either. He recalls his father talking about the Depression and the importance of the New Deal in giving all people an opportunity for achieving affluence in society. Waxman may have been even more influenced by his grandparents, immigrants from Russia who fled czarist persecution in 1905. "My grandparents would tell me about how the anti-Semites would come into town and destroy property, beat people up, threaten their lives, and they just felt they could no longer stay," he said. His grandparents eventually settled in the Boyle Heights section of Los Angeles. Waxman continues to honor his faith by going to temple, keeping kosher, and not working Saturdays.

As a teenager, Waxman attended Fremont High School, where he not only did well but made his first bid into politics by winning several student body posts. By 1961, Waxman enrolled at the University of California at Los Angeles (UCLA) as a political science major. He graduated with a B.A. in 1964. The civil rights movement fueled his already strong interest in politics and led Waxman to join the California Young Democrats. He was eventu-

ally elected statewide president of the organization, becoming more savvy in politics and meeting key friends who would later hold office and play a role in his political success.

In 1968, the 29-year-old Waxman graduated from UCLA Law School, and after a brief stint in law, he took the plunge into the political arena that same year. He ran for the state assembly against a veteran who seemed to be losing support among voters in his southwestern coastal district. Despite spending only $30,000 on the campaign and running as an outsider, Waxman won with 64 percent of the vote. In the state assembly, Waxman distinguished himself as an expert on health care issues and a leading advocate for the elderly. He served as chairman of the assembly Health Committee and on the Committee on Elections and Reapportionment and the Select Committee on Medical Malpractice during his three terms in the assembly.

In 1974, court-ordered reapportionment created a U.S. House of Representatives district seat in West Los Angeles, the very area where Waxman had lived since college. Because of the large concentration of elderly and Jewish voters as well as immigrants and gays, Waxman knew he had a good chance of winning. It was during the height of the Watergate era that Waxman and many other Democrats were swept into Congress. Immediately, he sought a place on the Interstate and Foreign Commerce Committee (now called Energy and Commerce) because health was an issue addressed by that committee. After just four years he challenged and beat out a senior colleague for the chairmanship of the Health and Environment Subcommittee. He held the chairmanship for 16 years, until the Republican sweep of Congress of 1994 stripped him of the title.

With health and the environment hot topics at the end of the 1970s, Waxman was the right person at the right time to bring about major policy changes. He became a major player in important amendments to legislation protecting the nation's air and water. Throughout the 1980s, he worked on strengthening amendments to the Clean Air Act of 1970. Although Waxman was one of the primary authors of the amendments to the act, he actually held up weaker versions of it for a decade until provisions addressing acid rain and emission controls were strengthened. The acid control programs required major reductions in sulfur dioxide emissions from electric power plants. The hazardous emissions control program replaced what had been an ineffective section of the Clean Air Act with a list of 190 toxic chemicals along with instructions for the Environmental Protection Agency on how to control them. He was able to block a key vote on a weaker version of the bill in the early 1980s by presenting 600 amendments wheeled into the committee room in a shopping cart. The motor vehicle emission standards enacted in 1990 have reduced allowable emissions from new cars by 75 percent.

In other environmental legislation, Waxman also sponsored the 1986 and 1996 Safe Drinking Water Act amendments, the 1996 Food Quality Act that regulates pesticides, the Radon Abatement Act, and the Lead Contamination Control Act. He is continuing to take the lead on environmental issues by pursuing legislation to address global warming. On March 1, 1999, Waxman released the first

report in the country to analyze the levels of hazardous air using current monitoring standards. The study was done in Los Angeles and showed that its residents may be exposed to hazardous air pollutants at levels far higher than the goals of the Clean Air Act.

Environmental health issues also have captured Waxman's attention. He has supported Pres. Bill Clinton in attempts to give the Food and Drug Administration the power to regulate tobacco as a way to cut down on teen smoking. A former smoker himself, Waxman grilled tobacco company chief executive officers in a series of widely publicized hearings in 1994 that brought to the public's attention the fact that secondhand smoke was a carcinogen. Waxman also exposed secret tobacco reports on the addictive properties of nicotine and the extent to which the industry targeted teens.

Waxman has staved off attempts to cap Social Security and Medicare benefits and has consistently supported programs to improve long-term nursing care as well as housing, nutrition, and prescription drug coverage for the elderly. In women's issues, Waxman has fought for the right of women to a safe and legal abortion, including the extension of this right to lower-income women dependent on Medicare. He has been a strong defender of Medicaid and has fought for the Nutrition Labeling and Education Act, the Breast and Cervical Cancer Mortality Act, the Safe Medical Devices Act, and the Orphan Drug Act. In 1996, Waxman was a key sponsor of the Ryan White Care Act amendments that increased spending on AIDS research. Throughout his political career, Waxman has remained a stanch supporter of Israel and has gained a reputation as an expert on Middle East Policy.

Waxman and his wife, Janet, maintain residences in Bethesda, Maryland, and Los Angeles, California. They have two children—daughter Carol and son Michael-David—and three grandchildren.

BIBLIOGRAPHY

Getlin, Josh, "What Makes Henry Tick," *Los Angeles Times*, 1990; Kosterlitz, Julie, "Watch Out for Waxman," *National Journal*, 1989; Meyerson, Harold, "The Liberal Lion in Winter," *Los Angeles Times Magazine*, 1994; "U.S. Rep. Henry Waxman," http//www.house.gov/waxman; Waxman, Henry, "False Alarms on Clean Air," *Washington Post*, 1997.

Werbach, Adam

(January 15, 1973–)
President of the Sierra Club, Television Producer

As the youngest-ever president of the Sierra Club, Adam Werbach led the club for two years, from 1996 to 1998, rejuvenating the membership by attracting a younger crowd, focusing on environmental justice issues, and shifting the club's activities from its former focus on lobbying to more outreach and grassroots organizing. With his company Act Now Productions, Werbach currently produces the environmental newsmagazine *The Thin Green Line*, broadcast on cable television.

Adam Werbach was born on January 15, 1973, in Tarzana, California, a suburb of Los Angeles. Although both his parents worked in Los Angeles proper, they chose to raise their two sons on a rustic two-acre ranch. Avid members of the Sierra Club, the Werbachs spent their vacations camping and hiking in national parks. Werbach's first activist episode came at the age of eight, when he recruited 200 of his classmates to sign a petition to dump President Reagan's antien-

vironmentalist interior secretary, James Watt. As a high school student, he spent one semester at the Mountain School in Vershire, Vermont, a branch of the Milton Academy. His time there was mostly dedicated to environmental education and work on the school's 300-acre organic farm. After living in pristine rural Vermont, he recounts in his 1997 book *Act Now, Apologize Later*, he was shocked by the level of environmental degradation in southern California, especially the choking air pollution that made asthma a virtual epidemic. He immediately got involved in California's Big Green campaign for a broad-based 1990 environmental referendum that would have tightened clean air bills, increased open space and forest preservation, and required greater fuel economy. Werbach's contribution was to organize high school students to campaign for the referendum. The extraction industries outspent Big Green supporters ten to one, however, and Big Green was defeated.

Adam Werbach at a press conference in San Francisco, 25 April 1998 (AP Photo/Paul Sakuma)

Undaunted by Big Green's defeat, Werbach and the teenagers he had worked with decided to continue to work together as a student group. They persuaded the Sierra Club to fund a summer camp for teenagers that would provide environmental education and rally participants for environmental activism.

Werbach entered Brown University in 1991, where his activism continued and expanded. By this time, he had obtained official endorsement of the Sierra Student Coalition, which grew to 30,000 members and registered thousands of new student voters. They lent critical support to the 1992 California Desert Protection Act, via "dorm storm," in which dormitory residents throughout the country's colleges and universities called their congresspeople to demand their support for the act. The Sierra Student Coalition's creative methods also included selling black snow cones at concerts and fairs to dramatize the effects of drilling for oil in the Arctic National Wildlife Refuge.

After graduating from Brown in 1995 with degrees in political science and media, Werbach returned to California. A member of the national board of the Sierra Club since 1994, Werbach was elected president of the club in 1996 and reelected in 1997, with endorsements from such Sierra Club luminaries as DAVE FOREMAN and DAVID BROWER. Werbach's mission as president was to focus more effort on grassroots organizing, especially around issues of environmental justice, and to attract more young members by spreading the message through such popular media vehicles as MTV and the Internet. He was president in 1998 when the club members voted against a controversial and well-publicized proposal for the club to take a stance against immigration to the United States as a response to overpopulation pressure in this country. Instead, the club vowed to continue work on global population stabilization. During Werbach's term, the Sierra Club also asked President Clinton to cease logging in national forests, a position the club had never advocated previously. He helped lower the average age of U.S. voters by a decade (from 47 to 37 years). He also helped pass the strongest clean air standards in national history. Werbach has gained much attention, both supportive and incredulous, for his proposal to drain Lake Powell by demolishing the Glen Canyon Dam. The dam was built in 1963 after Sierra Club president David Brower agreed that the Sierra Club would accept that dam in return for the federal government's not damming the Green and Yampa Rivers at Echo Park in Dinosaur National Monument. Brower had never seen Glen Canyon, but later when he did have a chance to tour the magnificent site before it was submerged, he realized his mistake. Brower has regretted that compromise ever since. Werbach's proposal to drain Lake Powell has garnered the official support of the Sierra Club, which works with the Glen Canyon Institute to further the dam's demolition.

Werbach currently heads Act Now Productions, which produces *The Thin Green Line*, a newsmagazine that profiles environmental activists and their struggles. The *Thin Green Line* is broadcast monthly on the Outdoor Life Network cable channel. Werbach described his interest in the medium of television to Jennifer Hattam of *Sierra*, "TV is signal-rich, but content-poor, while the environmental movement is content-rich, but signal-poor. Our goal is to take the great stories that need to be told to a medium that needs substance."

Werbach resides in San Francisco.

BIBLIOGRAPHY

"Babe in the Woods," *People Magazine*, 1996; Chetwynd, Josh, "'Splatter-casting' the Sierra Club's Message," *U.S. News & World Report*, 1997; Hattam, Jennifer, "Werbach Walks 'Thin Green Line,'" *Sierra*, 1999; King, Patricia, "A Sprout for Sierra," *Newsweek*, 1996; McManus, Reed, "Pitchman for the Planet," *Sierra*, 1996; Werbach, Adam, *Act Now, Apologize Later*, 1997; Wilke, Anne W., "Adam Werbach: The Youngest Sierra Club President Is Aiming for the Grassroots—MTV," *E Magazine*, 1997.

Whealy, Diane, and Kent Whealy

(January 1, 1950– ; April 27, 1946–)
Cofounders of Seed Savers Exchange

Kent and Diane Whealy work to conserve the genetic diversity of garden crops through Seed Savers Exchange (SSE), a private nonprofit organization they founded in 1975. The Whealys have long recognized the need to conserve food crop diversity because of its importance to sustainable agriculture, human nutrition, cultural richness, and evolution—and doing so is becoming ever more critical as biotechnology and profit-based agriculture threaten to intensify genetic erosion. The thousands of members of Seed Savers Exchange are dedicated to systematically collecting, maintaining, and distributing seeds from nonhybrid vegetables, fruits, and grains, thereby saving them from extinction. Seed Savers Exchange operates a 170-acre farm to maintain its collection of endangered food crops and currently preserves more than 18,000 rare heirloom (traditional) varieties there.

Diane Ott Whealy was born in New Hampton, Iowa, on January 1, 1950, and raised in the small town of Festina, Iowa. She attended Winona State University in Winona, Minnesota, but never graduated. Kent Whealy was born in Sioux Falls, South Dakota, on April 27, 1946, and grew up in Wellington, Kansas. He studied at the University of Kansas and received his bachelor's degree in journalism in 1969. In 1971, shortly after their wedding, the Whealys started building a house in the woods of Missouri. Before they finished, however, Diane's grandfather, Baptist John Ott, became ill. Knowing that Grandpa Ott did not want to leave his family farm near St. Lucas, Iowa, they moved to nearby Decorah, Iowa, to care for him. That summer when they started their first garden, Grandpa Ott gave them tips and, more important, seeds from two plants that his parents brought from Bavaria in the 1870s; one was a large, pink tomato and the other an exquisite deep purple morning glory.

The Whealys learned a lot from Grandpa Ott, and when he died the next winter they wanted to preserve the legacy he had left them. They began seek-

Diane and Kent Whealy (Courtesy of Anne Ronan Picture Library)

ing out other gardeners using heirloom varieties—varieties that have been passed down through generations—and became more and more aware of the extensive genetic riches of isolated rural gardens. The diversity of these heirloom varieties vastly outnumbers the entire selection offered by the garden seed industry in North America, yet they had never before been systematically collected or preserved. The Whealys also realized that few gardeners comprehended the scope of their garden heritage and how much was in danger of being lost. Each variety has a unique genetic makeup that manifests differences in plant size, taste, drought or disease resistance, adaptability, and other traits. This diversity is the raw material for evolution, which is a

prerequisite for survival. The varieties of food plants now in existence represent all of the breeding material that will ever be available for crops of the future; and in order to deal with diseases and pests, as much genetic diversity as possible is needed. When a strain becomes extinct, particular genetic characteristics are lost forever. What made the situation even more urgent was that many of the living heirlooms they found were being maintained by elderly gardeners—and as rural economic conditions deteriorated, younger generations were forced into more urban settings, leaving noone to replant these unique seeds.

In 1975 the Whealys took action and founded Seed Savers Exchange (SSE). By writing to editors of backwoods newspapers and gardening magazines, they continued to search for gardeners using seeds that had been passed down through generations. At the end of its first year SSE had six members who traded seeds by mail. The next year, Kent Whealy took a job in a print shop to make ends meet, while SSE grew to 29 members. In order to keep track of what seeds were available to other members, and who had them, the Whealys put together a newsletter; this was the start of the SSE yearbook, which has since grown to 500 pages listing 11,500 varieties of seeds. Starting in 1981, SSE also began compiling and publishing an inventory of all nonhybrid seeds available from seed company catalogs and over the years found an alarming rate of genetic erosion. Hundreds of varieties were dropped from catalogs each year, so SSE began to buy up samples of any variety offered only by a single source, establishing a mechanism for saving threatened varieties before they were lost.

By 1981, Kent Whealy was able to quit his printing job, and he and Diane Whealy began working for SSE full-time, although at the time their operation was not grossing more than $3,000 a year. But it continued to expand, and in 1986 SSE purchased Heritage Farm, a 170-acre tract of limestone bluffs and burr oak woods that now houses their headquarters, their Historic Orchard, and their Preservation Gardens. They have spent the past 13 years developing Heritage Farm into an extensive professional facility.

In the Historic Orchard, the most diverse public orchard in the United States, 700 varieties of 19th-century apples and 200 kinds of grapes are now maintained and displayed. These orchards are a hedge against the genetic erosion that has been exhausting apple crops—at the turn of the twentieth century there were 7,000 varieties in North America, but fewer than 1,000 of those remain and are steadily dying out. On another part of the farm, the Whealys keep a herd of extremely rare Ancient White Park cattle, a 2,000-year-old breed from the British Isles.

Currently Heritage Farm preserves a vast, ever-increasing collection of seeds that recently exceeded 18,000 varieties, including 4,100 varieties of tomatoes, 850 varieties of lettuce, 200 varieties of watermelons, and 3,500 varieties of beans, among others. The seeds are kept in cold storage vaults, and each year up to 2,000 endangered varieties are planted in the organic Preservation Gardens to multiply seeds, which are then processed and heat-sealed into foil packets and stored in seed vaults. Any remaining seeds are then available to SSE's members, who now number 8,000. Distributing and planting heirloom seeds as widely as possible helps ensure their survival, and the Whealys' mission is to get these heirlooms into as many gardens as possible across North America.

In 1989 Diane founded the Flower and Herb Exchange (FHE), a separate organization patterned after SSE. Many of the 3,000 varieties of heirloom flowers and herbs offered through FHE have never been available through seed catalogs, having always been passed down from generation to generation within families. FHE now has 2,000 members.

In recognition of their work, Kent and Diane Whealy received honorary doctorates from Luther College in 1991. Kent Whealy was also awarded a MacArthur Fellowship in 1990 for his efforts to conserve genetic resources. With part of the award money, he expanded the scope of the search for heirloom seeds to other countries. In 1992 he started developing Seed Savers International, a network of plant collectors rescuing food crops in foreign countries. Many Eastern European countries have an extremely rich variety of traditional food crops, with seeds still being produced by gardeners and farmers. But Western agricultural technology and seeds are currently encroaching on even the remotest areas, threatening these fragile genetic resources. Seed Savers International has hosted 12 trips to Eastern Europe and the former Soviet Union to gather seeds, an effort that brought 4,000 seed varieties from 30 countries to Heritage Farm.

SSE and Heritage Farm have served as a model for dozens of similar organizations in the United States and overseas. The Whealys' efforts to raise national awareness of heirloom varieties and the dangers of genetic erosion are paying off, as new varieties are beginning to show up in seed company catalogs. Kent and Diane Whealy are currently continuing their work as executive director and codirector of Seed Savers Exchange, respectively.

BIBLIOGRAPHY

Eddison, Sydney, "Saving Seeds for Future Generations," *Organic Gardening*, 1999; Fowler, Cary, and Pat Mooney, *Shattering: Food, Politics, and the Loss of Genetic Diversity*, 1990; Geeslin, Ned, "Kent Whealy's Seedy Operation Provides Garden Variety Veggies from Centuries Past," *People Weekly*, 1987.

White, Gilbert F.

(November 26, 1911–)
Geographer, Natural Hazards Specialist

Geographer Gilbert White has devoted his career to understanding—and helping change—how people deal with natural hazards, especially flooding. Starting as a young man during Pres. FRANKLIN D. ROOSEVELT's New Deal administration and continuing throughout his 70-year career, White has proposed creative responses to flooding that were much less environmentally harmful than previous approaches. White has also made major contributions to other twentieth-century challenges, including water management, water pollution, the environmental effects of dams and reservoirs, desertification, nuclear war, nuclear waste disposal, and global climate change. In all of his efforts, White is recognized by colleagues as a leader in coordinating interdisciplinary work and steering it to influence public policy decisions.

Gilbert Fowler White was born on November 26, 1911, in Chicago to Mary (Guthrie) and Arthur White. Neither of them had attended college, yet both felt it would be of value to their children, so they settled close to the University of Chicago so that their children would be able to attend the university and its Lab School. White spent summers from the time he was ten years old at a ranch of which his family was a part owner, on the Upper Tongue River in Wyoming. It was at the ranch that White first became aware of the environmental consequences of land use; in his biographical essay for *Geographical Voices* (2000), he recalled community debates that asked "When and how was the forest land over-grazed? When was there inadequate drainage leading to soil degradation? What systems of land use made for prosperity or poverty in the local community?"

White graduated from the University of Chicago with an S.B. in 1932 and an M.S. in 1934, both in geography. While at the university he was especially influenced by his professor Harlan Barrows, who had described geography as human ecology in 1923 and who urged White to take classes in plant ecology and urban ecology. As a Quaker, White had already decided by this time to guide his professional life according to the same constructive, nonviolent principles that he wanted to follow in his personal life. He was attracted to geography for its potential to help people live better, in greater harmony with the earth's natural processes.

In 1934 White left Chicago at the invitation of Barrows to work with the New Deal administration on natural resource management issues. Working on reports for the Mississippi Valley Committee, the National Resources Planning Board, and the National Resources Committee, White focused on flood and drought mitigation, searching for a systematic way to analyze social costs and benefits of different uses of the land, especially land vulnerable to floods and droughts. Far ahead of his time, White proposed that federal financing of flood-control dams in California be contingent on state regulation of human occupation of floodplains. His work resulted in his Ph.D. dissertation, "Human Adjustment to Floods," which suggested that single structural solutions, such as dams, levees, and seawalls, be supplemented with multiple solutions adapted to local and geographic context. White continued to expand on these recommendations throughout his career, suggesting such damage-reducing measures as better forecasting and warning, evacuation planning, flood-proofing of buildings, land use planning and floodplain zoning, and flood insurance. He has insisted throughout his career that intensive human occupation of floodplains only invites disaster and that better land use planning could help prevent flood catastrophes while making better use of the natural resources of the floodplain.

White worked nights to complete his dissertation with the help of his fiancée, Anne Underwood, while working during the day from 1940 to 1942 for the executive office of Pres. Franklin D. Roosevelt, handling all legislative matters that pertained to land and water issues. White was granted his Ph.D. from the University of Chicago in 1942. When the United States entered World War II after the bombing of Pearl Harbor, White, who was a pacifist conscientious objector, decided to leave government service so as not to be contributing to the war effort. He traveled to France in 1942 to work with a Quaker refugee relief effort until 1943, when he and 133 other Americans were detained by the Germans. They were held for 13 months in Baden-Baden, during which time White and the other detainees formed their own informal academy. White taught geography to the children, studied German and Russian, and gave a seminar on contemporary geographic theories. When he returned to the United States, he married Anne Underwood, with whom he would enjoy a long marriage and partnership.

Working through the Quaker-run American Friends Service Committee during the war and immediately afterwards, White helped with relief efforts in India, China, and Germany, before being offered in 1946 the presidency of Haverford College in Haverford, Pennsylvania, a position he would hold until 1955. At Haverford, a Quaker college, he increased the endowment, doubled the faculty, and decreased the size of the student body so that everyone could know each other. He taught one course per year in natural resources conservation and continued to serve on government commissions, including the Hoover Commission Task Force on Natural Resources, the United Nations Committee on Integrated River Development, and the new United Nations Educational, Scientific, and Cultural Organization (UNESCO) Advisory Committee on Arid Zone Research. He enjoyed the small college atmosphere but decided to leave the field of administration after Margaret Mead, an anthropology professor at the nearby University of Pennsylvania, warned him that by continuing to climb the administrative ladder, he would never have time to further his research. White realized that she was right and accepted a position as chairman of the Geography Department of the University of Chicago in 1955.

White remained at Chicago for 14 years, supervising the doctoral dissertations of students on groundbreaking topics, ranging from floodplain management problems to waste disposal. Although he continued to participate in high-level commissions dealing with international water planning and won awards for his teaching at Chicago, one of his most satisfying experiences during this period was to study the household use of water in East Africa with his wife, Anne, and medical researcher David Bradley. During this 1966 research project, they studied how more than 700 households in 34 different locations decided how much water to draw from the local water source and how it would be used. They measured the costs to the families in time, calories, and money; the health implications; and the way in which the family determined which source of water to use. This was the first study of its type and was eminently useful to water resource managers. More than 30 years later, an international team studied water use at the same sites to see how conditions and practices had changed.

After 14 years on the faculty at Chicago, the Whites moved to Boulder, Colorado, where he taught geography and headed the Institute for Behavioral Science (IBS) for the University of Colorado. Through the IBS, White and other geographers dealing with natural hazards collaborated with people from other fields, broadening both the scope and the potential audience of their work. White joined with a colleague in recommending the establishment of the Natural Hazards Research Applications and Information Center at the IBS; he served as its first director from 1976 to 1984 and then again from 1992 to 1994.

While working with the Scientific Committee on Problems of the Environment (SCOPE) during the 1970s to evaluate the problems of man-made lakes, White became concerned about the consequences of these and other global changes caused by humans. In 1979, he joined with Mostafa Tolba, director of the United Nations Environmental Programme, in issuing a statement about the dangers facing global life support systems. He was a member of an advisory group on greenhouse gases from 1986 to 1990.

White's complete list of committee memberships and affiliations is too numerous to be mentioned here, but those that stand out include his chairing of the American Friends Service Committee from 1963 to 1969, the Advisory Panel on Reducing Earthquake Losses for the Office of Technology Assessment in 1994, and the Committee on Sustainable Water Supplies for the Middle East of the National Research Council in 1996, through which he helped obtain a consensus on water management issues from Israeli, Palestinian, and Jordanian water experts.

Additional work of environmental importance has been through his work with the Lower Mekong Coordinating Committee in 1969, the SCOPE Steering Committee on the Environmental Consequences of Nuclear War from 1983 to 1988, and the Aral Sea Basin Diagnostic Panel in 1993. White has received numerous honorary degrees and awards for his efforts, including the United Nations Sasakawa International Environmental Prize (1985), the Tyler Prize for Environmental Achievement (1987), the Volvo Environmental Prize (1995), the National Academy of Science Public Service Medal (2000), and the 2000 Millennium Award from the International Water Resources Association.

Anne U. White died in 1989; Gilbert White is still working regularly and still actively involved in most of the areas that he has pioneered and developed over the past 60 years. He lives in Boulder, Colorado.

BIBLIOGRAPHY

Platt, Rutherford H., Tim O'Riordan, and Gilbert F. White, "Classics in Human Geography Revisited: White G.F. 1945: *Human adjustment to floods,*" *Progress in Human Geography*, 1997; White, Gilbert, F., "Geographer's Autobiographical Essay," *Geographical Voices*, ed. Peter Gould and Forrest Pitts, 2000; White, Gilbert, F., *Human Adjustment to Floods*, University of Chicago Research Paper 29, 1942; White, Gilbert, F., David J. Bradley, and Anne U. White, *Drawers of Water: Domestic Water Use in East Africa*, 1972.

White, Lynn, Jr.

(April 29, 1907–March 30, 1987)
Medievalist

Author of the 1967 essay "The Historical Roots of Our Ecological Crisis," Lynn White Jr.'s ideas have profoundly shaped the debate about the relation between religion and the environment. White traced the roots of Western attitudes about nature to the medieval period, particularly the way medieval Christian thought was based on a belief in humankind's "dominion" over nature. White was both praised and attacked for this article, which was seen as anti-Christian by some theologians and laypersons and as asking fundamentally necessary questions by others. By providing a rigorous examination of science, technology, and theology in medieval Europe, White reshaped twentieth-century ideas about ecology.

Lynn Townsend White Jr. was born on April 29, 1907, in San Francisco. White's father was a clergyman, and White followed in his footsteps to Stanford University, earning a B.A. in 1928, and to Union Theological Seminary for an M.A. in 1929. He went on to Harvard University to earn a second M.A. in 1930 and a Ph.D. in 1934. His first academic job was as an instructor in history at Princeton University from 1933 to 1937. He was hired at Stanford University in 1937, where he earned tenure and stayed until

1943, at which time he was hired as president of Mills College in California. In 1958 he became a professor of history at the University of California at Los Angeles (UCLA), where he remained until the end of his career.

When he resumed his teaching career at UCLA, White began his work on medieval technology, which resulted in the publication in 1962 of *Medieval Technology and Social Change*. This work challenges two basic assumptions about medieval history. First, White argues against the belief that the Middle Ages were the "dark" ages, devoid of science and culture. White shows that a number of fundamental technological changes began in the medieval period, including the development of improved plow technology, widespread use of wind and water power, and major advances in machine design. Second, White argues that medieval Europe was much more interactive with the far and near East than had previously been thought. Studies that place greatest importance on the Renaissance period in the development of present-day ideas, White argues, underestimate the cultural and economic exchanges between medieval Europe, Asia, and Africa. White makes a strong case for taking seriously the Middle Ages, rather than relegating them to a footnote between Rome and the Renaissance.

In 1967 White followed up on these ideas with "The Historical Roots of Our Ecological Crisis," published first as an article in *Science*. Here White made an even stronger case for viewing the medieval period as the source of modern ideas about science and technology. The Middle Ages, for White, provided the "psychic foundations" of the spirit of modern technology. White locates this spirit in a number of trends within medieval culture. One of these was the church's assault on paganism, with its animistic view of nature; this assault ensured that reverence toward the natural world would assume an air of blasphemy. Another important element of medieval thought was its emphasis on the dichotomy between spirit and matter; matter comes to be seen as inert, devoid of soul, and hence available for use without ethical qualm. Third, the medieval church emphasized those elements of the Genesis story that give humankind dominion over nature, characterizing the earth as created solely for human use. Finally, White argues that medieval monasticism prepared Western culture to embrace technological development with great energy. European monks in this period celebrated labor as integral to worship, a tenet of faith seen in White's readings of medieval manuscripts and cathedrals. Cathedrals celebrate mechanical clocks and organs, two of the most complex machines operating in premodern Europe. White argues that this reflects and helps to create an attitude that sanctioned technology as virtuous and reverent. These trends added up to a tendency to exploit and abuse nature, he believes, and if we are to come to terms with present-day ecological crisis, we need to reenvision a less anthropocentric view of the natural world. White closes with an examination of the ideas of St. Francis of Assisi, in whom White finds a way to recover a respect for nature within a Christian framework. St. Francis proposed equality among all creatures, including humankind, and White urges a return to a radical Franciscan ecology.

"The Historical Roots of Our Ecological Crisis" evoked a storm of controversy.

Many theologians took up White's call for a new Christian ecological theology, while some Christians expressed outrage at what they saw as White's anti-Christian bigotry. Today the perspective that there is something inherently negative in Christian beliefs about the environment is called the Lynn White Thesis, and it is still generating articles and books. White continued to develop his own views in this area, many of which are collected in two volumes, *Dynamo and Virgin Reconsidered* (1971) and *Medieval Religion and Technology* (1978). In 1964 White founded UCLA's Center for Medieval and Renaissance Studies, which he directed until 1970. White retired from UCLA in 1974 and continued to write, speak, and teach as an emeritus member of the faculty until shortly before his death. Lynn White Jr. suffered a heart attack at his home and died on March 30, 1987.

BIBLIOGRAPHY

Bennett, John, "On Responding to Lynn White: Ecology and Christianity," *Ohio Journal of Religious Studies*, 1977; Ruether, Rosemary, "Biblical Vision of the Ecological Crisis," *Christian Century*, 1978; Whitney, Elspeth, "Lynn White, Ecotheology, and History," *Environmental Ethics*, 1993; Wolkomir, Michelle, Michael Futreal, Eric Woodrum, and Thomas Hoban, "Substantive Belief and Environmentalism," *Social Science Quarterly*, 1997.

Whitman, Walt

(May 31, 1819–March 26, 1892)
Poet

W alt Whitman is the grandfather of American poetry. What Dante is for Italy, Shakespeare to England, Walt Whitman is to the United States. The Good Gray Poet is best known for his *Leaves of Grass* (1855). The most quoted book of American poetry, *Leaves of Grass* is a declaration of personal independence and the interdependence of all things. Poet Walt Whitman believed that the wild places in the United States were demonstrative of the national character, and he promoted the American landscape as the primary agent of American-ness. Environmentalists, naturalists, and general readers alike have been nourished by the enormous optimism of *Leaves of Grass*, and Whitman's pioneering spirit led many to take to the open road and explore this country's wilderness.

Walter Whitman was born at West Hills, Long Island, New York, on May 31, 1819, to Walter Whitman Sr. and Louisa Van Velsor Whitman. The second of eight children, Whitman came from long-established native stock, landowners, farmers, builders, and horse breeders. Falling on hard times, the Whitmans moved from West Hills in 1823 to Brooklyn, where Walter Sr. worked as a carpenter. During a Fourth of July parade through the streets of Brooklyn in 1825, Revolutionary War hero General Lafayette stooped down to kiss six-year-old Walt, who was watching the procession.

He was educated by his attentive parents and in public schools until he was 11. He showed an early knack for journalism; by the age of 12, he was already writing what he called "sentimental bits" for the *Long Island Patriot* (edited by Samuel E. Clements), and during his teens he published a couple of pieces in the *New York Mirror*. At 20, Whitman had saved enough money from teaching in country schools to purchase a press and type, and he briefly published the weekly *The Long Islander* of Huntington. Whitman did almost all the work, including deliveries throughout Long Island on horseback. Recognizing it as an unsustainable venture, he found work at *The Aurora*, *The Tattler*, and, later, a good position editing *The Brooklyn Eagle*. He took an interest in Democratic Party politics, wrote an early account of a baseball game in Brooklyn, and also covered a series of lectures by New England transcendentalist RALPH WALDO EMERSON. In 1848, Whitman was relieved of his editorship, as his political positions were contrary to those of the publisher, and he took the opportunity to travel for the remainder of the year. With his brother Jeff, Whitman went to New Orleans for a brief stint at a newspaper and traveled by boat to St. Louis, LaSalle, Chicago, Milwaukee, and Buffalo before returning to New York for more freelance journalism.

Walt Whitman (Corbis)

The constraints of editorial and newspaper reporting clearly could not contain Whitman's poetic voice. In his early thirties he declared: "I will also be a master after my own kind, making the poems of emotions, as they pass or stay, the poems of freedom, and the exposé of personality—singing in high tones democracy and the New World of it through These States." Whitman set out to write the poetry of America in the American language. He rejected conventional literary themes, rhyme, and formalism, anything that reflected the old world poetic traditions and social orders in favor of something new and bold, which he felt embodied the American landscape. The time between 1850 and 1855 was an incubation period for Whitman the poet. He read a lot and took long walks on the beach before penning the first versions of *Leaves of Grass*, which made its print debut on July 6, 1855, selling for two dollars. He would continue to revise and expand the work in subsequent editions his entire life.

The original edition did not include titles for the poems, but once they were added in the 1860 edition, Americans would know the best of them: "Song of Myself," "The Sleepers," "I Sing the Body Electric." Although the book's principal focus is the self, it also includes Whitman's vision of the United States in all its wild, bucolic, and industrious glory. From "Song of Myself," verse 31:

> I believe a leaf of grass is no less than the journeywork of stars,
> And the pismire is equally perfect, and a grain of sand, and the egg of the wren
> And the tree-toad is a chef-d'ouevre for the highest,
> And the running blackberry would adorn the parlors of heaven,
> And the narrowest hinge in my hand puts to scorn all machinery,
> And the cow crunching with depressed head surpasses any statue,
> And a mouse is a miracle enough to stagger sextillions of infidels,
> And I could come every afternoon of my life to look at the farmer's girl boiling her iron tea-kettle and baking shortcake.

Whitman's glorification of nature and the human spirit satisfied Ralph Waldo Emerson's criteria for what an American poet ought to bring to the national culture, and he said so in a 1856 letter to the poet. Whitman also gave a copy to HENRY DAVID THOREAU, and his response was also laudatory, but not without reservations about the more sensual passages in the work. Thoreau wrote in a letter to Harrison Blake that he thought Whitman "has spoken more truth than any American or modern that I know. I have found his poem exhilarating, encouraging. . . . Though rude and sometimes ineffectual, it is a great primitive poem—an alarm or trumpet-note ringing through the American camp." Whitman consulted naturalist

and nature writer JOHN BURROUGHS on the hermit thrush, the symbolic bird in his elegy to Abraham Lincoln, "When Lilacs Last in the Dooryard Bloom'd" (*Drumtaps*), and generally drew as much information from naturalists and other nature writers as he could. In kind, Whitman's raw exuberance and love for life found in his "Song of the Open Road," "Song of the Broad Axe," or "Song of the Rolling Earth" had a profound influence on Burroughs and JOHN MUIR and more recent generations of naturalists and nature writers. Many took Whitman's lead, who, upon seeing "the white-topped mountains point up in the distance," flung out his fancies toward them.

During the Civil War years, Whitman worked in Union army hospitals, a friend and comrade of the sick and wounded. He received a clerkship in the Department of the Interior from President Lincoln, but was later removed by Secretary Harland on account of the unorthodox character of poetical writings found in his desk. *Drumtaps* (1865) recounted the war years and included his elegy to Abraham Lincoln, whose assassination affected the poet greatly. Whitman worked at the attorney general's office until 1873, when, owing to a paralytic shock, he retired to his brother's home in Camden, New Jersey. The sudden death of his mother, a few months later, led to a relapse and partial paralysis. It was after this stroke that Whitman sought peace in the wild. Weak from illness and still depressed from the experience of the Civil War, Whitman retreated to the woods and creeks where he found solitude from "the whole cast-iron civilized life." His later writings, *Democratic Vistas* (1871) and *Specimen Days* (1882), were more reflective. Whitman's vision for the United States was being corrupted by ambition and urban industrialism. The pastoral idyll—a reflection of the Jeffersonian farmer and the source, Whitman believed, of American identity—was rapidly disappearing. Whitman also intimated in his poetry that the American was becoming overcivilized. "Without enough wilderness America will change," he asserted. "Democracy, with its myriad personalities and increasing sophistication, must be fibred and vitalized by regular contact with outdoor growths—animals, trees, sun warmth and free skies—or it will dwindle and pale."

The ecstatic spirit of *Leaves of Grass* that earned Whitman his national and international renown eventually evolved to a somewhat melancholic but empathetic view. Whitman died on March 26, 1892, in Camden, New Jersey, where he is buried. Although Walt Whitman was never awarded literary prizes and accolades during his lifetime, there are schools, bridges, and parks all over the United States bearing his name.

BIBLIOGRAPHY

Allen, Gay Wilson, *The Solitary Singer: A Critical Biography of Walt Whitman*, 1967; Canby, Henry S., *Walt Whitman, An American: A Study in Biography*, 1943; Kaplan, Justin, *Walt Whitman: A Life*, 1982; Whitman, Walt, *The Collected Writings of Walt Whitman*, ed. Gay Wilson Allen and Sculley Bradley, 1963; Woodress, James, *Critical Essays on Walt Whitman*, 1983.

Willcox, Louisa

(May 9, 1955–)
Program Director of Greater Yellowstone Coalition, Project Coordinator of Sierra Club Grizzly Bear Ecosystems Project

Louisa Willcox's assertiveness and resolve have helped to make her one of the most highly effective and recognized environmental leaders in the Rocky Mountain West. She has held leadership positions in several conservation organizations in the region, beginning with her ten-year directorship of the Greater Yellowstone Coalition (GYC)—during which time she fought to protect the unique ecosystem of the area surrounding Yellowstone National Park. She also became closely involved in the restoration of the grizzly bear, a threatened species that is integral to a healthy balance in Rocky Mountain ecological systems. To this end, she served as director of Wild Forever, a grizzly bear recovery project based in Bozeman, Montana, and now holds the position of project coordinator for the Bozeman-based Sierra Club Grizzly Bear Ecosystems Project.

Louisa Willcox was born on May 9, 1955, and raised in a Quaker family in Newton Square, Pennsylvania. Her father, John, was a mechanical engineer who died when she was 12, leaving her mother, Joyce, to raise Louisa and her two siblings alone. During her teenage years, Willcox spent summers working on ranches in Wyoming and Montana and fell in love with the West. After graduating from high school in Pennsylvania, she returned to Wyoming and became an expedition leader for the National Outdoor Leadership School (NOLS), teaching mountaineering, kayaking, skiing, and desert hiking. For the next nine years, in between attending Williams College in Massachusetts, Willcox continued conducting NOLS expeditions. In 1980 she finished her bachelor's degree at Williams, majoring in English with a concentration in environmental studies and graduating *magna cum laude*. She then attended the School of Forestry at Yale University, earning a master's degree in forestry in 1984, with an emphasis on natural resource policy and ecosystem dynamics. From 1984 to 1986, she was field studies director of the Teton Science School in Jackson, Wyoming, organizing, planning, and instructing all residential field ecology programs.

In 1986 Willcox was named program director for the Greater Yellowstone Coalition, an alliance of environmental groups founded in 1983 to conserve and protect the greater Yellowstone ecosystem, an area of about 18 million square miles centered around Yellowstone and Grand Teton National Parks. Willcox, with her knowledge of the region's ecology and her political savvy, was uniquely suited to the position. Her aggressive style also served her well as she coordinated the advocacy efforts of the group's 140 member organizations, lobbied agencies and public officials involved in Yellowstone conservation issues, and did fundraising and public speaking.

As part of her work for the GYC, Willcox fought a 1990 request by Crown Butte Mines Inc., the U.S. subsidiary of a giant Canadian conglomerate, Noranda Inc., to mine for gold on Henderson Mountain in Montana. The proposed site is near Yellowstone National Park, and Willcox believed pollution from the mine would harm Yellowstone's intricate ecosystem. Though Crown Butte officials insisted their mine would be environmentally safe, their plan included dumping five million cubic tons of toxic residue into a mile-wide pit located only two miles from Yellowstone's northern boundary. Willcox went into high gear to fight the mine, speaking at informal gatherings of Wyoming and Montana residents and making sure government officials reviewed the GYC's scientific studies of the region's aquatic ecology, including the potential for mine runoff to leach into groundwater. Well aware that politics would also influence the decision, Willcox rounded up 12 Montana residents who opposed the mine and flew with them to lobby Montana congressman Pat Williams in Washington, D.C. The issue had not been resolved by the time Willcox left GYC, but eventually an agreement was reached among the Clinton administration, Crown Butte Mines, Inc., and GYC for a buyout of the proposed mining project.

Willcox and the GYC also battled the Church Universal and Triumphant (CUT), a controversial New Age–oriented religious group headquartered near the north entrance of Yellowstone Park. The church planned a series of developments up and down the 50-mile-long Paradise Valley in Montana, close to the Wyoming border, and wanted to tap a hot spring for a swimming pool and to heat an office building. Fearing that the drilling could potentially damage the area's elaborate underground geothermal structure, the GYC loudly opposed the development, and thanks in part to its efforts, the U.S. Senate placed a moratorium on all drilling in geothermal areas until proper studies could be carried out. In 1994 the moratorium was lifted after the U.S. Geological Survey concluded that the plans would not affect geothermal features in the park; GYC along with three other local environmental groups then filed suit to try to block development on the 33,000 acres of church land, citing environmental concerns. Several years later, after running into financial troubles, CUT agreed to relinquish its geothermal water rights and to sell, trade, or place conservation easements on 7,850 acres of land on the north edge of Yellowstone, opening up public access and protecting valuable wildlife habitat.

Willcox left GYC in 1995 to become director of Wild Forever, a newly formed collaborative grizzly bear project based in Bozeman, Montana, that includes the Sierra Club, the National Audubon Society, Earthjustice Legal Defense Fund, and the Greater Yellowstone Coalition. Listed as threatened under the Endangered Species Act, grizzlies have been eliminated from 99 percent of their original range over the past 120 years. They once numbered as many as 100,000 and ranged throughout the mountains and plains of the West, but they have been reduced to about 1,000 bears, most of them in Yellowstone and Glacier National Parks and Montana's Bob Marshall Wilderness. Wild Forever develops and coordinates grassroots efforts to establish a strong grizzly bear recovery program that guarantees adequate habitat protection—which entails setting aside large roadless areas and recreating corridors to link isolated bear populations. In addition, Willcox and Wild Forever stood up to the U.S. Fish and Wildlife Service to challenge its 1993 Grizzly Bear Recovery Plan, a proposal to delist the Yellowstone population of grizzly bears from its threatened species status and remove existing protections.

In 1998 the Wild Forever effort folded into the Bozeman-based Sierra Club Grizzly Bear Ecosystems Project, a coalition of about 60 regional, local, and national conservation groups working to restore and recover grizzly bears in the lower 48 states. Willcox became the project coordinator for the new group, helping coordinate the advocacy and communication efforts of the groups and working closely with members of the press. From her new post she continues to dispute the Fish and Wildlife Service's recovery plan, voicing her objections to the fact that the draft does not address the biggest threat to grizzly bear habitat: encroaching human development. In March 2000, the Fish and Wildlife Service released its final environmental impact statement outlining a plan to reintroduce 25 bears over the course of five years into the Selway-Bitterroot and Frank Church wilderness areas of Idaho and Montana. Part of the plan involves an innovative proposal to establish a citizens' management committee, comprising Idaho and Montana citizens, representatives of state and federal agencies, and a member of the Nez Perce tribe, that would make crucial decisions about bear management issues such as road construction in bear habitat. Willcox opposes the plan. She is concerned about a political committee with little or no background in grizzly conservation making management decisions and would like to see the reintroduction occur under full endangered species status—as opposed to the "non-essential experimental" designation assigned by the plan, which allows for the destruction of bears that frequent areas of high human use or act aggressively toward humans.

In addition to her work with the Sierra Club, Willcox serves on the board of directors of the Wildlands Project and the Rocky Mountain Ecosystems Coalition. She also serves on the board of advisers of Wildlands CPR, a conservation organization working to raise awareness of the impacts of roads on ecosystem integrity, and on the board of advisers of *Wildlife Tracks*, a publication of the Humane Society of the United States.

In the summer of 1992 Willcox married Douglas Honnold, a lawyer with the Sierra Club Legal Defense Fund, who helps with her crusade to preserve wilderness. They live in Livingston, Montana.

BIBLIOGRAPHY

Reed, Susan, "Dig They Mustn't: A Yellowstone Guardian Opposes a Gold Mine," *People Weekly*, 1993; Willcox, Louisa, "The Last Grizzlies of the American West: The Long Hard Road to Recovery," *Endangered Species Update*, 1997; Willcox, Louisa, "The Yellowstone Experience: The Use of Science, with Humility, in Public Policy," *BioScience*, 1995.

Wille, Chris

(February 23, 1947–)
Conservation Project Leader and Public Information Specialist

Working with Central and South American conservation organizations, conservation broker Chris Wille has convinced the tropical fruit–growing industry that it pays to use environmentally friendly cultivation methods that work with nature rather than against it and to treat workers well so that they may lead healthy and dignified lives. Wille and his partners in the Rainforest Alliance have designed guidelines for responsible agroproduction, and award growers who follow these guidelines with the ECO-O.K. Wille's work is highly effective and unique because he works with an industry commonly cited for environmental damage with the specific goal of helping growers become more humanitarian and less damaging to the environment.

Chris Wille was born in Porterville, California, on February 23, 1947. He credits his dedication to the environment to a very intense early interest in butterflies. He received a B.S. in wildlife science from Oregon State University and an M.S. in English from Oklahoma State University. Wille was primed for his work in ecocertification by nearly 20 years of service to various conservation entities, including the National Audubon Society, the National Wildlife Federation, the Guam Department of Fisheries and Wildlife, and the Oklahoma Department of Wildlife Conservation. In each of these jobs, Wille specialized in disseminating information on conservation: editing publications, producing television and radio programs, and developing environmental education curricula.

Wille and his wife, DIANE JUKOFSKY, were two of the founding members of the Rainforest Alliance, an international conservation organization whose mission is to "develop and promote economically viable and socially desirable alternatives to the destruction of this endangered, biologically diverse natural resource." The Rainforest Alliance grew out of the world's first major rain forest conference ("Tropical Forests, Interdependence, and Responsibility," held in 1988 in New York City). At a time when most rain forest activists were boycotting rain forest wood products, the Rainforest Alliance united environmentalists, scientists, government officials, and industry representatives to design practical yet environmentally stringent guidelines for loggers. The result was a program called Smartwood that stimulated genuine change in the logging industry by certifying operators that meet certain standards.

Observing the success of Smartwood and the destruction wrought by banana companies in Costa Rica, Wille and Jukofsky organized multidisciplinary teams to study the problems associated with fruit farming and to recommend solutions. After nearly two years of work, the conservationists, scientists, community leaders, and banana industry representatives had found a middle ground, and standards were written for socially and environmentally responsible banana production. Specially trained teams of agronomists and biologists began inspecting farms, using the standards as guidelines. The banana industry was reluctant at first, and only small, independent farmers opened their farms to the inspectors. But eventually, Chiquita Brands—the industry leader—joined the Better Banana Program and enrolled farms throughout the region. Now, the banana industry is changing rapidly, controlling and reducing use of agrochemicals, improving conditions for workers, reforesting, protecting rivers, managing wastes, recycling, and seeking better relations with local communities. The program prohibits further deforestation for banana farms.

Meanwhile, Wille was brainstorming with a group of biologists in Guatemala called the Interamerican Foundation for Tropical Research (Fundación Interamericana de Investigación Tropical or FIIT). These naturalists had been studying wildlife populations in coffee farms, one of the few crops that can be grown in harmony with the rain forest. Unfortunately, many coffee farmers throughout the Americas are bulldozing their beautiful old forested farms and replacing them with new, high-tech, "factory farms" that leave no place for wildlife. Wille brought FIIT and the Rainforest Alliance together to create another certification program, ECO-O.K., which certifies ecofriendly coffee farms that meet strict standards for wildlife habitat and worker welfare.

Wille guided another group of biologists, the Conservation and Development Corporation (Corporación de Conservación y Desarrollo or CCD) in Ecuador, in using the ECO-O.K. model to create standards for forest-friendly cocoa production. In Brazil, another affiliate group, Instituto de Manejo e Certificacão Florestal e Agrícola (IMAFLORA), directed a two-year, consensus-building effort among sugarcane farmers, workers, and environmentalists to negotiate standards for low-impact sugar production.

In 1998, these groups banded together as the Conservation Agriculture Network. Although the specific requirements for certification vary from crop to crop and from country to country, they are always based on nine common principles and the three pillars of sustainable agriculture: community well-being, environmental protection, and economic vitality. The standards guide farmers in using nature as an ally on the farm and serving as stewards of soils, waters, wildlife, and other natural assets. On certified farms, the people involved in cultivation and harvest enjoy a safe, dignified, and economically just life, and farmers work in concert with local conservationists to promote a healthy habitat for native wildlife on the farm and in neighboring parks and refuges.

Farmers are interested both for reasons of conscience and profit; they know that many consumers prefer certified products and will pay more for them. Since the program originated in 1990, it has certified more than 100 banana plantations and five coffee farms in Ecuador and Colombia and throughout Central America. With support from the World Bank, El Salvador's leading conservation group, SalvaNATURA, is using the ECO-O.K. standards to evaluate farms in that country and plans to certify 200 farms in two years. The Rainforest Alliance certification programs in 1995 received the coveted Peter F. Drucker Award for Nonprofit Innovation.

BIBLIOGRAPHY

"Rainforest Alliance," http://www.rainforest-alliance.org; Rainforest Alliance, *Rainforest Alliance, Celebrating Ten Years*, 1997; "Rainforest Alliance Staff Biographies," http://www.rainforest-alliance.org/about/biographies.html.

Williams, Terry Tempest

(September 8, 1955–)
Writer, Naturalist

Terry Tempest Williams is a celebrated and accomplished writer of natural history, whose books expand on her observations as a naturalist and remind readers to pay attention to the natural world. A native of Utah, she brings to her writing a profound appreciation for the land in which she grew up—and through her example she presents a means for people to locate their own connection to nature. She also conveys a unique perspective on one of the tragic consequences of disregarding the environment—while growing up in Utah, she and her family were exposed to radioactive fallout from nuclear testing, and nine female members of her family eventually contracted cancer. After her mother died from cancer, Williams grew even more active in the environmental movement in her state and has fought for protection of Utah's wild areas.

Terry Tempest was born on September 8, 1955, in Salt Lake City, Utah, and grew up within sight of the Great Salt Lake. Her parents, Diane Dixon and John Henry Tempest III, raised Terry and her three younger brothers in the Mormon faith, which Terry later described as bestowing her with a strong spiritual appreciation of the land. At the age of five, her grandmother gave her a field guide to western birds, and one of her favorite activities was accompanying her grandmother on bird-watching expeditions to Bear River National Wildlife Refuge. When she was ten her grandmother decided she was old enough to join the Audubon Society on a birding outing to the wetlands surrounding Great Salt Lake. In 1971, when Tempest was 15, the family had a scare when her mother was diagnosed with breast cancer. After a mastectomy she completely recovered (though 12 years later the cancer would reappear in her ovaries).

After high school Tempest attended the University of Utah, where she studied English. She worked part-time in a bookstore in Salt Lake City and was pleasantly surprised one day when a customer approached the register, his arms filled with her favorite nature books. The customer, Brooke Williams, asked her for a date, and six months later, in June 1975, they were married. In 1978, she received her bachelor's degree in English.

Williams's first book, coauthored with Ted Major, was a work for children entitled *The Secret Language of Snow* (1984). It examines over a dozen different types of snow and explains how snow interacts with the environment. The book won a Children's Science Book Award from the New York Academy of Sciences. Also in 1984 she received a master's degree in environmental education from the University of Utah. After graduating, Williams taught Navajo children at the Navajo Reservation in Montezuma Creek, Utah. She found that the children taught her more about environ-

mental education than she had known beforehand, and the experience led to her next book, *Pieces of White Shell: A Journey to Navajoland* (1984). Inspired by the children's' stories and by Navajo legends and rituals, Williams used her gift as a storyteller to describe aspects of Navajo culture that exemplify a sense of respect for the earth. The next year she published *Between Cattails* (1985), a free-verse children's book about the lives of herons, muskrats, grebes, ducks, and other marsh life.

In 1991, with the publication of *Refuge: An Unnatural History of Family and Place*, Williams began to gain significant recognition. In the book, she tells the story of her mother's poignant second battle with cancer, this time ovarian, which claimed her life in 1987. Interwoven with this family history, Williams describes how rising water levels in the Great Salt Lake during this period threatened the fragile wetlands surrounding the lake, important habitat for birds and other marsh species. After her mother's death Williams learned that her family had been exposed to radioactive fallout when the United States performed above-ground nuclear testing at Yucca Flats, Nevada. These tests continued from 1951 to 1962, exposing many people in the region to radiation. She attributes the cancer of her mother, six aunts, and her grandmother to their living downwind of nuclear test sites and describes her anger and frustration at the governmental deceit surrounding the radioactive contamination with which her family and many others in the region grew up. By paralleling the story of her mother's death with descriptions of changes in the Great Salt Lake wetland ecosystems, Williams captures the significance of change in the environment and shows that people can change as a result of their environment too. She also reveals how she drew comfort and even a greater understanding of death from the landscape during the difficult time of her mother's illness, illustrating the importance of the natural world in people's lives.

Three years after *Refuge* came out, Williams published a collection of essays called *An Unspoken Hunger: Stories from the Field* (1994). Though the essays are connected by their central theme of women as intermediaries between human actions and the earth, they cover a wide range of subject matter—from nuclear protesting to life on the African Serengeti. She also focuses on the importance of family and reminds readers of the fragility of life on the planet.

Having become increasingly active in environmental causes since the death of her mother, in the summer of 1995 Williams helped lead a fight against the Utah Public Lands Management Act, a congressional proposal to open 22 million acres of southern Utah's buttes and canyons to commercial development— including mining, logging, oil-drilling, and dam building. She spoke out in opposition to the act before a Senate subcommittee hearing. In addition, she and fellow Utah writer Stephen Trimble persuaded 18 prominent western writers to contribute to a book of essays in defense of Utah's wilderness. These essays were combined into a limited-edition book called *Testimony: Writers in Defense of the Wilderness*, which was handed out to every senator and representative. It became a landmark document in the environmental movement and may have played a role in the eventual defeat of the Utah Public Lands Management Act.

Williams, who has had two biopsies for breast cancer and surgery for a tumor between her ribs diagnosed as borderline malignant, continues to write about the great beauty of the western landscape— and also its misuse. Her connection with her homeland is so great that it pains her to go on trips to make public appearances. She has been the naturalist at the Utah Museum of Natural History and is currently a visiting professor of English at the University of Utah. She and her husband, an environmental consultant, live in Emigration Canyon, near Salt Lake City, Utah.

BIBLIOGRAPHY

"Center for Environmental Literacy," http://www.mtholyoke.edu/proj/cel/williams.html; Lassila, Kathrin Day, *"Testimony: Writers of the West Speak on Behalf of Utah Wilderness," The Amicus Journal*, 1997; Reed, Susan, "Friend of the Earth," *People Weekly*, 1996; Williams, Terry Tempest, *Pieces of White Shell: A Journey to Navajoland*, 1983; Williams, Terry Tempest, *Refuge: An Unnatural History of Family and Place*, 1991; Williams, Terry Tempest, *An Unspoken Hunger*, 1994.

Wilson, Edward O.

(June 10, 1929–)
Entomologist

Harvard University entomologist Edward O. Wilson can be credited with inserting "biodiversity" into the public vernacular and has been instrumental in raising the alarm about the accelerated loss of planetary biodiversity during the 1980s and 1990s. During his long and productive career, he has advanced the study of his specialty, social insects (especially ants); pioneered new fields of biology, such as population biology, island biogeography, and sociobiology; and written prolifically for both specialists and the general public. His eloquence has made him the most popular American spokesperson for the preservation of biodiversity.

Edward Osborne Wilson was born in Birmingham, Alabama, on June 10, 1929, the only child of Inez and Edward O. Wilson. Growing up as a lonely, solitary boy in the Deep South of the 1930s, Wilson immersed himself in the fantastic ecology of the Gulf Coast. From the age of seven, the year his parents divorced, he was allowed to wander alone through the beaches, marshes, and mangroves of the area, observing wildlife and collecting specimens. According to Wilson's autobiography *The Naturalist* (1994), this freedom came with certain risks. He suffered four serious accidents: one while fishing that blinded his right eye, another in which he slashed himself to the bone with a machete, and two more involving venomous snakes.

Wilson's childhood interest in Gulf Coast wildlife became focused during his adolescence. His choice to specialize in ants—slow, silent insects—was a result of two disabilities: a congenital hearing defect and his partial blindness. As a Boy Scout and a student in a military high school, Wilson adopted a work ethic that has allowed him tremendous productivity in his professional life. While still in high school he performed a survey of the ants of Alabama, which he continued at the University of Alabama. His professors at Alabama treated him as a colleague and gave him space in their laboratories. Wilson graduated from Alabama in 1949 with a B.S. and earned an M.S. there in 1950. He gained admission to Harvard University in 1951, earned his Ph.D. in 1955, and immediately joined the faculty as assistant professor of biology. He has moved through the ranks and since 1976 has been the Frank B. Baird Jr. Professor of Science, as well as the entomology curator at Harvard's Museum of Comparative Zoology.

Wilson performed the first inventories of ants through most of the South Pacific, and during these explorations of tropical islands, he began to suspect that the chaotic display of life in the tropics concealed what was actually an orderly distribution of species. This suspicion led him to found a new field of science with mathematician and ecologist Robert MacArthur in 1962. "Island biogeography" relates the number of species on an island to its area and its distance from the mainland. This theory has been applied more recently to a different type of "island," shrinking islands of natural habitat in seas of deforested pastures or suburban residential expansion. Wilson's and MacArthur's work was expanded by

Edward O. Wilson (Courtesy Harvard University)

tropical ecologist THOMAS LOVEJOY in his forest fragment project in Brazil, which studies species composition in forest patches of different sizes and distances from one another within heavily deforested Amazonian cattle ranches. The results from Wilson's and MacArthur's work and from Lovejoy's on-going experiment give conservationists more accurate data about minimum sizes for potential wildlife reserves and about how species numbers will decrease if reserves are deforested.

At the same time he was charting ant populations and articulating island biogeography, Wilson made the landmark discovery that ants are able to communicate with pheromones, chemicals they secrete to influence others. Wilson's work with social insects led him to extrapolate and develop a field called sociobiology, which studies the biological basis of animal behavior. Scientists largely praised the book, but Wilson fell under widespread attack for the chapters of *Sociobiology* (1975) that suggested that human behavior was also controlled in part by our struggle for the survival of our genes. Critics felt that such an assertion denied the effect of environment on our behavior and could be used to promote racism. Wilson was hurt by what he felt was a misunderstanding of his proposal and retreated from sociobiology. In the years since he wrote the book, however, sociobiology has become a bona fide field of study.

Biophilia (1984) represented a new turn in Wilson's career. It was a more personal book than his previous works and explored why humans are drawn to other forms of life. "We learn to distinguish life from the inanimate and move toward it like moths to a porchlight," he claimed. The book attracted attention from a variety of fields, and a later volume (*The Biophilia Hypothesis*, 1993), edited by Wilson and social ecologist STEPHEN KELLERT, includes essays from 15 writers of different professions about how biophilia fits into their fields.

From the beginning of his career, Wilson was aware of habitat destruction and the danger it posed to the forms of life he adored and depended upon professionally. His shy personality and the embarrassment he suffered after *Sociobiology* prevented him from becoming a public advocate of environmental conservation until 1979. In that year, British ecologist Norman Myers published data that suggested that deforestation was claiming global forests at the rate of approximately 1 percent per year and that they were taking with them one-quarter of 1 percent of the world's species each year. Myers predicted that this rate would grow, owing to population increases and deep poverty in most tropical countries. Wilson's conscience pushed him to action, and he joined an informal alliance of biologists (including Myers, Jared Diamond, PAUL EHRLICH, THOMAS EISNER, DANIEL JANZEN, and Thomas Lovejoy) committed to conservation. Soon, Wilson became the science adviser to the World Wildlife Fund–U.S., joined its board of directors, and began to publish papers and give talks on the biological diversity crisis.

In 1986, the National Academy of Sciences (NAS) and the Smithsonian sponsored an international conference on biodiversity, and Wilson edited the proceedings, a collection of 56 essays entitled *Biodiversity* that became one of the best-selling books ever published by the National Academy Press. (With his characteristic generosity and respect for his colleagues, Wilson reminds his readers in *The Naturalist* that he should not be credited for inventing the term *biodiversity*, because it was proposed first by NAS administrator Walter Rosen.)

Wilson continues to advocate eloquently for a deeper appreciation of biodiversity and for a stronger international commitment to conserve it. He has been honored with two Pulitzer Prizes (for *On Human Nature*, 1979, and *The Ants*, 1990, which he cowrote with Bert Hölldobler and which was the first strictly scientific book ever to win the prize) and numerous awards, including the National Medal of Science (1977), the Royal Swedish Academy of Science's Craaford Prize (1990, shared with Paul Ehrlich), the WWF Gold Medal in 1990, and in 1993 the Japanese International Prize for Biology.

Wilson has been married to Renee Kelley since 1955, and together they have one daughter, Catherine, born in 1963. He resides near Boston, Massachusetts.

BIBLIOGRAPHY

Diamond, Jared, "Portrait of the Biologist as a Young Man," *New York Review of Books*, 1995; McKibben, Bill, "More Than a Naturalist," *Audubon*, 1996; Wilson, E. O., *The Diversity of Life*, 1992; Wilson, E. O., *In Search of Nature*, 1996.

Winter, Paul

(August 31, 1939–)
Musician, Composer

Saxophonist Paul Winter, who pioneered "earth music," the genre that incorporates the sounds of wildlife such as humpback whales, wolves, and bird song into musical meditations, became famous by fusing his two passions: music and preserving wildlife. Over the course of his career, which has so far yielded 32 albums, he has visited 35 countries, both to perform and to record wildlife sounds for his music. He continually works to promote music as an expression that can reconnect people with nature. He has donated the royalties from some of his albums to conservation organizations that protect wilderness and wildlife species and has given benefit concerts for a variety of environmental groups. Independent and innovative, Winter established his own record label, Living Music Records, which produces his work and reflects his respect for the musical traditions and natural environments of the earth.

Paul Theodore Winter was born on August 31, 1939, in Altoona, Pennsylvania, the son of Paul Theodore and Beulah (Harnish) Winter. He grew up in a family bound together by music. His grandfather, at 17, had been the youngest bandleader in the Civil War and had later owned the local music store. His father helped run the store and worked as a piano tuner. Winter dove into music lessons young, playing drums at age five, piano at six, and clarinet at eight, but he chafed under the academic rigidity of the classical music he was forced to practice. Even so, music had gotten under his skin, and by the time he was 12 he had picked up on jazz and latched on to the saxophone. He played in the local symphony and at 17 toured with the Ringling Brothers Circus Band.

He attended Northwestern University, where he again shied away from structured musical training and majored in English composition, obtaining his B.A. in 1961. That year he assembled some college student musicians and formed the Paul Winter Sextet, which went on to win first prize in an intercollegiate jazz competition and was signed by Columbia Records. The following year the group recorded its first album, *The Paul Winter Sextet*, and then toured 23 countries in Latin America, a liberating experience that added new influences to Winter's musical style. At about that time, Winter reached a turning point: He had been accepted for admission to the University of Virginia Law School but decided against a career as an attorney in favor of a career in music. Between 1962 and 1965, he and his group toured Latin America twice more, performed at jazz festivals and on television shows, and cut seven albums for Columbia Records. Sales of the group's recordings in the mid-1960s did not meet the expectations of executives at Columbia Records, however, and the Paul Winter Sextet lost its contract and disbanded.

In 1967 Winter formed the Paul Winter Consort, an ensemble that fused together a broad range of musical elements such as rock, classical, jazz, folk, and Brazilian sounds. He lived and worked out of a stone cottage in rural Connecticut, recording three albums with the Paul Winter Consort for A&M Records in the next few years. One of these, *Road* (1971), was so well loved by astronauts on Apollo 15 that they left a cassette of it on the moon and named two lunar craters after two of its songs. Meanwhile, Winter had begun pursuing another avenue, one that he later described as having opened the doors of nature to him and that eventually led to the creation of his distinctive "earth music." It had begun in 1968 when he first heard a recording of the sounds of humpback whales and had been amazed by their musical intelligence and shocked to learn that they were being hunted nearly to extinction. Over the next several years, Winter sailed the shores of such places as

El Salvador, Scotland's Inner Hebrides, Newfoundland, and British Columbia, seeking whales and other sea mammals and playing his saxophone to them. To his delight, off Baja California, he found he could call whales to his raft with his music.

Appreciating and understanding wildlife became a new passion for Winter, who also developed a lifelong admiration for timber wolves at around this time. In 1973 he visited Minnesota to see the last wild population of timber wolves in the lower 48 states. Two years later at a wildlife research center in the Sierra Nevada of California, he played his saxophone to captive wolves and heard their howling response as a kind of celebration. Expanding on these ideas, Winter began creating musical compositions that incorporated the actual voices of animals and sounds of nature, pioneering a genre of New Age music that would become abundantly popular. In 1977 the first of these albums, *Common Ground*, appeared, including whale songs, wolf howls, and the cry of an African fish eagle on various music tracks. Winter shared royalties from the album with environmental groups that supported whales, wolves, and eagles. His next release, *Callings* (1980), included the sounds of 13 different sea mammals.

Winter grew tired of working with big-label record companies, feeling that they did not know how to define his music and tended to just ignore it, so in 1980 he created his own label, Living Music Records. Also that year, Winter was named artist in residence at the Cathedral of St. John the Divine in New York City. While there he wrote an ecumenical mass for life on earth, *Missa Gaia/Earth Mass* (1982), which the Paul Winter Consort recorded partly at the cathedral and partly at the Grand Canyon. Beginning in the summer of 1984, Winter embarked on a series of visits to the wilderness areas of the Soviet Union, especially attracted to Lake Baikal, the world's deepest and largest freshwater lake—home to 1,200 species of wildlife found nowhere else on earth. Over the next ten years he collaborated with various Russian artists to produce a series of albums, hoping to spread through the universal language of music the message that love and respect for the earth can be a common ground for peace. In 1985 an album celebrating the Grand Canyon was released. Titled *Canyon*, it was recorded by the Paul Winter Consort during four rafting expeditions down the 279-mile length of the Colorado River, using the side canyons and natural amphitheaters as a "studio of the earth." *Canyon* was nominated for a Grammy in 1987 and made it to fourth place on *Billboard*'s jazz chart. This was followed by *Whales Alive* (1987), which used melodies "composed" by the whales themselves. Winter donated royalties from *Whales Alive* to the World Wildlife Fund.

By this point, Winter had achieved international recognition and had defined a whole new style of music that many describe as New Age, a label to which he objects. He prefers "earth music" or "living music," to remind himself that the purpose of his music is to honor the earth and its tapestry of life. Over the years, in addition to recording 32 albums (four of which have won Grammy Awards), he has performed benefit concerts for the Sierra Club, the Wilderness Society, Greenpeace, the ceremonial environmental observances of the United Nations, and the 1992 Earth Summit in Rio de Janeiro,

Brazil. At the 1985 United Nations General Assembly on World Environment Day, Winter was given a World Environment Day Award in recognition of his work in promoting environmental causes; he has also received the Humane Society of the United States' Joseph Wood Krutch Medal for service to animals (1982), the Peace Abbey's Courage of Conscience Award (1991), and the Promise to the Earth Award from the National Arbor Day Foundation (1996). He lives with his wife, Chez Liley, whom he married in 1991, and daughter, Keetu, on a 77-acre farm near Litchfield, Connecticut.

BIBLIOGRAPHY

Jerome, Jim, "Paul Winter; From Canyon to Cathedral, His Soaring Sax Calls Out to Seekers of a New Age," *People Weekly*, 1986; Sullivan, Karin Horgan, "Ode to the Wilderness: Paul Winter Honors Animals with His Music," *Vegetarian Times*, 1995; Verna, Paul, "Winter Takes His Studio Outside," *Billboard*, 1997.

Wolf, Hazel

(March 10, 1898–January 19, 2000)
Environmental Activist, Secretary of Seattle Audubon Society

A sharply intelligent and witty charmer, known for her tendency to defy rules, Hazel Wolf was one of the Pacific Northwest's most admired activists for environmentalism and social reform. As secretary of the Seattle chapter of the National Audubon Society, she helped found more than 20 chapters in Washington State and pressured the society and other environmental organizations to broaden their focus to include more ethnic and low-income members. She cofounded the Community Coalition for Environmental Justice, a social and environmental organization committed to improving the quality of life in low-income, inner-city neighborhoods. Although her environmental activism did not start until she was in her sixties, she more than made up for lost time through her fervent involvement: She organized and attended protests and rallies, recruited other supporters, lobbied office-holders, and built coalitions. For all this she was recognized with more than a dozen conservation awards.

Hazel Anna Cummings Anderson was born March 10, 1898, in Victoria, British Columbia, to a U.S. mother and a British seaman father who died when she was ten. She grew up in poverty and learned to be a fighter at an early age. When warned by her mother to be obedient or face the "boogey man," who her mother said was out on the porch waiting for her, Hazel eventually grew exasperated and one night threw open the front door to confront him, only to find that he did not exist. She also balked at having to do the dishes unless her brother had to do them too. As she got older, she dreamed of becoming a doctor, even though it was virtually unheard of for a woman in that era. She married early, but divorced soon after, and moved to the United States in 1923 with her daughter Nydia. (A later

Hazel Wolf (Natalie Fobes/Corbis)

marriage to Herbert Wolf also ended in divorce.) Settling in Seattle, Washington, she took odd jobs and struggled to make ends meet as a single mother. During the Depression she relied on welfare and lived in a Catholic boardinghouse in downtown Seattle, all the while following politics closely. Ever the reformer and activist, Wolf was fired from a job at the Works Progress Administration for trying to organize a union. She joined the Communist Party for a period because they crusaded for a system of unemployment assistance and other social programs.

In 1949 Wolf took a job as a legal secretary for noted Seattle civil-rights attorney John Caughlan, and she helped him promote civil rights and social reform.

Meanwhile, her continuing activism had caught the attention of immigration officers and the Federal Bureau of Investigation, who threatened her with deportation to Canada. In addition, her former ties to the Communist Party caused her trouble during the McCarthy era, and in 1955, 13 years after she had left the party, she was arrested and charged with "attempting to overthrow the government by force of violence." She spent half a day in jail working on a jigsaw puzzle before friends bailed her out. Eventually her employer helped to get the charges dropped, and her case, which had reached the Supreme Court, was dismissed. Finally, in 1970 she became a U.S. citizen.

At the age of 62, shortly before she retired from her secretarial job, Wolf humored a friend by joining the National Audubon Society and participating in a bird-watching field trip. She became enchanted watching a brown creeper, a small, inconspicuous bird that works its way meticulously up tree trunks in search of insects, and felt protective of it. Soon after, she became secretary of the Audubon Society's Seattle chapter, a position she held for 37 years—during which time she helped organize more chapters than anyone in the national organization's history—23 of the 26 chapters in Washington State and one in her birthplace, Victoria, British Columbia. She recruited members wherever she went, prompting a friend to comment that there probably is not a person who has sat next to her on an airplane who has not joined Audubon.

Wolf, whose personal philosophy was "everything connects," went beyond her work for the Audubon Society to become a skilled coalition builder among other groups. In 1979 she organized a conference of American Indian tribes from the Pacific Northwest together with environmental organizations, forming associations that have lasted many years. The purpose of the meeting was to strategize ways to protect the Columbia River from further dam building and to protest a proposal for an expanded irrigation project. The government eventually dropped the expansion plans, and Wolf later spoke of the conference as the most valuable and interesting thing she ever did. She also joined the Federation of Western Outdoors Clubs, serving as president from 1978 to 1980 and serving as editor of their environmental newsletter, *Outdoors West*, until she died. During the 1980s she made several trips to Nicaragua to study the Sandinistas' environmental record and endeared herself to the Nicaraguan people, who came to consider her their special saint.

With her infectious spark and wit, Wolf became a much-loved public speaker on environmental issues and lectured at many conferences, conventions, and schools. When asked by children if she had a boyfriend, Wolf would reply no, but she was searching for one who could cook. While attending environmental conferences all around the country, she often succeeded in confiscating the microphone during a plenary session to campaign for a particular cause. Possessed of seemingly endless energy, whenever Wolf was not doing Audubon Society work or editing for *Outdoor West*, she was testifying at hearings, lobbying officials, participating in rallies and protests, and urging others to get involved and to register to vote. She also pressured Seattle Audubon and other environmental organizations to reach beyond their largely White, middle- or upper-class membership. In 1993 she cofounded the first environmental justice group in the Seattle area, the Community Coalition for Environmental Justice (CCEJ), a multiethnic nonprofit organization that deals with social, economic, and environmental health issues, such as industrial pollution, that disproportionately affect people of color, women, children, and low-income people.

Wolf received numerous conservation awards over the years, including the Washington State Department of Game's Award for her work in wildlife protection (1978), the State of Washington Environmental Excellence Award (1978), the National Audubon Society's Conservationist of the Year Award (1978), and the Association of Biologists and Ecologists of Nicaragua's Award for nature conservation (1988). On March 10, 1996, on her 98th birthday, the governor of Washington issued a proclamation declaring her birthday Hazel Wolf Day. She received the Audubon Medal for Excellence in Environmental Achievement in 1997, and in June of that year Seattle University granted her, at 99, an honorary doctorate in humanities, upon which she asked to

be called "Doc." For her 100th birthday, King County honored her by renaming the 166-acre Saddle Swamp on the Sammamish River the Hazel Wolf Wetlands. Also in honor of her 100th birthday, the Audubon Society created a Kids for the Environment fund to foster environmental appreciation among young people. By January 2000 Wolf had accomplished the goal of having her life span touch three centuries. She asked friends to help plan her memorial service to make it into a fund raiser for Kids for the Environment, and she requested that if anyone showed up at the service who was not registered to vote, that they be registered on the spot. On January 19, 2000, she died in Port Angeles, Washington. She is survived by her daughter, five grandchildren, five great-grandchildren, and four great-great-grandchildren.

BIBLIOGRAPHY

Breton, Mary Joy, *Women Pioneers for the Environment*, 1998; Broom, Jack, "Honoring Hazel Wolf: Seattle Environmentalist Turns 100," *Seattle Times*, 1998; "Hazel Wolf," http://members.tripod.com/~HazelWolf/; Robin, Joshua, "Environmentalist Hazel Wolf Dies at 101," *Seattle Times*, 2000.

Wolke, Howie

(June 17, 1952–)
Wildlands Conservationist, Cofounder of Earth First!

In 1980, wildlands conservationist Howie Wolke, along with DAVE FOREMAN and three other environmental activists founded the radical environmental activist group, Earth First!, a group that has become well known for its dramatic and extreme methods of defending the ecological integrity of the earth. Wolke left Earth First! in 1990 and in recent years has been working through his nonprofit Big Wild Advocates on such wildland issues as the reintroduction of grizzly bears, the protection of the Greater Salmon–Selway Ecosystem (in northern Idaho and western Montana), and the proposed Northern Rockies Ecosystem Protection Act. Most recently, he has been working to build public support for the Roadless Area Initiative of the U.S. Forest Service (USFS), which has the potential to protect up to 50 million acres of Forest Service land as roadless wilderness.

The older of Arthur and Beverly Wolke's two children, Howie Wolke was born on June 17, 1952, in Brooklyn, New York. When Wolke was six years old, his family moved from New York to Nashville, Tennessee, where he discovered his love of wild things and places. Wolke claims that he was born with the "Neanderthal gene," a recessive trait that crops up in some individuals, making them fanatical lovers of the wild. This gene first manifested itself in Wolke when he was a child in Tennessee. He explored and enjoyed the woods, lakes, streams, and animals and decided that he wanted to be a forest ranger when he grew up. After two more moves with his family (to Pennsylvania and New Jersey), Wolke enrolled in college at the University of New Hampshire Forestry School, where he learned what foresters actually do and decided that he did not want to be one after all. He changed his field of study and graduated in 1974 with a B.S. in environmental conservation and wildlife ecology.

After graduation, Wolke worked until he had saved up enough money to head west. He loaded his car with all of his belongings and drove to Wyoming, where he lived, mostly in Jackson, from 1975 to

1986. When he first arrived, he volunteered with the Sierra Club and the Wilderness Society and soon became the Wyoming representative for Friends of the Earth. He supported his activism by working as a ranch hand and a bouncer at the Cowboy Bar in Jackson. During this time, Wolke was cataloging roadless areas and advocating the protection of large tracts of wilderness through the USFS's second national review of unprotected roadless lands (the Roadless Area Review Evaluation [RARE] II process). Through RARE II, the USFS was seeking public participation in determining how many acres of these roadless areas should be preserved as such. Ultimately, the USFS recommended the protection of only 15 million acres out of the 62 million under consideration (Wolke felt that a total of 80 million roadless acres should have been considered, but the USFS excluded 18 million of them). In his first book, *Wilderness on the Rocks*, Wolke writes, "RARE II was the grandiose defeat for the modern wilderness movement."

This defeat led directly to the formation of Earth First!, formed in 1980 by the disgruntled Wolke along with Dave Foreman, Mike Roselle, Bart Koehler, and Susan Morgan. Earth First! evolved out of the idea that a system of multi-million-acre wildland reserves should be developed in every major ecoregion of the United States. This was a controversial proposition in 1980 but has since become more mainstream, as conservation biologists today assert that this is just the type of program that needs to be implemented if we are to maintain an acceptable level of biological diversity. Earth First! also advocated dam removal, an idea so unthinkable in 1980 that it was literally dismissed outright. Twenty years later, in

2000, the U.S government was holding hearings to consider the removal of dams on the Snake River.

Earth First! was originally intended to act as the strategic arm of the wildland conservation movement. This meant that while Earth First!'s official position on the use of sabotage and other forms of civil disobedience was to "neither condemn nor advocate it," many individual members of the organization did employ such tactics to prevent the loss of wild habitat to industry. A few members of Earth First! engaged in tree spiking (hammering nails into trees to ruin saw blades), desurveying roads, and tree sitting. (In later years, after a California sawmill worker was injured when his saw blade hit a spike in the trunk he was hewing—likely put there in a labor dispute—most activists with Earth First! and other direct action environmental groups renounced tree spiking and other potentially injurious activities.) Creating quite a media stir in 1981, the organization unfurled a 300-foot-long fake polyethylene "crack" down the face of Colorado River's Glen Canyon Dam in protest of the dam as a representation of wilderness destruction in the western United States. Earth First! was attempting to make a fairly simple point: remove the dam and restore the wilderness.

In 1985 Wolke was caught pulling survey stakes out of the ground in the Grayback Ridge Roadless Area in Wyoming. He was attempting to prevent the Forest Service from bulldozing a new logging road and an oil field. He received a sentence of six months in jail, during which time he wrote a draft of *Wilderness on The Rocks* (1991), an extended critique of the U.S. land management bureaucracies.

He dedicated the book to the grizzly and to "all that is wild and pure, cyclical and free, and diverse." Wolke later cowrote *The Big Outside* (1992) with Dave Foreman, a catalog of 385 wilderness areas in the lower 48 states.

In 1990, Wolke quit Earth First! after becoming dissatisfied with the direction in which the organization was moving. "People wanted to talk about tree spiking and bombing, not eco-systems," he is quoted as saying in the March/April 2000 issue of *Sierra* magazine. He felt that he had little in common with the "militant vegan anti-hunting eco-witches for social justice," as he teasingly refers to the Earth First! culture of the early 1990s.

In recent years, Wolke has spent his time working on various wildland issues, such as the protection of the Greater Salmon–Selway Ecosystem and the reintroduction of grizzly bears there and the proposed Northern Rockies Ecosystem Protection Act, always emphasizing roadless area protection. Since October 1999, Wolke has been spending the majority of his time working on building public support for the Roadless Area Initiative, a USFS administrative action that, like RARE II, has the potential to provide a significant degree of protection for 50 million acres of USFS roadless areas. Wolke's recent activism has been done primarily through his non-profit advocacy organization, Big Wild Advocates. Wolke sees this organization as filling a niche in the wildlands conservation community; it addresses issues that other organizations overlook and provides tactical support to other non-profits that share his interest in protecting public wildlands, such as the Alliance for the Wild Rockies and the Friends of the Bitterroot.

In 1986, Wolke moved to Montana with Marilyn Olsen (whom he married in 1989) and her son, Josh. They run Big Wild Advocates as well as a wilderness guiding operation, Big Wild Adventures.

BIBLIOGRAPHY

Foreman, Dave, and Howie Wolke, *The Big Outside*, 1992; Kane, Joe, "Mother Nature's Army," *Esquire*, 1987; Kane, Joe, "One Man's Wilderness," *Sierra*, 2000; Wolke, Howie, *Wilderness on the Rocks*, 1991.

Woodwell, George

(October 23, 1928–)
Ecologist, Founder of the Woods Hole Research Center

George Woodwell, a life scientist with broad interests in global environmental issues, has studied terrestrial and marine ecosystems for many years, concentrating on how pesticides, nutrients, radioactive isotopes, and organic compounds are cycled through the environment. And despite criticism from some colleagues for abandoning scientific discretion, Woodwell has always been very vocal about the impact that humans are having on the global ecosystem. His research and testimony were instrumental in a case that ultimately led to the ban on dichlordiphenyltrichlor (DDT) in the United States in 1972, and he has never hesitated to spread the message about the effects of global warming and the need for rational policy responses. He founded and directs the Woods Hole Research Center and has been active in the founding of several environmental organizations, including Environmental Defense, the National Resources Defense Council, and the World Resources Institute.

George Masters Woodwell was born on October 23, 1928, in Cambridge, Massachusetts, the son of Philip and Virginia (Sellers) Woodwell. He grew up in Cambridge, spending summers in rural Maine, and came to love the changing landscape, the diversity, the water, and the history of the New England coast. He received a bachelor's degree in zoology from Dartmouth College, graduating with distinction in 1950. After three years of service in the United States Navy, he returned to school, this time to Duke University, where he earned a master's degree in 1956, and then a Ph.D. in botany in 1958. He began teaching in 1957 at the University of Maine at Orono, first as assistant professor, later as associate professor of botany. In 1961 he started working at the Brookhaven National Laboratory in Upton, New York, as an assistant scientist, eventually becoming senior scientist and building a program of basic ecological research.

Woodwell's scope of research has always been broad. He studies the planet's biosphere as a whole and addresses questions of how the global environment functions as a single system and what impact human activities have on it. In the 1960s, he became one of the first scientists to study the ecological effects of

George Woodwell (Courtesy of Anne Ronan Picture Library)

chronic exposure to ionizing radiation. He also published pioneering investigations on the circulation and effects of chemical toxins in different ecosystems and closely studied forests and estuaries in North America. In the 1950s, he had studied the negative impact of DDT, which was being sprayed on forests in Maine to battle spruce budworm, and was consulted by RACHEL CARSON during her research for *Silent Spring*. He studied DDT again in the late 1960s, this time in Michigan, where it was contaminating groundwater and harming wildlife. Not content to merely publish his findings and move on, Woodwell became one of the first to take legal action against the producers of DDT, and thanks in part to his urging and in part to the lobbying ef-

forts of the Environmental Defense Fund (which used Woodwell's research to back them up), the Environmental Protection Agency banned the pesticide in 1972.

In 1975 he founded and became the first director of the Ecosystems Center at the Marine Biological Laboratory in Woods Hole, Massachusetts, while also working as assistant director for education there. For ten years, Woodwell continued his studies of the earth's carbon, nitrogen, and sulfur cycles from the Ecosystems Center, before branching out and founding and becoming director of the Woods Hole Research Center in 1985. He founded the center to pursue a broader range of scientific studies and to explore the public policy implications of that research. The Woods Hole Research Center has come to be well known for its ecological research and studies of global climate change.

By approaching his work with a constant awareness of its ramifications, Woodwell set a new standard for science. In the traditional model of good science, scientists conduct high-quality research and then publish it for the benefit of colleagues. But Woodwell insists that another step is now necessary. Since they study the processes that support life and the impact of human activities, ecologists have a responsibility to inform the general public of the implications of their work. Woodwell has publicly articulated his own findings on a wide range of subjects. For example, he frequently speaks out about the controlling influence of global forests on climate and urges an immediate halt to deforestation around the world in order to counteract the impoverishing effect it has on biodiversity and the escalating effect it has on global warming. He advocates an international protocol that would regulate deforestation and implement a program to increase forested areas worldwide in order to store more carbon. In recent years, Woodwell's public testimony and policy recommendations regarding his studies of global climate change have placed him at the forefront of the issue, and he is often consulted by other scientists, the U.S. Congress, and foreign governments. In 1996 he joined a panel of experts in a news briefing that highlighted evidence that human activities are leading to the destabilization of the earth's climate. Concentrations of carbon dioxide in the atmosphere are now 20 percent higher than they have been in the last 260,000 years, explained Woodwell, a situation that is contributing to a rise in global temperatures that may lead to more severe weather conditions, rising sea levels, and outbreaks of disease-bearing mosquitoes and other pests. Throughout his years of research, Woodwell has achieved distinction in elucidating these kinds of biotic interactions associated with the warming of the earth. And he has proven his commitment to increasing the influence of scholarship on public policy, showing that scientists can take part in and even lead public policy debates.

In addition to the research and advocacy work he conducts from Woods Hole Research Center, Woodwell has been active in many other organizations. From 1981 to 1984 he served as chairperson of the World Wildlife Fund, he has served as vice president (1976–1977) and president (1977–1978) of the Ecological Society of America, and he has been on the board of trustees at the Sea Education Association (1980–1985). From 1982 to 1983, he was chairperson of the Conference on Long Term Biological Consequences of Nuclear War. Since 1969 he has been a lecturer at the School of Forestry at Yale University and has published over 300 major papers and books in ecology.

He has also been instrumental in the founding of several environmental organizations, including the Environmental Defense Fund (1967), the National Resources Defense Council (1970), and the World Resources Institute (1982). He is a member of the National Academy of Sciences and a fellow of the American Academy of Arts and Sciences. He has been recognized with the Green World Award from the New York Botanical Garden (1975), the Distinguished Service Award from the American Institute of Biological Sciences (1982), the Silver Bowl Award from the Connecticut River Watershed Council (1984), the Hutchinson Medal from Garden Clubs of America (1993), and the Heinz Environment Award (1996), plus numerous honorary doctorates.

He is married to Katharine, administrator of the Woods Hole Research Center, and has four children. He lives in Woods Hole, Massachusetts.

Bibliography

Woodwell, George M., "Ecological Science and the Human Predicament," *Science*, 1998; Woodwell, George, ed., *Biotic Feedback in the Global Climatic System: Will the Warming Feed the Wakening?* 1993; Woodwell, George, ed., *The Earth in Transition: Patterns and Processes of Biotic Impoverishment*, 1991.

Worster, Donald

(November 14, 1941–)
Environmental Historian

One of the leading scholars and teachers of environmental history, Donald Worster is a catalyst of this rapidly growing discipline that examines the relationships between nature and culture over time. One of Worster's particular emphases has been to give nature agency in environmental history. Too often, nature has played a subsidiary role in environmental history rather than that of a leading character. Such a model is flawed if we are to properly appreciate the human place in the natural environment. Worster's work in the environmental history of the American West has sought to correct that flaw and push for more enlightened perspectives of the natural world.

Donald Eugene Worster was born November 14, 1941, in Needles, California, to Bonnie Pauline (Ball) and Winfred Delbert Worster, a railroad worker. He completed his B.A. at the University of Kansas in 1963 and earned an M.A. the following year, both degrees in English and communications. During the summer of 1965, he studied American history at Harvard University, before going to Yale for American studies. He received an M.Phil. in 1970 and was awarded his doctorate in 1971.

Worster started his teaching career at Brandeis University in 1971. He spent three years there before moving to the University of Hawai'i at Manoa, where he spent nine years in American studies before returning to Brandeis University to become the Meyerhoff Professor of American Environmental Studies in 1984. Worster left Brandeis University in 1989 to become Hall Distinguished Professor of American History at the University of Kansas.

In 1979, Worster produced *Dust Bowl: The Southern Plains in the 1930s*. More than two decades after its publication, *Dust Bowl* is still regarded as the definitive work on the history of the ecological disaster that wrought havoc during the Depression. Worster argues that capitalist economic culture—which led to ecologically unsound farming practices in the interest of short-term financial gain— played as significant a role in the formation of the massive dust storms as did drought and high winds. The following year this book won the Bancroft Prize for the best book in American history.

In 1981, Worster was awarded a Guggenheim fellowship and became the president of the American Society for Environmental History for two years. He spent the summer of 1984 as a Humanities Research Centre Fellow at the Australian National University in Canberra and was elected to the Society of American Historians in 1988. In 1995, he was awarded the Balfour Jeffrey Achievement Award for research in the humanities and social sciences from the University of Kansas and received a Distinguished Achievement Award from the Society for Conservation Biology in 1997.

Worster has also published eight other books and collections of essays, some of which have been translated into five languages. Two of Worster's books, *Rivers of Empire: Water, Aridity, and the Growth of the American West* (1986) and

The Wealth of Nature: Environmental History and the Ecological Imagination (1993), have been nominated for the Pulitzer Prize. His work is constantly seeking to push the boundaries of environmental history. In *The Wealth of Nature*, Worster presents a dozen essays on topics that range from a review of the religious traditions of American environmentalism, to a brief history of the U.S. Soil Conservation Service, to promoting the idea of a Darwinian approach to history with an emphasis on chaos theory and its impact on ecology. Worster's central theme in the book is the contention that crucial aspects of a society's past, present, and future can be best explained by an examination of the manner in which it organizes itself and reorganizes nature in order to extract from the land those resources necessary for survival. Environmental history, he argues, gains its power from a three-level analysis. The first of its three layers is the character of the natural environment in the time and place under study. Such a discussion requires that environmental historians explore other fields of academic study such as ecology, physics, archaeology, genetics, and so on. The next level is the environmental historian's exploration of the role of technology and the modes of production of a society. Worster defends this near-Marxian concept by arguing that recent history is all about capitalism and its expansion to be the globally dominant form of social organization. He describes modes of production as cultural mechanisms that determine how the available technologies are employed to relate to the landscape by a particular people at a particular time and in a particular place. But before going so far as to grant technology an autonomous moral existence, Worster emphasizes the importance of the third level of analysis, which he identifies as the importance of culture and the specific manner in which a whole human society has perceived and valued nature. Combined, Worster argues, these three levels make environmental history a valuable and distinct field that bridges the divide between the social and the natural sciences.

Worster currently teaches at the University of Kansas. He is the major professor for 17 graduate students in history and one in American studies as well as serving on other thesis and dissertation committees. His courses include "Environmental History of North America," "The American West in the 20th Century," and "Agriculture in World History." Outside of his academic work, Worster is a member of the Board of Directors of the Land Institute in Salina, Kansas, and has served on the boards of the Kansas Land Trust and the Thoreau Society.

Worster is married to Beverley Marshall Worster. They have two children.

BIBLIOGRAPHY

Worster, Donald, *Nature's Economy: The Roots of Ecology,* 1977; Worster, Donald, *A River Running West: The Life of John Wesley Powell,* forthcoming; Worster, Donald, "Transformations of the Earth: Toward an Agroecological Perspective on History," *Journal of American History,* 1990; Worster, Donald, *Under Western Skies: Nature and History in the American West,* 1992.

Yard, Robert Sterling

(February 1, 1861–May 17, 1945)
Editor, Founder of National Parks Association, Cofounder and Secretary of the Wilderness Society

Editor and publicist Robert Sterling Yard produced much of the early promotional material about the U.S. National Park Service, lifting this country's national parks from relative obscurity to worldwide fame. He worked as chief educational secretary for the National Park Service under director STEPHEN MATHER from 1915 to 1918 and then founded the National Parks Association (NPA), a nonprofit organization that raises private funds for the national parks. Yard was one of the seven original founders of the Wilderness Society in 1935 and served as its president and permanent secretary until his death in 1945.

Born on February 1, 1861, in Haverstraw, New York, Robert Sterling Yard graduated in 1883 from Princeton University. For the first 30 years of his professional life he lived and worked in the publishing world, mostly in New York City, in such capacities as head of foreign cables and correspondence at the W. R. Grace and Co. shipping firm, reporter at the New York *Sun*, Sunday editor at the New York *Herald*, book advertising manager at Charles Scribner's Sons, and editor in chief at the influential *Century Magazine*.

It was not until Yard was 53 years old that he entered the field for which he is principally remembered, that of wilderness preservation. He was invited to Washington, D.C., in 1915 to assist his former colleague and best friend from the *Sun*, Stephen Mather, who had just become assistant secretary of the interior in charge of national parks. Yard's job paid only $30 per month, but millionaire Mather supplemented his salary so that he was earning about $5,000 per year. Throughout the rest of Mather's life, while Yard was working for the government and later for the nonprofit National Parks Association, Mather continued to bankroll Yard's work.

The more than 1,000 articles that Yard wrote, edited, or inspired with his press releases during his four years with the national parks helped convince Congress to pass the National Park Service Act, the law that created the National Park Service to manage all U.S. national parks. Yard's articles also raised public awareness of the parks and critical public support for them. His promotional articles could be tailored to any segment of society. He convinced the automotive industry that supporting national parks would help their business. (It did. Automobiles were just becoming more affordable to the middle class, and families were using them to drive cross-country to visit such wonders as Mesa Verde and Glacier National Parks.) Railroad companies, which built special spurs to Yellowstone, Zion, Bryce, Grand Canyon, and Mount Rainier National Parks, among others, financed widely distributed picture books about the parks. Yard distributed almost 350,000 feet of movies free to schools.

Despite his productivity and success, Mather had to dismiss Yard in 1918, because a new law prohibited private payment to government employees. With

Mather's financial backing, Yard went on to found the National Parks Association, whose mission, according to its charter, was "to defend the National Parks and National Monuments fearlessly against the assaults of private interests and aggressive commercialism." The NPA applied vital pressure at key moments in history. In 1919, shortly after its founding, Yard and the NPA lobbied Congress to kill the Falls-Bechler bill, which would have allowed dams to be built within national parks. Yard continued as executive secretary of the NPA until 1933, when he stepped down because the organization lacked the funds to pay him. At that time he was named editor of publications, a post he kept until he was succeeded by Devereux Butcher in 1942.

In 1935, Yard's friend Robert Marshall, the director of forestry at the Bureau of Indian Affairs, invited him to cofound the Wilderness Society (TWS) along with himself and six others (Benton MacKaye, Harvey Broome, Harold Anderson, Bernard Frank, Aldo Leopold, and Ernest Oberholtzer). Marshall was a staunch defender of wilderness, opposing the same private assaults on public wild lands as Yard and the NPA. Marshall paid Yard's salary, personally contributed at least 80 percent of TWS's operating budget, and controlled TWS membership so that its work could be focused on wildlands preservation, unhampered by internal debate. For 30 years, TWS pushed for legislation that would declare certain wilderness areas inviolable by roads, vehicles, buildings, or any other artificial development. Its work finally bore fruit with the 1964 passage of the Wilderness Act, which signed nine million acres into untouchable wilderness and gave Congress the possibility of declaring other areas untouchable too. In the years since, 64 million more acres of wilderness have been set aside through the Wilderness Act.

TWS founder Marshall died suddenly in 1939, leaving Yard to run the TWS office from his Washington, D.C., apartment. Yard served as TWS's executive secretary, its president, and editor of its quarterly magazine *Living Wilderness* until his death on May 17, 1945, at the age of 84. Yard was survived by his wife, Mary, and one daughter, Margaret.

BIBLIOGRAPHY

Fox, Stephen, *John Muir and His Legacy: The American Conservation Movement*, 1981; Glover, James M., *A Wilderness Original: The Life of Bob Marshall*, 1986; Shankland, Robert, *Steve Mather of the National Parks*, 1970; "The Wilderness Society," http://www.wilderness.org/; Yard, Robert Sterling, *The Book of the National Parks*, 1919.

Zahniser, Howard

(February 25, 1906–May 5, 1964)
Executive Secretary and Executive Director of The Wilderness Society, Editor

As executive secretary and executive director of the Wilderness Society (TWS) from September 1945 to May 1964, Howard Zahniser led two major conservationist battles: the first to prevent a dam from being built in Dinosaur National Monument and the second to convince Congress to pass a wilderness bill that established the National Wilderness Preservation System. Both accomplishments required cooperation among many conservation and civic organizations that previously had not collaborated. Zahniser is remembered for his skill and diplomacy in building coalitions and developing local leadership for national conservation struggles.

Howard Clinton Zahniser was born on February 25, 1906, in Franklin, Pennsylvania, and grew up in Tionesta, Pennsylvania. His father was an energetic Free Methodist minister and, later, district elder whose diary records 227 pastoral calls in the first quarter of the year he died of heart trouble. His mother was descended from colonist Mary Jemison, who was captured by the Seneca tribe and chose to live with them even after Indian-White relations normalized. An early lover of the Allegheny landscape and its birds, Zahniser joined a Junior Audubon Society as a fifth grader. In college he concentrated on English literature and journalism, editing the student newspaper; he graduated with a B.A. in 1928 from Greenville College in Illinois. His first jobs were as reporter and editor for newspapers in Pittsburgh and Greenville and as English teacher for the Greenville

High School. In 1930 he went to work for the U.S. Department of Commerce's Division of Publications as editorial assistant. In 1931 he transferred to the U.S. Department of Agriculture's Bureau of Biological Survey, which later became the Department of the Interior's U.S. Fish and Wildlife Service. There he was mentored in an ecological view of the world by naturalists Edward Preble and OLAUS J. MURIE and by J. NORWOOD "DING" DARLING, Ira N. Gabrielson, and ALDO LEOPOLD and worked briefly with RACHEL CARSON. Zahniser wrote, edited, and produced broadcasts on topics of wildlife research, management, and conservation. In 1942 he became principal research writer for the Department of Agriculture's Bureau of Plant Industry, Soils, and Agricultural Engineering and directed publicity for the World War II Victory Gardens campaign. Beginning in 1935, Zahniser moonlighted as books editor for *Nature Magazine* and contributed entries on conservation and wilderness preservation to the *Encyclopaedia Britannica* yearbook series.

Zahniser was a charter member of the Wilderness Society, which had been founded in 1935. In 1945, when TWS executive secretary ROBERT STERLING YARD died, Zahniser was recruited as its new executive secretary and editor of its journal, *Living Wilderness*. He had started to question continued government service because of the U.S. use of atomic bombs. Zahniser created a news section for *Living Wilderness* that became a national clearinghouse for conservation and related civic issues. This work was key to

Howard Zahniser in 1955 (Wilderness Society)

his later coalition building. Zahniser first emerged into public view in the struggle to oppose the Echo Park dam proposed inside Dinosaur National Monument in the 1950s. He was instrumental in building and coordinating this first national coalition for a conservation cause and coleading the six-year struggle with DAVID BROWER, then head of the Sierra Club, who produced WALLACE STEGNER's illustrated book *This Is Dinosaur* and a film. Zahniser, based in Washington, D.C., testified, lobbied, and marshaled grassroots support. In 1956, the Colorado River Storage Project Act provided that "no dam or reservoir constructed under the authorization of the Act shall be within any National Park or Monument." Zahniser negotiated the historic final settlement on behalf of conservationists. It established the principle of the inviolability of National Park System lands, threatened since Hetch Hetchy Valley in Yosemite National Park was dammed earlier that century.

Beginning in 1946 and through the Dinosaur years, Zahniser brainstormed about a national program for preserving wilderness. He urged the idea at the biennial national wilderness conferences that TWS cosponsored with the Sierra Club and also at the 1955 National Citizens Planning Conference on Parks and Open Spaces for the American People. The Sierra Club in 1955 resolved to support federal protection of wilderness. Upon the Dinosaur victory, Zahniser drafted a wilderness bill, and he and Brower determined to turn the Echo Park coalition toward pursuing it. Zahniser circulated the four-page document to friends and allies in conservation and civic groups and convinced Sen. Hubert H. Humphrey and Rep. John P. Saylor to introduce the legislation in Congress.

In the eight-year battle for the Wilderness Act, Zahniser and the Wilderness Society worked closely with David Brower and the Sierra Club and with many national and regional conservationist organizations and civic, garden, and women's clubs. No legislation in history had generated so many letters to Congress. Despite overwhelming grassroots support, industry groups vehemently opposed the wilderness bill. Timber, oil, grazing, and mining interests and proponents of motorized access to wild lands lobbied against the legislation, saying it would "lock up" such areas for lesser numbers of hikers and campers. Zahniser patiently countered opposition with ecological, spiritual, and philosophical viewpoints derived from the conservation tradition of GEORGE PERKINS MARSH, HENRY DAVID THOREAU, and JOHN

MUIR and his intimate knowledge of the writings of the Book of Job, Dante Alighieri, and William Blake. Zahniser agreed with Wilderness Society cofounder ROBERT MARSHALL that wilderness offered social, mental, and spiritual health. He also believed that wilderness can help humans sense themselves as interdependent members of the whole community of life on earth. Zahniser articulated these ideas in his articles for *Nature, National Parks, American Forests, Living Wilderness,* and the *Britannica* yearbooks.

Zahniser devoted himself to assuring passage of a wilderness bill. He took his family—his wife, Alice Bernita Hayden, and their four children—on numerous backpack and canoe trips during these years to see areas proposed for wilderness preservation. Zahniser testified at every public hearing for the bill, in Washington, D.C., and in several western states. In his travels he made his own share of "pastoral calls" on congressmen, newspaper editors, other conservationists, and civic leaders. He wore suits his son Ed called "fabric filing cabinets," whose tailor-made oversized pockets held quantities of wilderness leaflets and usually a book by Thoreau, Dante, or Blake. Under Zahniser's leadership, TWS membership grew from 2,000 to 27,000.

During the final stretch toward passage of the Wilderness Act, Zahniser died of a heart attack at the age of 58 years, on May 5, 1964, at the family's home in Hyattsville, Maryland.

The fight for passage of the act and for wilderness preservation generally was carried on by Zahniser's long-time TWS associate Michael Nadel and by STEWART BRANDBORG, Zahniser's successor as executive director of TWS. The legislation was passed in August 1964 and signed by Pres. Lyndon B. Johnson on September 3, 1964. At 9 million acres, the initial National Wilderness Preservation System was far less than the 60 million acres Zahniser originally proposed. However, the Wilderness Act provided a mechanism that has, in the years since 1964, enabled wilderness preservationists to secure the preservation of the system's present 104 million acres.

BIBLIOGRAPHY

Fox, Stephen, *The American Conservation Movement: John Muir and His Legacy,* 1981; "Howard Clinton Zahniser, 1906–1964," *Living Wilderness,* 1964; Nash, Roderick, *Wilderness and the American Mind,* rev. ed., 1973; Zahniser, Ed, "Howard Zahniser, Father of the Wilderness Act," *National Parks,* 1984.

List of Leaders by Occupation or Work Focus

Activists and Organizers

Ansel Adams
Jane Addams
Dana Alston
Adrienne Anderson
Betty Ball and Gary Ball
Judi Bari
Peter Berg
Roberta Blackgoat
Stewart Brandborg
Walt Bresette
Robert Bullard
César Chávez
Benjamin Chavis
William Colby
Barry Commoner
Paul and Ellen Connett
Bill Devall
Marjory Stoneman Douglas
Rosalie Edge
Dave Foreman
Lou Gold
Robert Gottlieb
Lois Gibbs
Richard Grossman
Alice Hamilton
Dorothy Webster Harvey
Denis Hayes
Randy Hayes
Julia Butterfly Hill
William Temple Hornaday
Hazel Johnson
Claudia Alta (Lady Bird) Johnson
Jim Jontz
Owen Lammers
Alicia Littletree
Jackie Lockett
Oren Lyons
Mary McDowell
Donella Meadows
Enos Mills
Marion Moses
John Muir
Olaus and Mardy Murie
Steve Packard

Jane Perkins
Melissa Poe
Jeremy Rifkin
Karen Silkwood
Rocky Smith
Wilma Subra
Terri Swearingen
JoAnn Tall
Chief Tommy Kuni Thompson
Grace Thorpe
Brian Tokar
William Toor
D. Chet Tzchozewski
Hazel Wolf

Advertising Executive

Jerry Mander

Analysts

Gary Ball
Marion Clawson
Donella Meadows
Rocky Smith

Architects, Landscape Architects, and Planners

Carl Anthony
Catherine Baeur
Arthur Carhart
Buckminster Fuller
Robert Gottlieb
Benton MacKaye
Ian McHarg
William McDonough
Lewis Mumford
Frederick Law Olmsted, Jr.
Frederick Law Olmsted, Sr.
Frank Popper
Paolo Soleri

Artists

John James Audubon
William Bartram
Albert Bierstadt
George Catlin

Thomas Cole
Oren Lyons
Margaret Owings
Roger Tory Peterson
Ernest Thompson Seton

Attorneys

John Hamiton Adams
Richard Ayres
Michael Bean
Peter Berle
Harry Caudill
Chris Desser
Robert Golten
Ralph Nader
J. William Futrell
Robert F. Kennedy, Jr.
David Sive
Christopher Stone

Aviators

Anne Morrow and Charles
Lindbergh

Cartoonist

Jay Norwood "Ding" Darling

Consultants

Stewart Brand
Molly Harriss Olson
Marc Reisner
Peter Warshall

Editors

Elias Amidon and Elizabeth
Roberts
Stewart Brand
Devereux Butcher
Frank Chapman
Ellen Connett
Mark Dowie
David Ehrenfeld
Robert Underwood Johnson
Betsy Marston
Peter Montague

Peter Warshall
Robert Sterling Yard
Howard Zanheiser

Educators/Professors

Elias Amidon and Elizabeth
    Roberts
Adrienne Anderson
Albert Bartlett
Jack Collom
Paul Connett
Richard Dawson
Robert Golten
Robert Gottlieb
Aldo Leopold
Roderick Nash
Eugene Odum
David Orr
David Sive
James Gustave Speth
Oakleigh Thorne, II
Martin Walter
Barbara Warburton

Elected Officials

Congresspeople

John Chafee (Senator from
    Rhode Island)
John Dingell, Jr. (Representative
    from Michigan)
Henry Jackson (Senator from
    Washington)
Jim Jontz (Representative from
    Indiana)
George Mitchell (Senator from
    Maine)
Edmund Muskie (Senator from
    Maine)
Gaylord Nelson (Senator from
    Wisconsin)
Morris Udall (Representative
    from Arizona)
Stewart Udall (Representative
    from Arizona)
Henry Waxman (Representative
    from California)

Governors

Cecil Andrus (Idaho)

Bruce Babbit (Arizona)
Jimmy Carter (Georgia)
Gaylord Nelson (Wisconsin)
Russell Peterson (Delaware)

Vice Presidents of the United States

Albert Gore, Jr.

Presidents of the United States

John Quincy Adams
Jimmy Carter
Franklin D. Roosevelt
Theodore Roosevelt

Mayor

William Toor (Boulder,
    Colorado)

Entertainers

Actors

Ed Begley, Jr.
Woody Harrelson

Musicians

Alicia Littletree
Woody Guthrie
Pete Seeger
Paul Winter

Orators/Lecturers/Storytellers

Lou Gold
Enos Mills
Chief Sealth (Seattle)
Ernest Thompson Seton

Radio and Television Personalities

Peter Berle
Chris and Martin Kratt
Carl Sagan
Adam Werbach

Electrician

Debby Tewa

Engineer

George Waring

Entrepreneurs

Ray Anderson
Amos Bien
Kate and Tom Chappell
Anita Clark
Lisa Conte
William Drayton
Paul Hawken
Danny Seo
Ted Turner

Explorer

John Wesley Powell

Farmers/Gardeners

Wendell Berry
Helen and Scott Nearing
Robert Rodale
Cathrine Sneed

Foresters

Mollie Beattie
Jeff DeBonis
Jerry Franklin
Robert Marshall
Gifford Pinchot

Futurists

Buckminster Fuller
Hazel Henderson

Health Professionals

Nutritionists

Joan Dye Gussow

Physicians

Alice Hamilton
Marion Moses
Herbert Needleman
Irving Selicoff

Toxicologists

Eula Bingham

Historians

Henry Adams

William Cronon
Bernard Devoto
Samuel P. Hays
George Perkins Marsh
Edmond Meany
Carolyn Merchant
Roderick Nash
Francis Parkman
Wallace Stegner
Frederick Jackson Turner
Lynn White, Jr.
Donald Worster

Inventors
George Washington Carver
Buckminster Fuller
George Perkins Marsh

Leaders of Environmental Organizations and Institutions
John Hamilton Adams (Natural Resources Defense Council)
Carl Anthony (Urban Habitat)
Richard Ayres (Natural Resources Defense Council)
Peter Bahouth (Greenpeace U.S.A.; Turner Foundation)
Judi Bari (Earth First!)
Peter Berg (Planet Drum Foundation)
Peter Berle (National Audubon Society)
Barbara Bramble (National Wildlife Federation)
Stewart Brandborg (The Wilderness Society)
David Brower (Sierra Club, Friends of the Earth, League of Conservation Voters, Earth Island Institute)
Janet Brown (Environmental Defense Fund)
Lester Brown (Worldwatch Institute)
Devereux Butcher (National Parks Association)

Jasper Carlton (Biodiversity Legal Foundation)
Marjorie Carr (Florida Defenders of the Environment)
Aurora Castillo (Mothers of East Los Angeles)
César Chávez (United Farm Workers)
Benjamin Chavis (National Association for the Advancement of Colored People)
Marion Clawson (Resources for the Future)
William Colby (Sierra Club)
Barry Commoner (Center for the Biology of Natural Systems)
Jay Norwood "Ding" Darling (National Wildlife Foundation)
Jeff DeBonis (Association of Forest Service Employees for Environmental Ethics and Public Employees for Environmental Ethics)
Chris Desser (Muir Investment Trust, Migratory Species Project)
Will Dilg (Izaak Walton League of America)
Hank Dittmar (Surface Transportation Policy Project, Great American Station Foundation)
Richard Donovan (SmartWood)
William Drayton (Ashoka)
Alan Durning (Northwest Environmental Watch)
William Dutcher (National Audubon Society)
Dave Foreman (Earth First!)
Dian Fossey (Karisoke Research Center)
Kathryn Fuller (World Wildlife Fund)

J. William Futrell (Environmental Law Institute)
George Bird Grinnell (National Audubon Society, Boone and Crockett Club)
Lois Gibbs (Center for Health, Environment and Justice)
Richard Grossman (Environmentalists for Full Employment, Grogram on Corporations, Law and Democracy)
Juana Gutiérrez (Mothers of East Los Angeles)
Randy Hayes (Rainforest Action Network)
Jay Hair (National Wildlife Federation)
Woody Harrelson (Oasis Preserve International)
Debra Harry (Indigenous Peoples Council on Biololonialism)
Tim Hermach (Native Forest Council)
William Temple Hornaday (New York Zoological Society)
Dolores Huerta (United Farm Workers)
Helen Ingram (Udall Center for Studies in Public Policy)
Wes Jackson (The Land Institute)
Hazel Johnson (People for Community Recovery)
Diane Jukofsky (Conservation Media Center)
Jim Jontz (American Lands Alliance)
Hal Kane (Pacific Environment and Resources Center)
Daniel Katz (Rainforest Alliance)
Henry Kendall (Union of Concerned Scientists)
Danny Kennedy (Project Underground)

Fred Krupp (Environmental Defense)

Winona LaDuke (Indigenous Women's Network)

Owen Lammers (Glen Canyon Action Network)

Francis Lappé (Institute for Food and Development Policy, Center for Living Demcracy)

Aldo Leopold (The Wilderness Society)

Jackie Lockett (Border Information and Solutions Network)

Amory Lovins and Hunter Lovins (Rocky Mountain Institute)

Benton MacKaye (The Wilderness Society)

Robert Marshall (The Wilderness Society)

Michael McCloskey (Sierra Club)

Donella Meadows (The Sustainability Institute)

Edmond Meany (Washington Mountaineers)

Russell Mitermeier (Conservation International)

Peter Montague (Environmental Research Foundation)

Marion Moses (Pesticide Education Center)

John Muir (Sierra Club)

Mardy and Olaus Murie (The Wilderness Society)

Ralph Nader (too numerous to list here)

Carlos Nagel (Friends of Pronatura)

Reed Noss (Society for Conservation Biology)

Sigurd Olson (National Parks Association, The Wilderness Society)

Fairfield Osborn, Jr. (New York Zoological Society, Conservation Foundation)

Margaret Owings (Friends of the Sea Otter)

Paula Palmer (Global Response)

Jane Perkins (Friends of the Earth, American Federation of Labor-Congress of Industrial Organizations)

Russell Peterson (National Audubon Society)

Melissa Poe (Kids F.A.C.E.)

Sandra Postel (Global Water Policy Project)

Richard Pough (The Nature Conservancy)

Paul C. Pritchard (National Parks Trust, National Parks Conservation Association)

Peter Raven (Missouri Botanical Garden)

William Kane Reilly (World Wildlife Foundation, Conservation Foundation)

Jeremy Rifkin (Foundation on Economic Trends)

John Robbins (EarthSave International)

Vicki Robin (New Road Map Foundation)

Carl Safina (Living Oceans Program of the National Audubon Society)

Kirkpatrick Sale (E.F. Schumacher Society)

John Sawhill (The Nature Conservancy)

Rodger Schlickeisen (Defenders of Wildlife)

Danny Seo (Earth 2000)

Christopher Shuey (Southwest Research and Information Center)

David Sive (Natural Resources Defense Council)

Michael Soulé (Society for Conservation Biology, Wildlands Project)

James Gustave Speth (Natural Resources Defense Council, World Resources Institute, United Nations Development Program)

William Steel (Mazamas)

Kierán Suckling (Center for Biological Diversity)

JoAnn Tall (Native Resource Coalition, Indigenous Environmental Network, Seventh Generation Fund)

Oakleigh Thorne, II (Thorne Ecological Institute)

Grace Thorpe (National Environmental Council of Native Americans)

Brian Tokar (Vermont Greens)

D. Chet Tzchozewski (Global Greengrants Fund)

Adam Werbach (Sierra Club)

Louisa Willcox (Greater Yellowstone Coalition, Sierra Club Grizzly Bear Ecosystems Project)

Chris Wille (ECO-OK Certification Program)

Diane Whealy and Kent Whealy (Seed Savers Exchange)

Hazel Wolf (Seattle Audubon Society)

Howie Wolke (Earth First!)

George Woodwell (Woods Hole Research Center)

Howard Zanheiser (The Wilderness Society)

Robert Sterling Yard (National Parks Association, The Wilderness Society)

Naturalists

John Burroughs

Aldo Leopold

Enos Mills

John Muir

Olaus Murie

Terry Tempest Williams

Philanthropists and Funders

Dana Alston
John D. Rockefeller, Jr.
Laurance Rockefeller
Ted Turner

Philosophers

Peter Berg
Murray Bookchin
J. Baird Callicott
Bill Devall
Paul Goodman
Stephanie Mills
Lewis Mumford
Scott and Helen Nearing
Bryan Norton
Holmes Rolston, III
Theodore Roszak
George Sessions

Photographers

Ansel Adams
Devereux Butcher

Publishers

Stewart Brand
Ellen and Paul Connett
Ron Mader
Stephanie Mills
Robert Rodale
Mark Dowie

Recyclers

Anita Clark
Pete Grogan

Scientists

Life and Physical Scientists

John James Audubon (ornithologist)
Albert Bartlett (nuclear physicist)
John Bartram and William Bartram (botanists)
Marston Bates (zoologist)
William Beebe (marine biologist and ornithologist)
Hugh Hammond Bennett (soil scientist)
Tom Cade (ornithologist)
Archie Carr (zoologist)
Marjorie Carr (biologist)
Rachel Carson (biologist)
George Washington Carver (agricultural scientist)
Frank Chapman (ornithologist)
Theo Colborn (zoologist)
Barry Commoner (biologist)
Paul Connett (chemist)
Henry Cowles (botanist, ecologist)
Paul Cox (ethnobotanist)
Frank and John Craighead (wildlife biologists)
Michael Dombeck (fisheries biologist)
René Dubos (microbiologist)
William Dutcher (ornithologist)
Thomas Eisner (entomologist)
Silvia Earle (marine biologist, oceanographer)
David Ehrenfeld (ecologist)
Anne and Paul Ehrlich (biological researcher; population biologist)
Dian Fossey (zoologist)
Stephen Jay Gould (paleontologist)
Asa Gray (botanist)
Garrett Hardin (biologist)
Donna House (ethnobotanist)
Wes Jackson (plant geneticist)
Dan Janzen (tropical ecologist)
Henry Kendall (physicist)
Thomas Lovejoy (tropical ecologist)
George Perkins Marsh (environmental scientist)
Dennis Martínez (restoration ecologist)
Russell Mittermeier (zoologist)
Olaus Murie (wildlife ecologist)
Gary Nabhan (ethnobiologist, agricultural and desert ecologist)
Reed Noss (conservation biologist, ecologist)
Eugene Odum (ecologist)
Roger Tory Peterson (ornithologist)
Steve Packard (restoration ecologist)
Mark Plotkin (ethnobotanist)
Ellen Swallow Richards (sanitary chemist)
Peter Raven (botanist)
Carl Safina (marine ecologist)
Carl Sagan (astronomer)
Charles Sprague Sargent (dendrologist)
Stephen Schneider (climatologist)
Richard Evans Schultes (ethnobotanist)
Michael Soulé (conservation biologist)
Sandra Steingraber (ecologist)
Wilma Subra (analytical chemist)
Oakleigh Thorne (ecologist)
William Vogt (ornithologist, ecologist)
Edward O. Wilson (entomologist)
George Woodwell (ecologist)

Social Scientists

Kenneth Boulding (economist)
Janet Brown (political scientist)
Lester Brown (economist)
Robert Bullard (sociologist)
Lynton Caldwell (political scientist)
Marion Clawson (agricultural economist)
Herman Daly (economist)
Garrett Hardin (human ecologist)
Hazel Henderson (economist)
Glenn S. Johnson (sociologist)
Stephen Kellert (social ecologist)
Winona LaDuke (economist)
Paula Palmer (sociologist)

Deborah Popper (geographer)
Laura Pulido (sociologist)
Gilbert White (geographer)

Theologians/ Religious Teachers

John B. Cobb, Jr.
Oren Lyons
Holmes Rolston, III
Rosemary Radford Ruether

Government Appointed Officials

Horace Albright (National Park Service Director)
Cecil Andrus (Secretary of the Interior)
Bruce Babbit (Secretary of the Interior)
Mollie Beattie (Director of the U.S. Fish and Wildlife Service)
Hugh Hammond Bennett (Director of the U.S. Soil Conservation Service)
Eula Bingham (Director of the Occupational Safety and Health Administration)
Carol Browner (Administrator of the U.S. Environmental Protection Agency)
Douglas Costle (Administrator of the U.S. Environmental Protection Agency)
Michael Dombeck (Chief of the U.S. Forest Service)
William O. Douglas (Supreme Court Justice)
Newton Drury (National Park Service Director)
Willie Fontenot (Community Liason Officer in the Louisiana Attorney General's Office)
Harold Ickes (Secretary of the Interior)
Stephen Mather (National Park Service Director)

George Mitchell (U.S. Attorney)
John Wesley Powell (U.S. Geological Survey Director)
Gifford Pinchot (Founder and First Chief of the U.S. Forest Service)
William K. Reilly (Administrator of the U.S. Environmental Protection Agency)
Carl Schurz (Secretary of the Interior)
William Steel (Commissioner of Crater Lake National Park)
Stewart Udall (Secretary of the Interior)

Whistle-blowers

Adrienne Anderson
Jeff DeBonis
Hugh Kaufman
William Sanjour

Writers

(These include authors, journalists, poets, nature writers, and novelists)
Edward Abbey
Diane Ackerman
Henry Adams
Mary Austin
Marston Bates
William Beebe
Wendell Berry
Murray Bookchin
Michael Brown
John Burroughs
Devereux Butcher
Archie Carr
Rachel Carson
Harry Caudill
Jack Collom
James Fenimore Cooper
Bernard Devoto
Marjory Stoneman Douglas
Mark Dowie
Ralph Waldo Emerson

Michael Frome
Theodor Geisel (Dr. Seuss)
Ross Gelbspan
Paul Goodman
Stephen Jay Gould
George Bird Grinnell
Dorothy Webster Harvey
Paul Hawken
Edward Hoagland
Diane Jukofsky
Hal Kane
Winona LaDuke
Frances Moore Lappé
Aldo Leopold
Anne Morrow and Charles Lindbergh
Barry Lopez
Ron Mader
Richard Manning
Peter Matthiesen
Bill McKibben
John McPhee
Donella Meadows
Stephanie Mills
John Muir
Gary Nabhan
Helen and Scott Nearing
Sigurd Olson
David Orr
Fairfield Osborn, Jr.
Roger Tory Peterson
Marc Reisner
Jeremy Rifkin
John Robbins
Robert Rodale
Theodore Roszak
Carl Safina
Kirkpatrick Sale
Gary Snyder
Luther Standing Bear
Wallace Stegner
Sandra Steingraber
Henry David Thoreau
Brian Tokar
William Vogt
Walt Whitman
Terry Tempest Williams

Index

About the Authors

ANNE BECHER teaches Spanish at the University of Colorado–Boulder, and writes on environmental matters.

KYLE MCCLURE has a degree in environmental studies from Pomona College. He is a freelance writer and environmental activist based in Missoula, Montana, where he is also doing graduate work at the University of Montana.

RACHEL WHITE SCHEUERING—a graduate of Oberlin College in biology and English—has worked as a field biologist in Montana, Oregon, and California, mainly studying bird ecology. She lives in Portland, Oregon, with her husband and son.

JULIA WILLIS has a Ph.D. in English from Rutgers University and has long been involved in issues of social justice. She teaches writing for the Student Academic Services Center at the University of Colorado–Boulder, a program that works to increase access to higher education for students from underrepresented groups.